Polymer Characterization

Santanu Chattopadhyay · Nikhil Kumar
Editors

Polymer Characterization

Structure to Applications

Springer

Editors
Santanu Chattopadhyay
Rubber Technology Centre (RTC)
Indian Institute of Technology Kharagpur
Kharagpur, West Bengal, India

Nikhil Kumar
Advanced Technology Development Centre
Indian Institute of Technology Kharagpur
Kharagpur, West Bengal, India

ISBN 978-981-95-2608-6 ISBN 978-981-95-2609-3 (eBook)
https://doi.org/10.1007/978-981-95-2609-3

This Springer imprint is published by the registered company Springer Nature Singapore Pte Ltd.
The registered company address is: 152 Beach Road, #21-01/04 Gateway East, Singapore 189721, Singapore

I dedicate this book to my father, Late Nihar Ranjan Chatterjee, a gifted lyricist who, at age of just 17, earned a prestigious Rank 7 at All India Radio, which was the nation's sole broadcasting platform of those days. I lost him three years ago when he was 82. He devoted his life to literature, crafting several thousand poems, songs, dramas, and dance dramas that showed his intense creativity and unwavering passion for the arts and literature. The smell of a newly published book in our own printing press fascinates me, and my aspiration to write a book myself started during my childhood days. However, I was destined for the world of research, particularly the diverse world of polymers. In my academic pursuit, I met one of the most incredible and knowledgeable personalities, Dr. J. Carson Meredith, Professor, School of Chemical and Biomolecular Engineering, Georgia Institute of Technology, USA who has been my postdoctoral supervisor for 2 years. His wisdom, perspective, and vision still inspire me. I devote this book to him, as my "Gurudakshina".

I, too, dedicate this book to my beloved mother, Late Lilabati Chatterjee, whom I lost in September 2016. I spent my playful childhood with two of my soul companions, my younger brother Santu, and my little sister Bubai. They motivate me in every single phase of my life and remind me of those days of companionship and the reason why I started this journey. Although they are my younger siblings, in the absence of our parents, they protect and nurture me as my guardian during any hardship. From the bottom of my heart, this book is dedicated to my childhood buddies, Santu and Bubai. Finally, I dedicate this book to the piece of my heart, my daughter, Sagnika, my beloved wife, Saoli, and my army of strength, my research group, SPACE.

—Santanu Chattopadhyay, Professor, Rubber Technology Centre, Indian Institute of Technology Kharagpur.

I dedicate this book to all those who have provided me with support in both my personal and professional life. This work is not merely a scientific foundation; it is a sincere expression of gratitude for the collective strength, motivation, and support that I have received over the years. First and foremost, I am grateful to my father (Shri. Raghunath Singh) and mother (Smt. Ranju Singh), sisters and family members, for their unwavering support. The sacrifices they have made and their faith in me to support my pursuit for knowledge and its dissemination. To my dear friends and colleagues Rajan, Priyanka, Roumita, Abhay bhai, Shailesh bhai, Homen, Nikhil Dhore, Manas, Shashi

Rastogi, Rahul, Anurag and Manas for their support, friendship, and faith. Each of you has contributed in your own special way to shaping my views, both in life and in my career.

To my research group, SPACE, I am indebted for the collaborative energy, thought-provoking conversations, and the motivation that fuels our collective hunger for knowledge.

I am deeply grateful to Dr. Biswajoy Ghosh, my research mentor, for his constant support, motivation, and scientific expertise. His guidance has been vital to my professional growth as a researcher.

I am also obliged to my Ph.D. supervisors, Profs. Santanu Chattopadhyay and Santanu Dhara, for their invaluable assistance. This book has been significantly influenced by their academic guidance and encouragement.

To everyone mentioned and many others whose silent support accompanied me, I owe my deepest gratitude. This work is as much yours as it is mine.

—Nikhil Kumar, Senior Research Fellow, Advanced Technology Development Centre, Indian Institute of Technology Kharagpur, India.

Foreword by Prof. Carson Meredith

Over the last century, polymers have emerged as one of the most versatile material classes. Now ubiquitous in applications like packaging, medicine, electronics, fabrics, and structural materials, the attraction of polymers stems from the ease of scalable processing and limitless variation in chemistry and structure. At the same time, most polymers used today are produced from non-renewable resources and have contributed to environmental burdens like microplastics.

For those involved in the design and production of polymers, the mastery of polymer characterization continues to be just as important today as it has been in the past. The current volume presented by Prof. Chattopadhyay will be invaluable as a timely guide to characterization, providing a modern and comprehensive introduction to classical as well as emerging methods. The book begins with a review of the molecular and mesoscale organization of polymers and composites. From there, the book presents in significant detail and clarity the concepts of chemical analysis in identification of polymers, additives, elastomers and composites. Spectroscopic techniques are discussed, including IR, UV, NMR, and their applications to polymers, elastomers, blends, composites and additives. Thermal analysis techniques including

TGA, DSC, DMA, and DETA are reviewed relative to degradation, thermal parameters as well for reverse engineering. Application of X-Ray diffraction techniques like SAXS and WAXS is covered as well as surface characterization via XPS, SIMS and MS. A detailed application of microscopy including optical, SEM, TEM and AFM is presented. In closing, I'd like to bring attention to several unique features that I think readers will find particularly valuable. These include sections on accelerated stability assessment, advanced multidimensional (e.g., 2D SEC-HPLC) and operando methods, and an intriguing chapter on reverse engineering that includes engaging real-life case studies.

Carson Meredith
Professor
School of Chemical and Biomolecular Engineering
Georgia Institute of Technology
Atlanta, GA, USA

Foreword by Prof. Muthupandian Ashokkumar

The world of polymer science is progressing at an incredible pace, driven by modern technology and the increasing demand for sustainable materials. In this exciting journey, polymer characterization has become a crucial foundation, allowing researchers to understand the structure-property relationships that dictate how polymeric materials perform. I am honored to introduce this well-timed and precisely organized book on polymer characterization, designed to be both a reference and a guide for students, researchers, and professionals engaged in the study and application of polymers. Presently, characterizing polymers goes far beyond imagination. It includes a wide variety of advanced methods, such as spectroscopic and chromatographic techniques, along with microscopic and mechanical tests, which examine the properties of polymers at molecular, mesoscopic, and macroscopic levels. This book emphasizes diversity by offering comprehensive coverage of both conventional and emerging techniques, with a focus on practical relevance.

The editors and authors have made sure that the chapters cover basic to advanced application-based concepts that make this book accessible to readers with different levels of expertise. All chapters are explained with their theoretical background, operational principles, and real-world applications in polymer science, which makes the content not just educational but also directly useful in both industrial and academic contexts.

As someone doing research in sonochemistry and materials science, I truly understand how crucial accurate characterization is when it comes to customizing materials for specific purposes. Whether we are talking about designing polymers for biomedical uses, electronics, coatings, or eco-friendly packaging, getting the precise and accurate characterization is key to achieving success.

I would like to extend my congratulations to the editors and authors for assembling a compact and insightful resource. This book is an example of excellent collaborative work and highlights the increasingly interdisciplinary nature of polymer science. I have no doubt it will motivate and assist the next generations of polymer scientists, engineers, and industry professionals.

Professor Muthupandian Ashokkumar
Director
Melbourne Global Centre Delhi
Deputy Pro Vice-Chancellor International (South Asia and Middle East)
University of Melbourne
Melbourne, VIC, Australia

Foreword by Prof. Sabu Thomas

Polymer characterization has emerged as a foundation for the advancement of materials engineering from structure to application. I introduce this book on *Polymer Characterization* with great enthusiasm, as it is well-organized and conceptualized. Students, researchers, and professionals who are active in the study and application of polymers across a wide variety of fields are the target audience for this volume, which was developed to satisfy their needs.

The book covers a wide range of methodologies, ranging from the most fundamental spectroscopic and chromatographic techniques to the most advanced, multifaceted approaches that define polymers at the molecular, mesoscopic, and macroscopic dimensions. It provides an in-depth investigation of both traditional and modern approaches to characterization, highlighting their practical significance. Chapters are well organized and crafted to integrate theoretical foundations, operational principles, and applied viewpoints that cross academic and industry contexts.

The purpose of this book is to provide a comprehensive foundation for understanding how characterization techniques stimulate innovation to support research and development across various industries. This book addresses vital needs of polymer science and engineering by offering a solid framework to appreciate how characterization helps innovation and ensures quality across diverse sectors.

I extend my heartfelt appreciation to the editors and authors for assembling such a comprehensive, insightful, and updated resource. In a world where materials innovation shapes our future, this book stands as both a guide and a source of inspiration. I wholeheartedly recommend it to all who are eager to explore the fascinating and essential domain of polymers.

Professor Sabu Thomas
Former Vice-Chancellor and Professor
Mahatma Gandhi University
Kottayam, India

Preface

Polymers are versatile and significant part of our daily lives, from commodity plastics to high-performance engineering materials in aerospace and biomedical devices. Despite their ubiquity, understanding the basic science and technology behind polymers, particularly structure-property relationships, can be a complex and challenging subject. This book, *Polymer Characterization* was written to elucidate structure to applications aspects for making the field of polymer more approachable for students, researchers, and industry professionals.

Our motivation for writing this book came from a humble but essential need, which is scripting a basic, comprehensive, and practical guide for polymer characterization. Over the years, the struggle with theoretical concepts, lack in basic understanding, and limited technology transfer have been noticed. We have also seen the highly skilled industry professionals who have expertise in their field but need multidisciplinary knowledge. This book represents our attempt to fill that knowledge gap.

We intend to build the basic foundation for any learner. We discuss the basic concepts in detail, including average molecular weight, chain architecture, crystallinity, and structural insights that influence polymer behavior. By discussing these fundamental principles, we set the stage for more advanced discussions throughout the book.

The main focus of this book is on various methods, including chemical analysis, spectroscopic techniques (like FTIR, NMR, and UV-Vis), thermal methods (such as DSC, TGA, and DMA), mechanical testing, processing, microscopy (like optical, SEM, TEM, and AFM), and special techniques like gel permeation chromatography (GPC) and X-ray diffraction (XRD). We explain each technique with its working principle, instrumentation, and case studies to highlight practical applications and common challenges.

One of the most evident messages we hope to convey is that polymer characterization is not just a collection of analytical methods. It is a way of thinking and executing. Each technique will give the readers a piece of the puzzle, and true understanding comes from putting those pieces together. For instance, a specific analysis reveals a specific property, but when combined with multidisciplinary information from various techniques, it presents a comprehensive picture of the material's performance in a specific application. This integrative approach is essential for both research and industrial development.

In the final chapter of the book, we turn our focus toward reverse engineering. We have also discussed the advanced techniques for characterizing polymers and reverse engineering. Industries ranging from packaging and automotive to electronics, healthcare, and environmental technologies utilize polymers extensively. This causes an environmental burden, which makes it difficult to achieve sustainability goals. We have provided an in-depth exploration on how characterization plays an essential role in the reverse engineering of waste management.

We have made a deliberate effort to keep this book conceptualized and practical. We have included illustrations, tables, and summaries to help explain key points related to mechanism and instrumentation. Our aim is not just providing information but to equip readers with the perception and confidence to apply what they have learned in their own work.

This book is designed for a diverse readership. Undergraduate and graduate students will find it beneficial for academic assignments and research endeavours. Scientists and engineers in various organizations can use it as a reference standard to better understand their materials and make data-driven decisions. Even experienced researchers may find value in the integrated approach and updated coverage of modern techniques.

As editors and authors, we have learned from different aspects of polymer science, whether it is in classrooms or laboratories or research across disciplines. Writing this book has been a chance to bring them together and share what we have found the most useful, valuable, and practical. We sincerely hope that you find the book not only informative but also encouraging as you explore the fascinating world of polymers.

Thank you for choosing this book. We hope it becomes a helpful companion in your studies, research, or professional practice.

Kharagpur, India

Santanu Chattopadhyay

Nikhil Kumar

Contents

Editors and Contributors

About the Editors

Dr. Santanu Chattopadhyay, Ph.D. Professor of Rubber Technology Centre and Ex-Associate Dean (SA) at Indian Institute of Technology Kharagpur, India. Prof. Santanu Chattopadhyay is a distinguished Professor of Rubber Technology Centre (ex-Head) and Chairman, Hostel Management Committee of Indian Institute of Technology Kharagpur, India. He has been serving there as a faculty since 2004. He has pursued his post-doctoral research work in Georgia Institute of Technology, USA from, the Chemical Engineering Department from 2002–2004. Prior to that, he did his post-doctoral studies at the University of Western Ontario, Canada, for one year. Prof. Chattopadhyay completed his Ph.D. from IIT Kharagpur in 2001 on electron beam irradiated thermoplastic elastomer blends as shape-memory materials. He completed his M.Tech. (Materials Science and Engineering) and M.Sc. (Chemistry) from IIT Bombay and IIT Kharagpur, respectively. He has published more than 200 research articles, along with 170 conference proceedings, 13 book Chapters and 5 patents. His research interests include green and sustainable soft materials, smart rubber composites, rubber blends, viscoelastic and rheological characteristics of rubbers, flexible biomaterials for health care, FEA and MD simulation. He has completed supervising 32 Ph.D. students and more than 60 master's students and is currently mentoring 11 Ph.D. students. He has been awarded with prestigious JSPS Fellowship in 2023–2024 and Elected as Fellow of West Bengal Academy of Science and Technology. He has developed two NPTEL courses on "Characterization of Polymers Elastomers and Composites" and "Rheology and Processing of Paints, Plastic, and Elastomer Based Composites" to profoundly encourage the students, academicians and relevant industrial delegates.

Nikhil Kumar is a senior research fellow the Advanced Technology Development Centre, Indian Institute of Technology Kharagpur. He is specialized in fabrication of smart polymeric materials for diagnosis and therapeutics. Nikhil completed his Bachelor of Pharmacy and Master of Pharmacy in Pharmacology. His masters research work was focused on developing dressings for diabetic foot ulcers. His current research revolves around the design and application of smart polymeric biomaterials for advanced diagnosis and therapeutic interventions for cutaneous squamous cell carcinoma. He has published 12 peer-reviewed research articles. He is actively engaged in research focused on polymer characterization, with particular interest in the structural, thermal, and rheological analysis of advanced polymeric materials. Mr. Kumar has been involved in research projects that explore the application of spectroscopic techniques (FTIR, NMR), thermal analysis (DSC, TGA), to evaluate polymer performance under varied environmental conditions. Mr. Kumar is also developing a novel diagnostic tool that integrates multiple characterization techniques FTIR, Raman spectroscopy, AFM and nanoindentation to probe and differentiate pathological tissue conditions such as fibrosis and hyperkeratosis.

Contributors

Komal Agarwal Department of Physics and Technology, UiT-The Arctic University of Norway, Tromsø, Norway

Abhijit Bera R&D Centre, ALP Overseas Pvt. Ltd., Alwar, Rajasthan, India

Santanu Chattopadhyay Rubber Technology Centre, Indian Institute of Technology, Kharagpur, India

Homen Dahal Department of Chemistry, Indian Institute of Technology Kharagpur, Kharagpur, India

Priya Devi Department of Chemical Sciences, Tezpur University, Tezpur, Assam, India

Nikhil Dhore Rubber Technology Centre, Indian Institute of Technology Kharagpur, Kharagpur, West Bengal, India

Biswajoy Ghosh Department of Physics and Technology, UiT-The Arctic University of Norway, Tromsø, Norway

Soumajyoti Ghosh Rubber Technology Centre, Indian Institute of Technology Kharagpur, Kharagpur, India

Roumita Hore Rubber Technology Centre, Indian Institute of Technology Kharagpur, Kharagpur, West Bengal, India

Manas Jain Department of Aerospace Engineering, Indian Institute of Technology Kharagpur, Kharagpur, India

Chandra Khatua Rubber Technology Centre, Indian Institute of Technology Kharagpur, Kharagpur, India

Abhay Kumar Apollo Tyres Ltd., Chennai, India

Nikhil Kumar Advanced Technology Development Centre, Indian Institute of Technology Kharagpur, Kharagpur, India

Shashi Kumar Department of Chemical Engineering, National Institute of Technology, Rourkela, Odisha, India

Payel Maity School of Nanoscience and Technology, Indian Institute of Technology Kharagpur, Kharagpur, India

Aparajita Pal Rubber Technology Centre, Indian Institute of Technology Kharagpur, Kharagpur, India

Chapter 1
Polymer Overview and Background

Nikhil Kumar, Roumita Hore, Soumajyoti Ghosh, and Nikhil Dhore

Abstract This chapter provides a comprehensive introduction to the fundamental principles of polymer science, offering a foundational understanding for students. It begins with an overview of the historical background and evolution of polymer science, outlining the development of synthetic and natural polymers and highlighting key milestones in the field. The classification of polymers based on their origin, structure, and physical properties is also discussed. Furthermore, the chapter explores various polymerization techniques and polymer processing methods such as extrusion, injection molding, blow molding, and compression molding, explaining how raw polymer materials are transformed into functional products. Special attention is given to the influence of processing conditions on the final properties of polymer materials. Additionally, the chapter outlines the wide range of applications of polymers across various industries.

Keywords Polymer science · History of polymer · Polymerization · Polymer processing

1.1 Introduction

The word "polymer" comes from Greek word "Poly" means many and "meros" parts or units [1]. The term was first coined in the early nineteenth century in the context of chemical substances having the same composition but different molecular weights. Jöns Jakob Berzelius, a Swedish chemist, coined the term "polymer" in 1833. The modern understanding of polymers as long chains of covalently bonded monomers was established by Hermann Staudinger in the 1920s [2].

N. Kumar (✉)
Advanced Technology Development Centre, Indian Institute of Technology Kharagpur, Kharagpur, India
e-mail: nikhilsingh25694@gmail.com

R. Hore · S. Ghosh · N. Dhore
Rubber Technology Centre, Indian Institute of Technology Kharagpur, Kharagpur, India

S. Chattopadhyay and N. Kumar (eds.), *Polymer Characterization*,
https://doi.org/10.1007/978-981-95-2609-3_1

In the era of material revolution beginning from the stone age to the modern day, lifestyle progress of the human mankind is closely related to the discovery and development of new polymers Fig. 1.1. It began with simple stones of varying shapes and sizes has progressed into the use of advanced materials like nano-sized metals, semiconductors, and polymers. Among these, polymers exhibited exceptional adaptability. Their wide-ranging properties have made them essential across various industries, including medicine, cosmetics, packaging, energy, healthcare, electronics, and aerospace [3]. Regardless of their widespread application, its molecular architecture and behavior remain complex and require detailed study, i.e., design, synthesis, and analysis. This complexity has driven the growth of polymer characterization, a multidisciplinary approach that utilizes techniques like spectroscopy, chromatography, microscopy, etc., to understand their characteristic physical, chemical, and mechanical properties.

1.1.1 Concept and Classification of Polymers

Polymers are large molecule consists of repeated small, simple chemical units in some cases this repetition is linear, branched or interconnected to form three-dimensional networks. This repeat unit of the polymer is equivalent to the monomer, or starting material from which, the polymer is formed. Different types of polymers along their monomer and repeat units are given in Table 1.1. Although the repeating unit and monomer contain the same atoms in similar relative positions, their structures may differ due to rearrangements of electrons during polymerization.

Degree of polymerization (DP) refers to the number of repeating units (monomers) present in a polymer chain. It is a key parameter that determines the length and size of the polymer. The molecular weight of a polymer is calculated by multiplying the molecular weight of the repeat unit by the DP (Fig. 1.2). For example, in the case of Polystyrene, if the repeat unit has a molecular weight of 36 and the polymer has a DP of 1000, then its molecular weight would be $36 \times 1000 = 36{,}000$.

Polymers with high DP typically exhibit better mechanical properties and are more suitable for industrial applications. Different polymers have varying degrees of polymerization and molecular weights depending on their structure and use. For example, polyethylene (PE) and polypropylene (PP) have molecular weights in the range of 100,000–500,000 and are widely used in packaging and household products. Polystyrene (PS), with a DP yielding molecular weights around 200,000, is used in disposable goods and insulation. Polyethylene terephthalate (PET), used in bottles and textiles, typically has molecular weights of 80,000–200,000. High-performance polymers like polytetrafluoroethylene (PTFE) can reach molecular weights up to 1,000,000, making them suitable for applications requiring extreme chemical resistance.

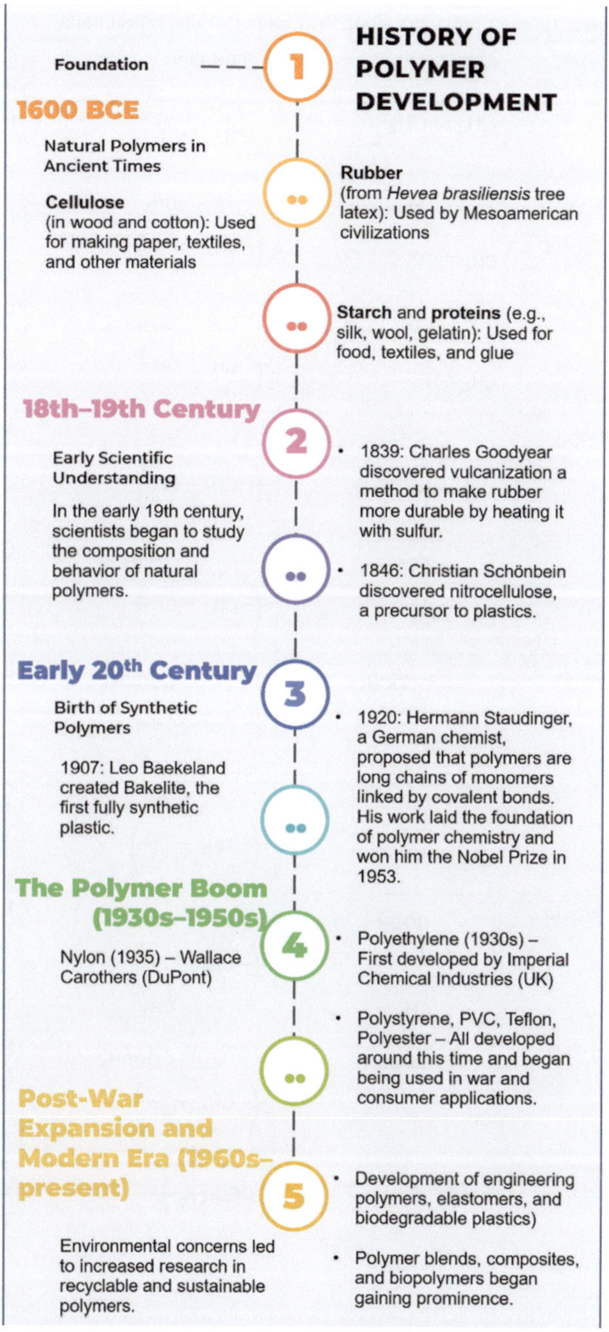

Fig. 1.1 History of polymer development

Table 1.1 Different types of polymers along their monomer and repeat units

Polymers	Monomer	Repeat unit	Commercial name
Polyethylene	$CH_2 = CH_2$	$-(CH_2-CH_2)_n-$	Polythene
Polypropylene	$CH_2=CH(CH_3)$	$-(CH_2-CH(CH_3))_n-$	Moplen, Propylux, Proteus, Cefor
Polystyrene	$CH_2=CH(C_6H_5)$	$-(CH_2-CH(C_6H_5))_n-$	Styrofoam
Polyvinyl chloride	$CH_2=CH(Cl)$	$-(CH_2-CH(Cl))_n-$	Vinyl
Polyacrylic acid	$CH_2=CH(COOH)$	$-(CH_2-CH(COOH))_n-$	Carbomer
Polyvinyl acetate	$CH_2=CH(OCOCH_3)$	$-(CH_2-CH(OCOCH_3))_n-$	Elmer's glue
Polyacrylonitrile	$CH_2=CH(CN)$	$-(CH_2-CH(CN))_n-$	Orlon, Acrilan
Polymethacrylic acid	$CH_2=C(COOH)(CH_3)$	$-(CH_2-C(COOH)(CH_3))_n-$	Eudragit
Poly(methyl acrylate)	$CH_2=CH(COOCH_3)$	$-(CH_2-C(COOCH_3)(H))_n-$	Plexiglas or Lucite
Poly(tetrafluoroethylene)	$CF_2=CF_2$	$(CF_2-CF_2)_n-$	Teflon, Fluon, Hostaflon, and Tedlar

(continued)

Table 1.1 (continued)

Polymers	Monomer	Repeat unit	Commercial name
1,4 Polyisoprene	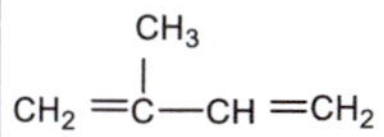	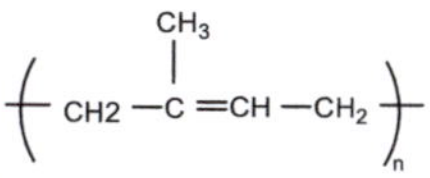	Natural rubber
Polycaprolactam	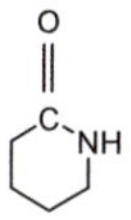	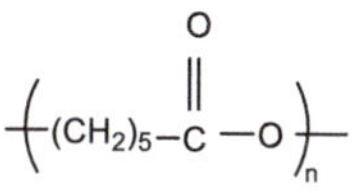	Re-form, shapelock, TechTack, friendly plastic

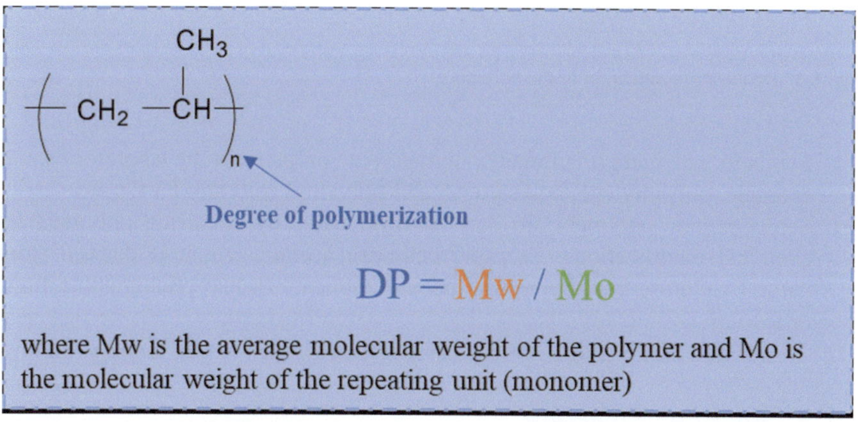

Fig. 1.2 Degree of polymerization

1.2 Classification

Polymers can be broadly classified in different ways Fig. 1.3. The classification is based on the origin of the polymer, polymer structure, thermal response, polymerization, conductivity, biodegradability, and the end-use applications [4, 5]

1.2.1 Based on Origin

Based on the source of availability polymers are classified into three categories, i.e., natural, synthetic, and semisynthetic.

(a) Natural polymers obtained from nature (plants and animals). These polymers are very essential for life. Example: starch, cellulose, proteins, nucleic acids (Biopolymers).

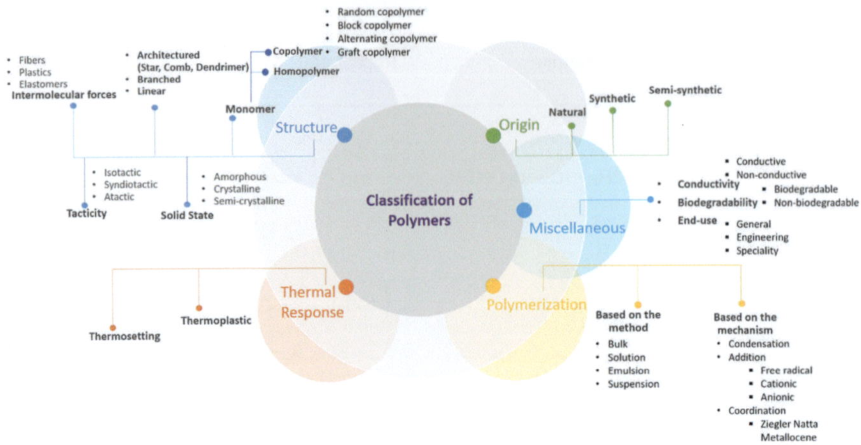

Fig. 1.3 Detailed classification of the polymers

(b) Synthetic polymers (man-made polymers) are prepared in the laboratories. For example, polyethylene, polystyrene, and polyvinyl chloride (PVC), etc.
(c) Semisynthetic polymers are derived from naturally occurring polymers by chemical modifications. Example cellulose acetate, which is derived from natural cellulose but altered for specific applications like photographic films or textiles.

1.2.2 Based on Structure

Depending upon the arrangement of structural units formed from monomers, polymers can be classified into linear, branched, crosslinked, and architecture types Fig. 1.4.

(a) In linear polymer structures the monomer units are arranged in a straight, chain-like structure. Example: high-density polyethylene (HDPE).
(b) A branched polymer is a polymer with side chains or branches extending from a main chain. These branches are typically made of the same repeating units as the main chain. Example: low-density polyethylene (LDPE).
(c) Cross-linked polymers are formed when polymer chains chemically bond with one another to form a three-dimensional network. Example: Bakelite.
(d) Architecture polymers are specific type of polymer structure where the arrangement and branching of the polymer chains are carefully designed to achieve tuneable properties such as mechanical, viscosity, and solubility. Example: Star (chains radiating out from a central core), Dendrimer (hyperbranched structures with a highly branched core), etc.

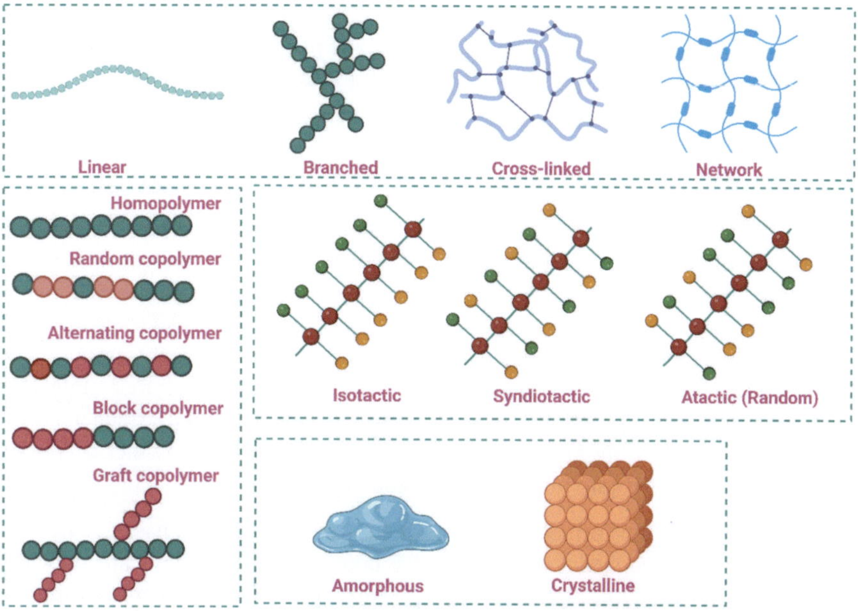

Fig. 1.4 Type of polymer based on structural units' arrangement, copolymers, and spatial arrangements

Based on the composition (nature of monomers), polymers can be classified into two types homopolymers and copolymers as shown in Fig. 1.4.

(a) Homopolymers are composed of one repeating unit of monomer. Examples: Polythene (from ethylene), Teflon (from tetra-fluoroethylene), Polystyrene (from styrene), and Nylon 6 (from caprolactam).
(b) Copolymers are composed of repeating structural units that are derived from two or more different types of monomer molecules. Examples: Buna-S (from styrene and butadiene), Buna-N (from butadiene and acrylonitrile), Nylon 6,6 (from hexamethylenediamine and adipic acid), Polyester (from various monomers depending on the type), and Melmac (from melamine and formaldehyde).

In copolymer systems, the repeating units on the polymer chain can be arranged into various degrees of order, further these systems classified into.

- Random copolymer: The repeating units are arranged randomly on the polymer chain. Example: Ethylene-vinyl acetate (EVA), Nitrile rubber, etc.
- Alternating copolymer: Two distinct monomers alternate in a repeating sequence. Example: Nylon 6,6.
- Block copolymers: Different blocks of identical monomer units alternate with each other. Example: polystyrene-b-poly (methyl methacrylate) (PS-b-PMMA) and Acrylonitrile Butadiene Styrene (ABS). These are synthesized using

controlled polymerization techniques like living cationic or anionic polymerization, or atom transfer radical polymerization (ATRP).

- Graft copolymer: Homopolymer branches of one monomer units are grafted (chemically bonded) on the homopolymer chains of another monomer units. Example: high-impact polystyrene (HIPS), where polybutadiene is grafted onto a polystyrene backbone.

The primary bonding within polymer chains is covalent, while secondary bonding arises from valence forces of attraction such as van der Waals forces, hydrogen bonds, and dipole interactions. These secondary forces cause cohesive aggregation between polymer chains. Since these interactions do not involve valence electrons, they are relatively weak though hydrogen and dipole interactions are generally stronger than van der Waals forces. For secondary bonding to be effective, polymer molecules must come into close proximity, which depends on the structural arrangement of the molecules. According to these intramolecular forces within polymer networks, polymers are classified into fibers, plastics, and elastomers.

(a) Fibers: Fibers are formed when polymer chains align closely, i.e., effective utilization of secondary intermolecular bonding forces. These are typically linear polymers with high symmetry and strong intermolecular forces (due to the presence of polar groups). Fibers are characterized by a high modulus, high tensile strength. Example: Nylon 66, Dacron, silk, etc.
(b) Elastomers have irregular molecular structure, weaker intermolecular attractive forces, and highly flexible polymer chains (significant chain mobility). In resting/normal state, molecules are in coiled configuration, at high stress show high extensibility and recover their shapes easily on removing stress. Usually, elastomers have low initial tensile modulus but become stiffer when stretched. Example: Natural rubber.
(c) Plastics are having structural and mechanical position between fibers and elastomers. Despite variations in chemical structure, the distinction between plastics and fibers is not always clear. For example, polymers such as polypropylene and polyamides can function as both fibers and plastics depending on the processing conditions.

The tetrahedral nature of carbon bonding has a significant impact that is not revealed by simple two-dimensional structural formulas. Atoms/groups attached to the carbon can occupy different spatial positions, and if all four substituents are different, these arrangements are not geometrically equivalent. Such carbons are termed as *chiral* and can exist in two different 3D forms known as *enantiomers*. In a polymer chain due to presence of multiple chiral carbon atoms give rise to multitude of enantiomeric configuration. In practice, these chiral polymers are classified into three types of stereoregularity (*Tacticity*).

Tacticity describes the spatial arrangement of repeating units in the polymer chain, specifically the orientation of side groups relative to each other. The tacticity influences properties of polymers [6]. For example, atactic polymers, being more disordered, cannot crystallize Fig. 1.4.

(a) Isotactic: Similar or identical spatial arrangements of the substituent in the repeat units. Crystalline in nature. Example: polypropylene
(b) Syndiotactic: Side groups alternate on opposite sides of the backbone. Semi-crystalline in nature. Example: Gutta-percha (dental fillings)
(c) Atactic: Side groups are randomly arranged, lacking a specific pattern. Amorphous in nature. Example: polystyrene and polyvinyl chloride (PVC).

– Structurally polymers in solid state can be classified into three types, i.e., Amorphous, Crystalline, and Semi-crystalline.

  (a) In crystalline polymers chains are highly ordered and tightly packed in an orderly fashion contributes to higher density, strength, and melting point. Crystallinity may be partial, as most polymers have both crystalline and amorphous regions (semi-crystalline). Examples: Polyethylene (HDPE), Nylon, Polypropylene.
  (b) Semi-crystalline polymers have both crystalline and amorphous regions, offering a balance of strength, flexibility, and processability. The degree of crystallinity affects mechanical and thermal properties. Examples: Polyethylene (LDPE, HDPE), Polyvinylidene fluoride (PVDF), Polyethylene terephthalate (PET).
  (c) Amorphous polymers have a random, disordered arrangement of their molecular chains with no long-range order. They are typically transparent, flexible, and soften gradually upon heating (do not have a sharp melting point). Examples: Polystyrene, Poly-(methyl methacrylate) (PMMA), and Polycarbonate.

1.2.3 Based on Thermal Response

Polymers can be classified based on their thermal response into the following two major categories:

(a) Thermoplastic Polymers: Polymers that soften upon heating and harden upon cooling in a reversible physical process. This property allows them to be melted, reshaped, and reused multiple times without significant chemical change. The chains in thermoplastic polymers are linear or slightly branched and held together by weak intermolecular forces (like van der Waals forces), allowing them to slide past one another when heated. Examples: Polyethylene (PE), Polypropylene (PP), Polyvinyl chloride (PVC), Polystyrene (PS), Polycarbonate (PC).
(b) Thermosetting Polymers: Polymers that undergoes chemical changes during initial heating (curing) to form a rigid, cross-linked structure. Once set, they cannot be remelted or reshaped. On further heating, they decompose rather than melt. This irreversible process makes them heat-resistant and dimensionally

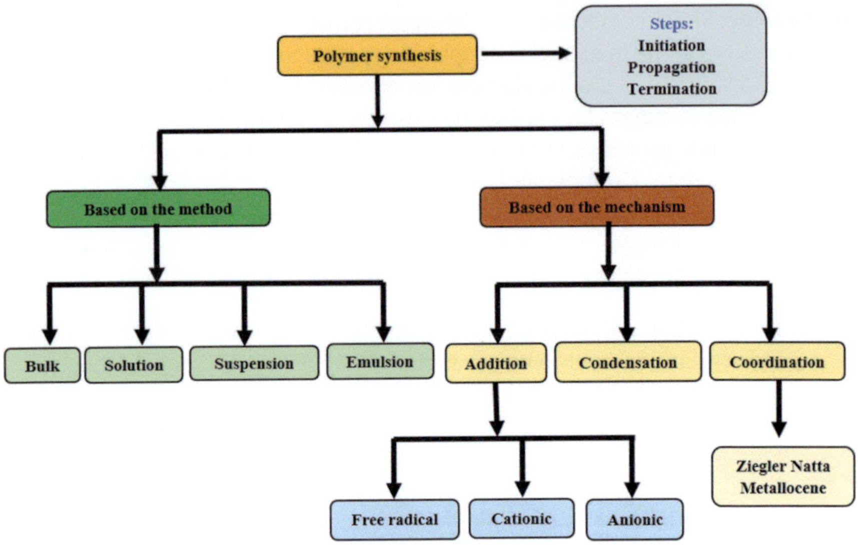

Fig. 1.5 Methods and mechanisms of polymer synthesis

stable. Example: Epoxy resins, Phenol–formaldehyde (Bakelite), Melamine–formaldehyde, Urea–formaldehyde

1.2.4 Based on Polymerization (Preparative Techniques)

The process in which the monomers are chemically bonded with each other and converted into polymers is called polymerization. It involves three significant steps: Initiation, propagation, and termination. Polymerization can be further segregated based on the mechanisms and methods as shown in Fig. 1.5.

Depending upon the synthesis mechanism, polymerization is divided into three major categories: addition, condensation, and coordination polymerization. Considering the methods, polymerization is again divided into four main kinds: Bulk, solution, suspension, and emulsion polymerization.

1.2.4.1 Polymer Synthesis and Mechanism

Addition polymerization: Addition polymerization, also known as chain growth polymerization, has three kinds: Free radical, cationic, and anionic polymerization. The pathway should be chosen based on the nature and stability of the propagating chain. In the following Fig. 1.6 X is the functional group, adjacent to the double bond. If X is a group with +I effect or electron pushing effect or –R effect, the polymerization should follow a cationic pathway. On the other hand, if X has an –I effect or

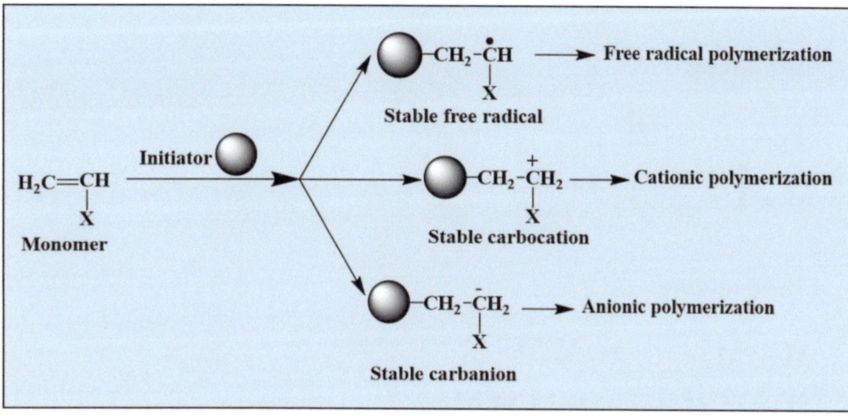

Fig. 1.6 Strategy of polymer synthesis

electron-withdrawing effect or +R effect, the ideal pathway should be anionic or free radical.

Free Radical Polymerization

In free radical polymerization, the polymer is formed by the successive addition of monomer radicals. The initiator (I_2) dissociates in the medium as (I•) and converts the monomers (M) into active radicals (M•). The chain grows as the active monomer radical subsequently reacts with another monomer, and so on (Fig. 1.7).

Being unstable, these active sites (radicals) cause uncontrolled and rapid chain growth. This is why, in less time, the molecular weight of the propagating chain attains a high value (Fig. 1.11). The termination of the growing chain can happen in two ways. If two propagating polymer chains collide at the active center, the termination happens by combination. Another way is by disproportionation, by which the propagating chain ends with an unsaturation [7].

Cationic Polymerization

Cationic polymerization involves a Lewis acid, such as boron trifluoride (BF3), as an initiator. Protonic acid can also be used to initiate the monomer. The acid dissociates and transfers the positive charge to the monomer to form a stable intermediate carbocation. Thereafter, the active carbocation reacts further with the monomers, and chain propagation takes place (Fig. 1.8). The termination predominantly happens by disproportionation, chain transfer, or solvent transfer. The typical monomers used for cationic polymerization are those that can form a stable carbocation, like styrene, vinyl ether, isobutene, cyclic acetals, etc. [8].

Anionic polymerization

Anionic polymerization is a chain growth mechanism where the intermediate propagating chain is a stable carbanion. As mentioned earlier, monomers with −I or +R effect are suitable for anionic polymerization. Sodamide ($NaNH_2$) and butyl

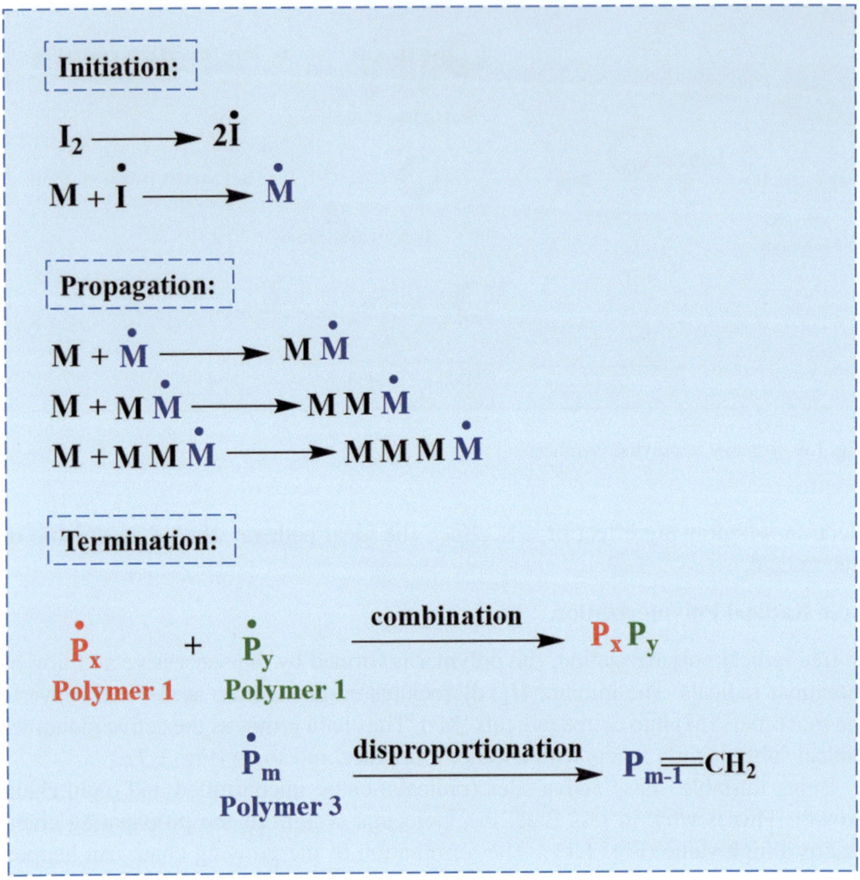

Fig. 1.7 Free radical mechanism

lithium (BuLi) are the commonly used initiators for this polymerization. In Fig. 1.9, the synthesis of polystyrene is shown via an anionic pathway. Anionic pathway is also referred as living polymerization, as the probability of chain termination is limited and can only be possible by the presence of some proton sources [9]. There are several other living polymerizations available, such as nitroxide-mediated radical polymerization (NMP), reversible addition fragmentation chain transfer polymerization (RAFT), and atom transfer radical polymerization (RAFT), which are also discussed in detail in the upcoming section.

Condensation Polymerization

Polycondensation reactions can be either externally or internally catalyzed by adding a mineral acid or using carboxylic acid within the medium, respectively. Condensation polymerization, also known as step-growth polymerization, involves two variants of monomers with at least bifunctionality [10]. Functionality is defined as

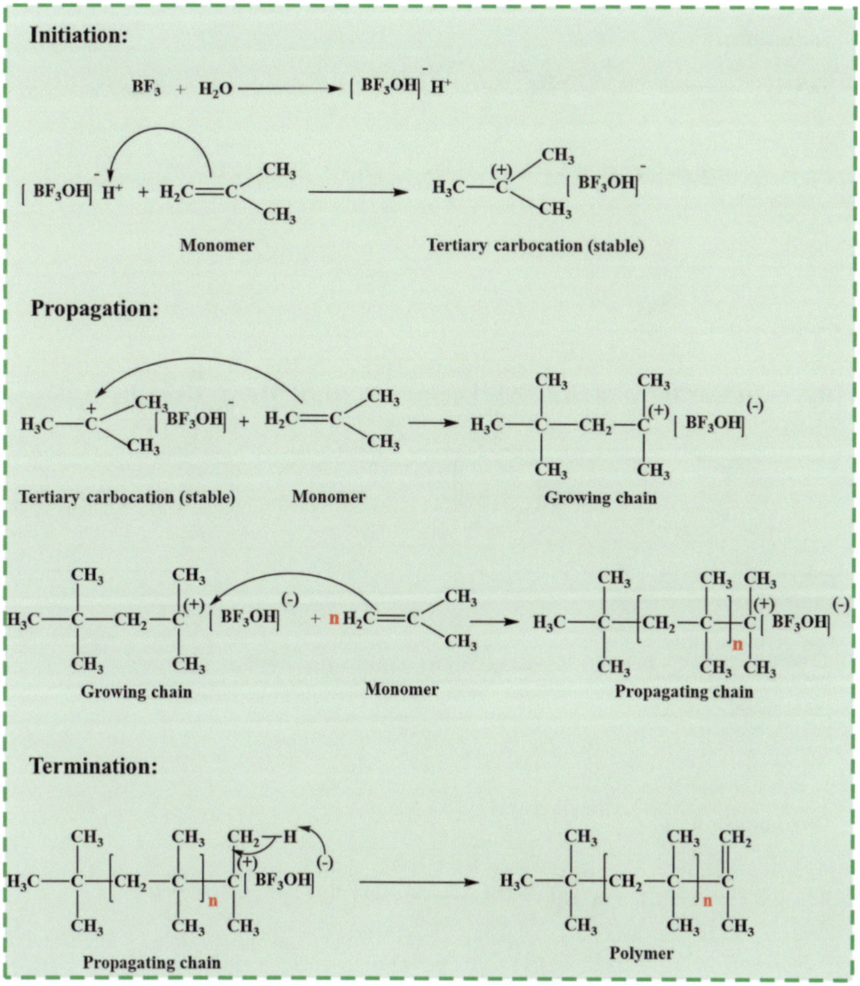

Fig. 1.8 Cationic polymerization

the number of active sites present in a monomer. The common polymers synthesized using condensation polymerization are polyamides, polyurea, polyurethane, and polyesters [11–13].

In general, condensation reaction facilitates the elimination of by-products, water (H_2O), or carbon dioxide (CO_2) according to the choice of monomer. It is interesting that the elimination of gas as a by-product, is used for creating a cellular foam structure using CO_2 as an inherent foaming agent (polyurethane foam).

Figure 1.10 shows that the polymerization of adipic acid and hexamethylene diamine results in the formation of Nylon 6,6, as a condensation polymer.

As mentioned above, in the addition or chain growth polymerization, the uncontrolled growth of the propagating chain happens at the very early stage of the

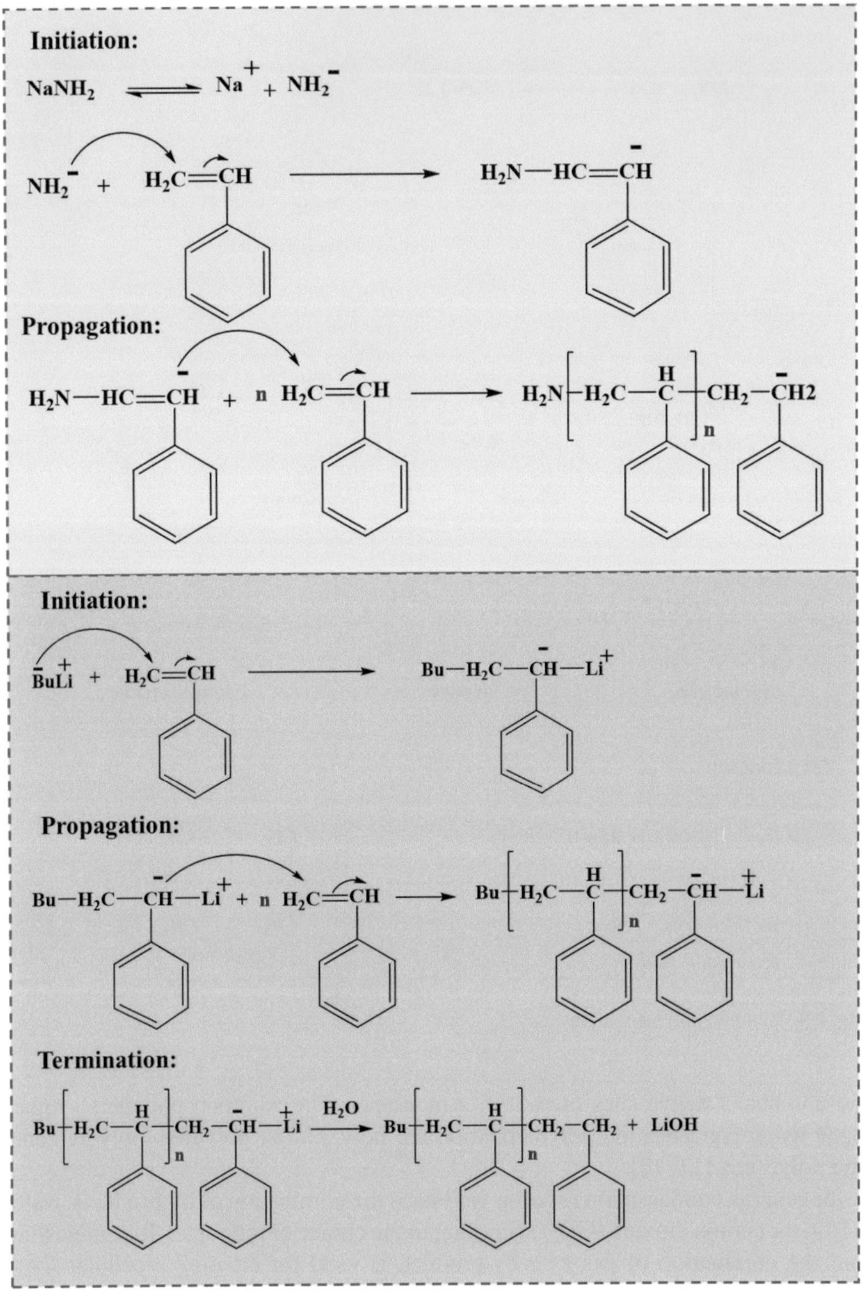

Fig. 1.9 Anionic polymerization

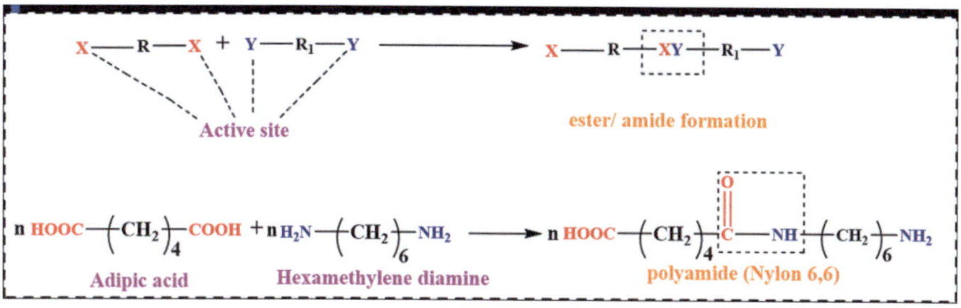

Fig. 1.10 Condensation polymerization of Nylon 6,6

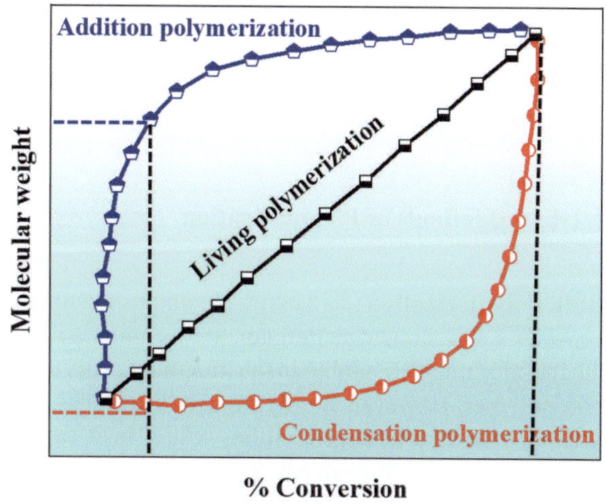

Fig. 1.11 Molecular weight versus % conversion for addition, condensation, and living polymerization

polymerization, causing a rapid growth in molecular weight at a very low conversion. Contrarywise, the condensation polymerization shows reverse analogy. At the end of polymerization, at more than 95% conversion, the rise in molecular weight occurs. Thus, polycondensation is named step-growth polymerization as the extent of increase of molecular weight happens stepwise.

Coordination Polymerization

Coordination polymerization is where the monomers are added to the growing polymer chain with a specific spatial arrangement by a metal complex catalyst. This polymerization technique is highly significant when a tacticity-specified polymer is manufactured. The primary metal complex catalysts are the Ziegler–Natta (Z-N) and the metallocene catalyst. There are several generations of Z-N catalysts available in the polymer industry [14] (Table 1.2). Nevertheless, the main constituents include titanium halide and metal alkyl. As per efficiency and overcoming drawbacks, the functional groups in the Z-N catalyst are modified. Figure 1.12 portrays

Table 1.2 Generations of Ziegler–Natta catalysts

Year	Generation	Chemical constituents
1954–1957	1st	$TiCl_3/Al(C_2H_5)_2Cl$
1968–1970	2nd	Porous $TiCl_3/Al(C_2H_5)_2Cl$
1971–1973	3rd	$TiCl_4/MgCl_2/Al(C_2H_5)_2Cl$
1980	4th	$TiCl_4/MgCl_2/Al(C_2H_5)_2Cl$, phthalate (internal donor), Silane (external donor)
1988–1991	5th	$TiCl_4/MgCl_2/Al(C_2H_5)_2Cl$, 1,3-diether (internal donor), no external donor
1999–2000	6th	$TiCl_4/MgCl_2/Al(C_2H_5)_2Cl$, p-ethoxy esters (internal donor), diether (internal donor)

a typical mechanism of monomer addition to create a stereoregular polymer using Z-N catalyst.

1.2.4.2 Methods of Polymerization

Bulk Polymerization: As a major chemical constituent, bulk polymerization [15, 16] consists of a monomer, an initiator, and a chain transfer agent (Fig. 1.13a). In this case, the initiator must be soluble in the monomer. This technique provides the purest form of a polymer. However, the significant drawbacks include a lack of heat dissipation and uncontrolled heat generation, which lead to gel formation. A gel is a three-dimensional cross-linked network structure; once formed during polymerization, the viscosity shoots up as the reaction progresses, and the end product becomes futile. This unrestrained increase in polymerization rate leading to gel formation is commonly known as auto-acceleration, the Trommsdorff effect, or the gel effect. To prevent gel formation, a chain transfer agent is generally added to terminate the reaction at approximately 65% conversion.

Solution Polymerization: Solution polymerization, on the contrary, consists of a solvent, monomer, initiator, and a chain transfer reagent (optional) (Fig. 1.13b). The solvent ensures better heat dissipation in the system but interferes with chain propagation. The propagating active chain undergoes solvent transfer, leading to immature chain termination. The end product is finally obtained using a non-solvent. Here, at the final stage, solvent evaporation and polymer purification become the challenge. Since most monomers and polymers are soluble in organic solvents, the process of purification and solvent evaporation augments the environment by bringing more volatile organic content (VOC). Henceforth, solution polymerization is not an ideal technique for polymer synthesis [17].

Suspension Polymerization: Suspension polymerization, also known as pearl polymerization or bead polymerization, involves a solvent, liquid monomer (insoluble), initiator, and suspending agent. Being insoluble, the monomer and the solvent

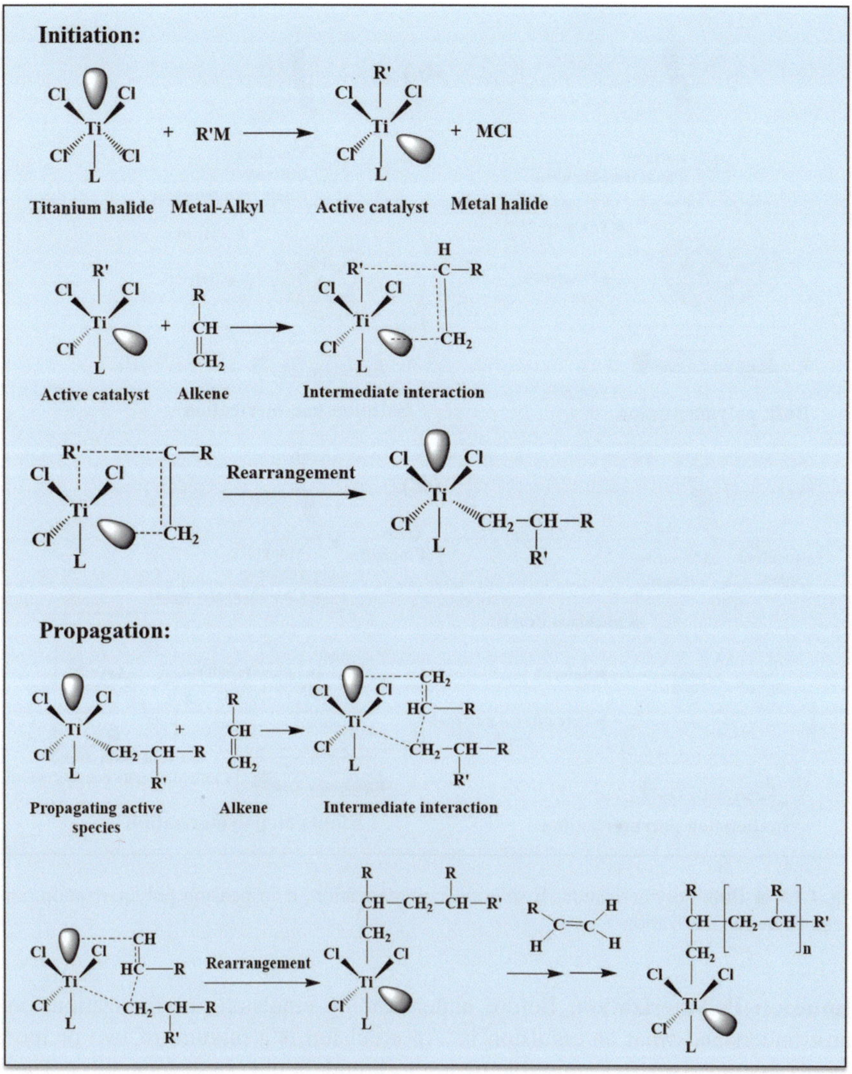

Fig. 1.12 Mechanism of coordination (Ziegler–Natta) polymerization

exist in two different phases. The vigorous agitation breaks the monomer into tiny droplets, and the suspending agent provides essential stability to the colloid-like system. The polymerization happens inside the monomer droplet, and as the reaction progresses, the molecular weight becomes higher, and the polymer beads get precipitated (Fig. 1.13c). The major disadvantage of this technique involves the impurity of polymers [18].

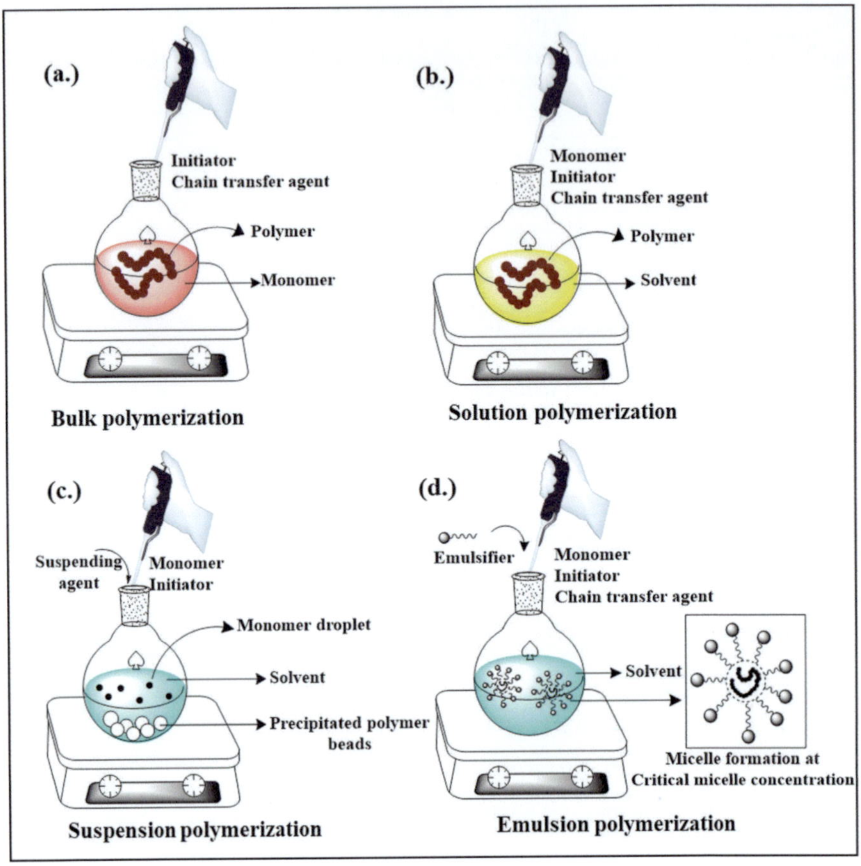

Fig. 1.13 **a** Bulk polymerization, **b** solution polymerization, **c** suspension polymerization, and **d** emulsion polymerization

Emulsion Polymerization: Before understanding emulsion polymerization, one must understand what an emulsion is. An emulsion is a mixture of two or more liquids that are usually immiscible owing to liquid–liquid phase separation. There are two types of emulsion: oil-in-water and water-in-oil. An emulsifier or surfactant is a molecule having both hydrophilic and hydrophobic parts. Thus, it acts as a bridge between two different phases and makes the emulsion stable. Examples of emulsions we use daily are milk, butter, mayonnaise, body lotions, etc. Emulsion polymerization requires water as the primary phase and a water-insoluble monomer as another phase. The surfactant consists of hydrophilic (head) and hydrophobic groups (long chain tail), which are arranged to form micelles (Fig. 1.13d) at a critical micelle concentration (CMC). The monomer, being hydrophobic in nature, enters the micelle, and the entire polymerization takes place inside the micelle. As the process remains uninterrupted from the solvent transfer, this technique provides a very high molecular weight polymer. In the end, the polymer is recovered by adding an electrolyte solution (aqueous solution of sodium or potassium chloride) and washing it with an adequate solvent. In inverse emulsion polymerization, organic solvent is used as the primary phase, and the water-soluble monomer is taken as the other phase. Polyacrylamide is synthesized using inverse emulsion polymerization [19, 20].

1.3 Architecture-Controlled Polymerization

The above-discussed polymerization methods give a broad molecular weight distribution in an uncontrolled way. To synthesize the precise control polymer or to architect a polymer like a star or a dumbbell, block copolymer, etc., as shown in Fig. 1.14, one should do the synthesis in an architecturally controlled polymerization manner. Several living or controlled polymerization techniques, such as ring-opening polymerization, controlled radical polymerization, reversible addition/fragmentation chain transfer polymerization, atom transfer radical polymerization, and nitroxide-mediated polymerization, are used to design the structure of a polymer precisely or to manipulate the growth of polymer chain [21]. The monomers, initiators, and reaction parameters must be properly selected to achieve the desired molecular design. It offers significant benefits such as low/narrow polydispersity, precise molecular weight control, tailored architecture design, functional group incorporation, wide monomer compatibility, and improved material properties.

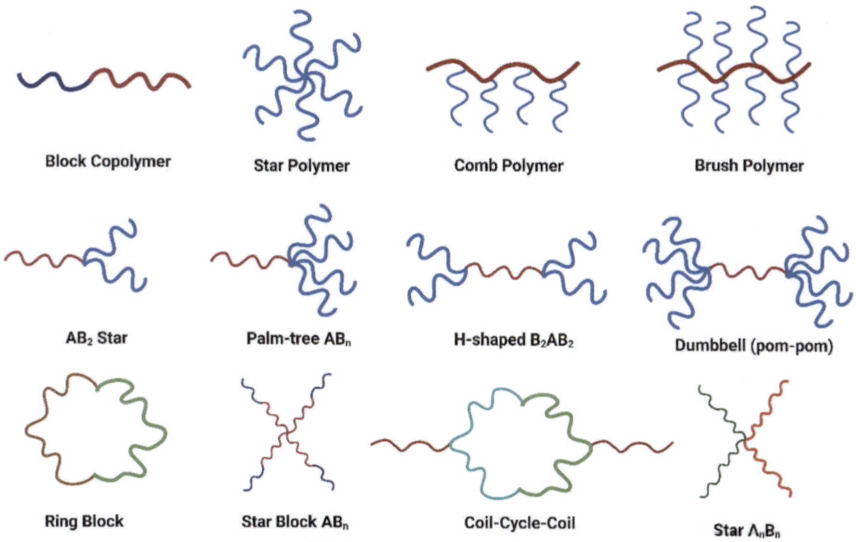

Fig. 1.14 Controlled architecture of polymers

1.3.1 Ring-Opening Polymerization

Ring-opening polymerization (ROP) comes under chain growth polymerization, which gives a longer polymer chain. The ring opening occurs due to the strain in ring (inherent instability or high energy state of a cyclic monomer due to distorted bond angles) of the monomers (Fig. 1.15) to form the reactive center. These activated monomers are added to the growing chain end to produce a long polymer chain [22]. The most favorable rings are typically 3-, 4-, and smaller-sized rings due to their high ring strain, while larger rings like 6-membered rings generally have minimal ring strain and therefore are less likely to polymerize via ring opening [23]. The thumb rule here is that a higher ring strain means easier polymerization. According to the present reactive center in the polymer chain, ROP has cationic ROP, anionic ROP, coordination ROP, radical ROP, and ring-opening metathesis polymerization (ROMP) types. These types are studied separately due to their unique mechanism during polymerization. The summary of different kinds of ROP with mechanisms, typical monomers, and key features is discussed in Table 1.3.

Mechanism of ε-caprolactam (Fig. 1.16)

Initiation. The water molecule initiates the reaction by attacking the carbonyl group of ε-caprolactam. It opens the ring and forms ω-aminocaproic acid. The formed ω-aminocaproic acid protonates another molecule of the ε-caprolactam, which generates a protonated monomer with carboxylate anion. This protonated monomer is more vulnerable to nucleophilic attack.

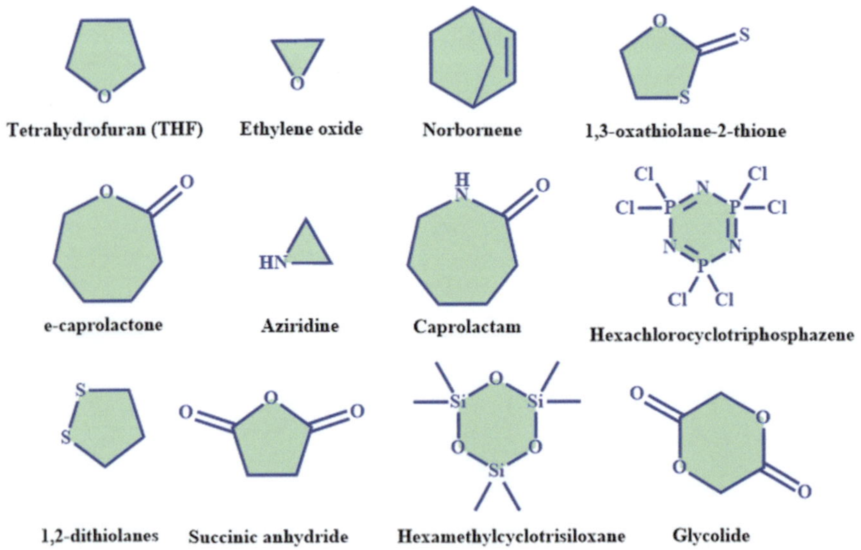

Fig. 1.15 Cyclic monomers

Table 1.3 Different types of ROP [25, 26]

Type of ROP	Mechanism	Typical monomers	Key features
Anionic ROP	Nucleophilic attack	Epoxides, aziridines, episulfides, lactones	– Initiated by nucleophilic reagents – Can be living polymerization – Example: ε-caprolactone initiated by alkoxide
Cationic ROP	Electrophilic attack	Lactones, lactams, amines, ethers	– Proceeds through SN1 or SN2 propagation – Affected by cationic species stability – Can be living polymerization – Terminated by nucleophilic reagents
Radical ROP	Free radical mechanism	Vinyl substituted cyclic monomers; methylene substituted cyclic monomers	– Produces polymers with functional groups in the backbone – Can compete with vinyl chain polymerization
Ring-Opening Metathesis Polymerization (ROMP)	Organometallic catalysis	Cycloalkenes, bicycloalkenes	– Produces unsaturated polymers – Requires organometallic catalysts – Follows olefin metathesis pathways
Coordination-insertion ROP	Metal-catalyzed	Lactones, lactides	– Often uses metal alkoxides or carboxylates as catalysts – Can provide control over molecular weight and stereochemistry

Propagation. The carboxylate anion attacks on the protonated monomer. This attack opens the ring of the monomer and forms a dimer. This formed dimer has two ends: amino group and carboxylate group. This process gets repeated, and carboxylate ends attacks on another protonated molecule, thus extending the chain by inserting one more repeating unit. In this process, the propagating chain can transfer a proton to the monomer.

Termination. Propagating chains can react with each other and terminate growth. This reaction specifically is the reaction between carboxyl group (–COOH) of one

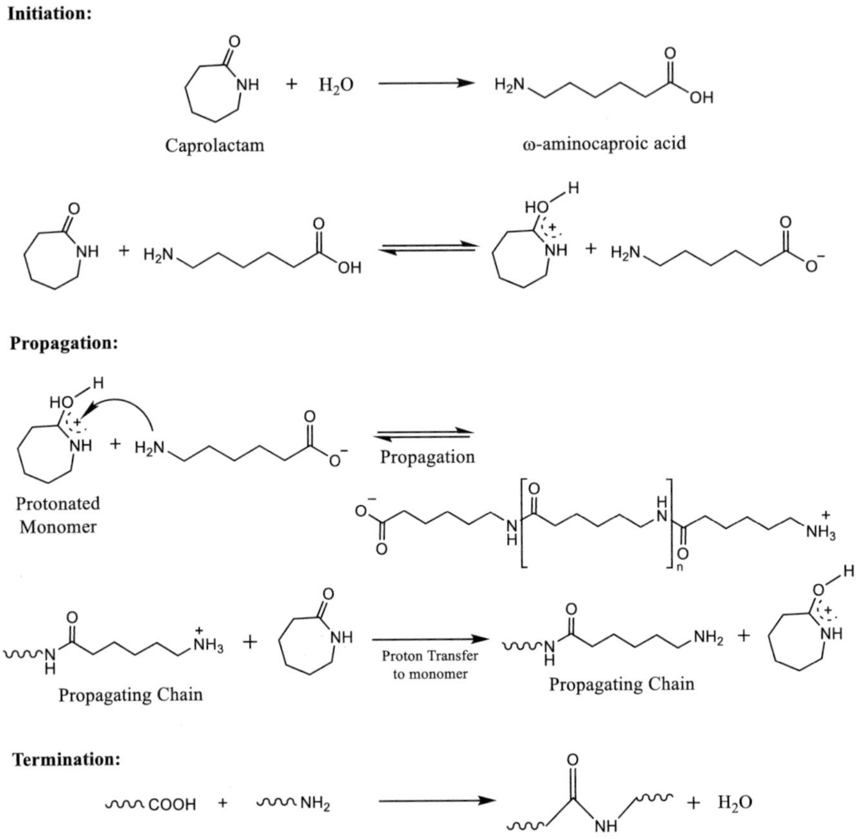

Fig. 1.16 Hydrolytic ring-opening polymerization mechanism of ε-caprolactam [26]

chain with an amino group (-NH_2) of another chain, forms amide bond and releases water molecule (H_2O) [24, 25].

ROP has number of advanced applications across biomedical materials, functional materials, advanced polymer architectures, sustainable materials and specialty applications. Biocompatible and biodegradable materials for medical applications can be developed using ROP for drug delivery systems, tissue engineering scaffolds and bioabsorbable sutures and implants. For developing eco-friendly polymers such as (bio)degradable materials to address environmental concerns. Specialty applications such as low shrinkage materials for dental and optical applications.

1.3.2 Atom Transfer Radical Polymerization

Atom Transfer Radical Polymerization (ATRP) technique is a type of controlled radical polymerization that utilizes a transition metal catalyst to reversibly activate and deactivate the propagation polymer chains (reversible electrochemical reaction) [27]. This activation and deactivation during the propagation helps us for precise control on the molecular weight of polymer, architecture and narrow distribution of molecular weight. This technique allows to regulate the active radical's concentration through a dynamic equilibrium between "dormant" and "active" species. The key techniques during synthesis in ATRP are appropriate selection of initiators, catalysts (generally copper complexes with ligands), optimizing reaction conditions, and monitoring the polymerization progress to ensure controlled polymer chains growth. The monitoring of the reaction can be done using NMR, GPC and UV techniques [27, 28]. ATRP is also considered as "living polymerization" where the ability of a growing polymer chain to terminate has been removed.

Mechanism

Initiation. The initiation step generates radical species to begin the polymerization process. It consists of activation and radical initiation of monomers' reactions.

- Activation: The initiator generally alkyl halide (R–X) reacts with a transition metal catalyst (Mt^n/L). The halogen group form the initiator get removed and forms highly active alkyl radical ($R^•$), while transition metal complex oxidized to a higher oxidation state ($X–Mt^{(n+1)}/L$) from lower oxidation **state**.

$$R{-}X + Mt^n/L \rightleftharpoons R^\bullet + X{-}Mt^{(n+1)}/L$$

- Radical Initiation of Monomers: The formed active alkyl radical ($R^•$) reacts with the first monomer (M) species in the reaction vessel and forms an activated propagating radical species ($P_n^\bullet$). A fast initiation step ensures that the polymer has narrow molecular weight distributions.

$$R^\bullet + M \rightarrow P_n^\bullet$$

Propagation. The growing polymer chain stays in its radical (active) form once the first monomer unit (M) is added, so enabling the addition of extra monomers in continuation. The repeatedly monomer addition during the propagation step grows chain with active radical activity.

$$P_n^\bullet + M \rightarrow P_{n+1}^\bullet$$

This process continues until the equilibrium between active and dormant chains.

Dynamic Equilibrium: Activation-Deactivation Cycle. This step helps to achieve controlled polymerization. Dynamic equilibrium occurs in between the active propagating radicals ($P_n^\bullet$) and dormant species (P_n–X).

- Deactivation (Reversible Halogen Transfer): The oxidized metal complex (X–$Mt^{(n+1)}$/L) combines with the developing polymer radical ($P_n^\bullet$). This returns the halogen (X) to the polymer radical such that it becomes a dormant polymer species (P_n–X). The transition metal simultaneously is lowered back to its lower oxidation state (Mt^n/L), so regaining its catalytic activity again.

$$P_n^\bullet + X{-}Mt^{(n+1)}/L \rightleftharpoons P_n{-}X + Mt^n/L$$

- Re-activation of Dormant Chains: Some present dormant chains (P_n–X) can be again reactivated and get back into the radical form ($P_n^\bullet$) back by reacting with the reduced metal catalyst (Mt^n/L) due to equilibrium nature of ATRP. This reactivation ensures that all growing polymer chains have the same rate of growing.
- Overall Equilibrium:

$$P_n{-}X + Mt^n/L \rightleftharpoons P_n^\bullet + X{-}Mt^{(n+1)}/L$$

Equilibrium constant ($K_{eq} = k_a/k_d$) controls the balance between activation and deactivation, and it has to be optimized for each polymer synthesis process.

Termination Phase. (Uncontrolled Side Reactions). A small fraction of permanently inactive polymer chains due to some radical terminations are still present in the ATRP although it is designed to minimize termination. Two primary ways are there to occur termination:

- Combination (Coupling): The formed two radical species ($P_n^\bullet$ and $P_m^\bullet$) combine with each other to form a single long polymer chain. This chain has double the polymer molecular weight than a normal ATRP chain.

$$P_n^\bullet + P_m^\bullet \rightarrow P_n{-}P_m$$

- Disproportionation: One formed polymer radical chain donates its hydrogen atom to another formed radical chain. This gives one saturated polymer chain (P_n–H) and one unsaturated polymer chain (P_m =)

$$P_n^\bullet + P_m^\bullet \rightarrow P_n{-}H + P_m =$$

Generally, termination reaction in the ATRP is not desired because they lower the polymerization control and generate dead chains [29–32]. The number-average degree of polymerization (DP_n) in ATRP follows the equation [33]:

$$DP_n = \frac{[\text{Monomer}]_0 - [\text{Monomer}]}{[\text{Initiator}]_0}$$

where $[\text{Monomer}]_0$—Initial monomer concentration, [Monomer]—Remaining monomer concentration and, $[\text{Initiator}]_0$—Initial initiator concentration.

ATRP has broad applications and is widely used in biomedicinc, nanotechnology, coatings, adhesives, and electronics sectors.

1.3.3 Reversible Addition/Fragmentation Chain Transfer Polymerization

Reversible addition/fragmentation chain transfer polymerization (RAFT) is also a controlled radical polymerization technique which utilize the utilizing a chain transfer agent (CTA) to regulate polymer growth. Chain transfer agent (RAFT agent) enables a reversible "on-and-off" cycle between active and dormant polymer chains during polymerization. The RAFT agent is a molecule containing a thiocarbonylthio group with substituents R and Z that reversibly interacts with the growing polymer chain, acting as a "mediator" for chain transfer. Due to the reversible nature of the chain transfer, RAFT polymerization can be considered "living" which means chains can be extended by adding more monomers later in the reaction. Monomers for reversible addition-fragmentation chain-transfer (raft) polymerization includes (meth)acrylates, (meth)acrylamides, acrylonitrile, styrenes, dienes, vinyl monomers and fatty acids. The classes of RAFT agents include dithioesters, dithiocarbamates, trithiocarbonates, and xanthates. Few raft agents are dithioesters, dithiocarbamates, xanthates, trithiocarbonates. The choice of RAFT agent depends on the monomer's reactivity. The general structure of RAFT agent and its classes are shown in Fig. 1.17 [34–36].

Mechanism

Initiation. The decomposition of an initiator (e.g., AIBN) to generate two primary radicals ($R^{\bullet}$) from 1 molecule of initiator at room temperature or on heating depends on the decomposition temperature of an initiator.

$$\text{AIBN} \rightarrow 2\text{R}^{\bullet}$$

This generated primary radical reacts with the monomer (M) to form monomer radical ($P_n^{\bullet}$) which is propagating radical.

$$\text{R}^{\bullet} + \text{M} \rightarrow \text{P}_1^{\bullet}$$

$$\text{P}_1^{\bullet} + \text{M} \rightarrow \text{P}_2^{\bullet}$$

$$\text{P}_n^{\bullet} + \text{M} \rightarrow \text{P}_{n+1}^{\bullet}$$

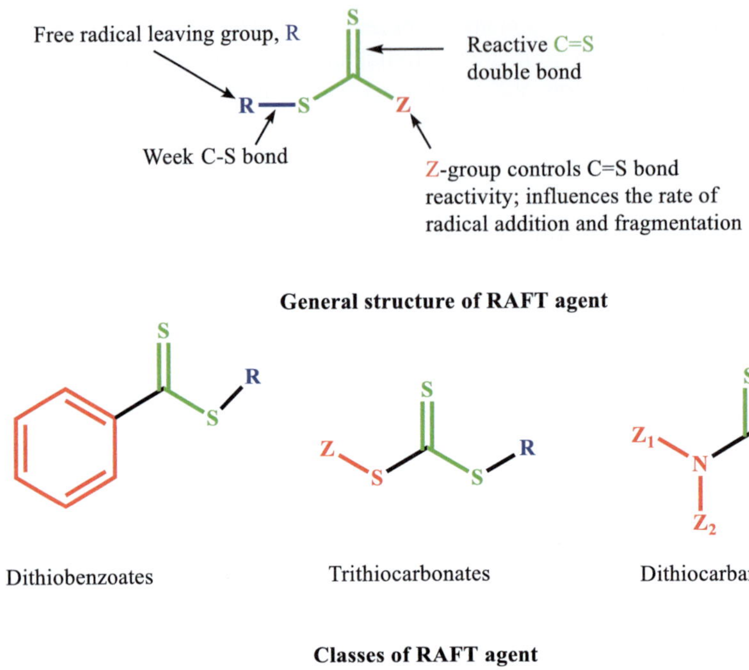

Fig. 1.17 General structure of RAFT agent and its classes

Addition to RAFT agent. The formed radical polymer chain ($P_n^{\bullet}$) gets added into the RAFT agent to form the RAFT intermediate radical. The Z-group in the RAFT agent controls radical stability, while the R-group is a good leaving group.

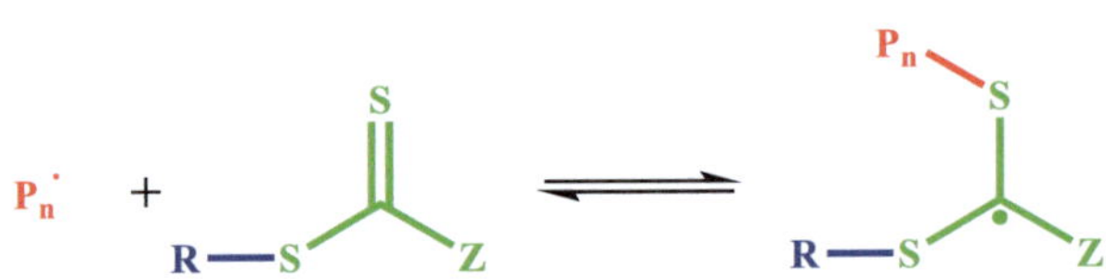

Fragmentation. The formed RAFT intermediate radical fragments transfer the polymer growing chain P_n to the leaving group- R. Thus, it forms a new propagating radical $R^{\bullet}$.

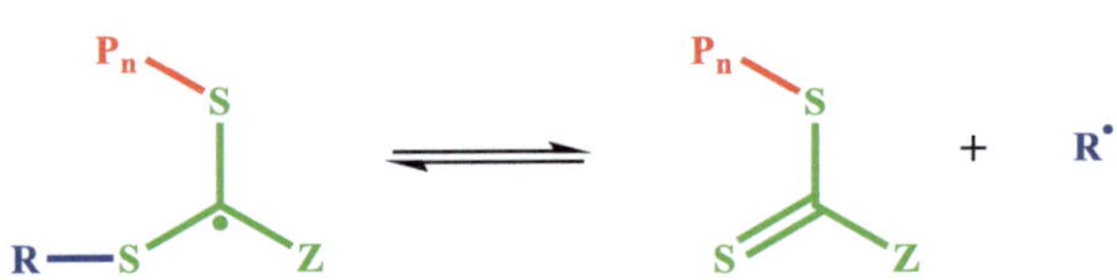

Now, the polymerization of fresh monomer molecules is initiated by $R^{\bullet}$.

$$R^{\bullet} + M \longrightarrow P_m^{\bullet}$$

Propagation. The new propagating radical $P_m^{\bullet}$ continues to react with the remaining monomer units. Also, at the same time, the RAFT equilibrium is maintained, i.e., the polymer chains are continuously transferred between the dormant and active states.

The dormant RAFT-capped polymer can undergo further chain transfer reactions, maintaining the equilibrium. This step is essential, ensuring controlled chain growth and preventing uncontrolled termination during the polymerization.

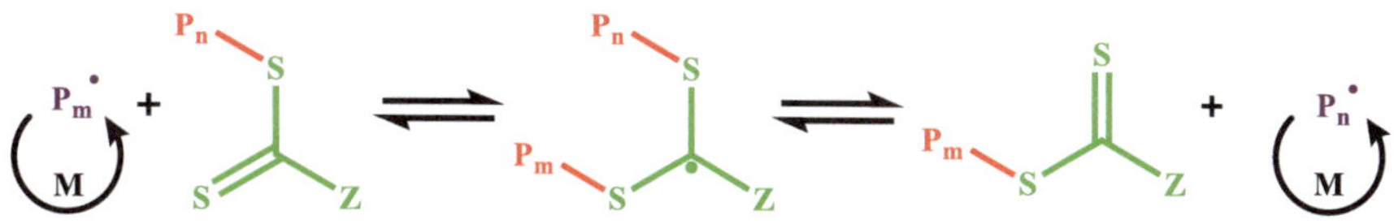

Termination. Though RAFT helps to minimize the termination, radical termination via combination or disproportionation can still occur [37].

$$R^{\bullet}\ R^{\bullet}\ R^{\bullet}\ R^{\bullet} \longrightarrow \textbf{Termination}$$

This technique is widely used to create complex polymer architectures like block copolymers, graft copolymers, and functionalized polymers having precise control over their properties.

1.3.4 Nitroxide-Mediated Polymerization

Nitroxide-mediated polymerization (NMP) is also a controlled/living radical polymerization technique that uses nitroxide radicals to synthesize polymers with well-controlled stereochemistry and narrow molecular weight distribution which helps to produce multi-block copolymers [38]. NMP is historically the first and represents perhaps the most straightforward controlled/living radical polymerization technology to apply among the other controlled techniques. In NMP, alkoxyamines are the initiators and controlling agents. It provides the reactive initiating radical and the stable, mediating nitroxide radical for controlled radical polymerization. The nitroxide radical acts as a controlling agent by reversibly trapping the growing polymer radicals, which helps prevent chain termination and allows for controlled chain growth. The most common example of alkoxyamine used in this technique is 2,2,6,6-tetramethyl-1-piperidinyloxy (TEMPO). The NMP technique is metal-free and not colored, and the polymer does not require any purification after synthesis.

Mechanism

Initiation. A radical initiator (I) decomposes upon heating and generates a reactive free radical ($R^{\bullet}$). This free radical reacts with the monomer (M), form a polymer radical RM•.

$$I \xrightarrow{\Delta} R^{\bullet} + R_2NO^{\bullet}$$

$$R^{\bullet} + M \longrightarrow RM^{\bullet}$$

Propagation. The growing polymer radical ($P_n^{\bullet}$) reacts with other monomers to form a polymer chain.

$$P_n^{\bullet} + M \longrightarrow P_{n+1}^{\bullet}$$

Reversible Termination. The propagating polymer radical ($P_n^{\bullet}$) can undergo a reversible reaction with the nitroxide radical ($R_2NO^{\bullet}$) to form a dormant alkoxyamine species and maintain equilibrium between dormant and active species. Thus, it prevents premature termination while allowing controlled growth.

$$P_n^{\bullet} + R_2NO^{\bullet} \rightleftharpoons P_n\text{-}ONR_2$$

Chain Extension and Block Copolymer Formation. The alkoxyamine can regenerate the active radical upon heating, which helps to introduce another monomer (M') into the polymer chain.

$$P_n\text{-}ONR_2 \xrightarrow{\Delta} P_n^{\bullet} + R_2NO^{\bullet}$$

$$P_n^{\bullet} + M' \longrightarrow P_{n+m}^{\bullet}$$

The main limitation of NMP is the availability of a range of monomers that can be effectively controlled. Some alkoxyamines and nitroxides can control most of the conjugated vinyl monomers. Some of these double-bonded polymers are styrene and derivatives, acrylates (including some functional acrylates), acrylamides, acrylonitrile, and methacrylates (with some limitations), and also some dienes such as isoprene [29, 39].

1.4 Miscellaneous

- Conductivity: Based on electrical conductivity polymers are divided into the following categories:
  (a) Non-conductive/Insulating Polymers: Polymers having very low electrical conductivity (in the range of 10^{-14} to 10^{-8} S/cm) and lack free electrons or charge carriers. Example: Polyethylene (PE), Polypropylene (PP), Polystyrene (PS), PVC, Teflon (PTFE).
  (b) Semi-conductive polymers: Polymers having conductivity within the range of 10^{-8} to 10^{2} S/cm, due to doping with electron donors or acceptors. Example: Polyaniline, Polypyrrole, Polyacetylene.

(c) Conductive polymers: Polymers with high electrical conductivity up to 10^3 S/cm, comparable to that of metals. Their structure includes conjugated double bonds which enable delocalization of electrons along the polymer backbone. Example: Doped Polyacetylene.

- Biodegradability: Polymers can be divided into two main categories based on their ability to degrade through biological processes [40]:

  (a) Biodegradable Polymers: These are non-toxic, eco-friendly polymers that can be broken down by natural biological processes, typically involving microorganisms such as bacteria, fungi, or enzymes. Examples: biopolymers, Polylactic acid, Polycaprolactone, Polyhydroxyalkanoates, Polyglycolic acid.

  (b) Non-biodegradable Polymers: These polymers resist degradation by biological agents and can survive in the environment for hundreds of years. They accumulate in landfills and ecosystems, contributing to long-term pollution. Examples: Polyethylene (PE), Polypropylene (PP), Polystyrene (PS).

- End-use: polymers can also be classified based on their end-use applications, where the classification is linked to the specific applications like general purpose, industries, engineering, etc. For example, diene polymers are primarily used in the rubber industry, Polyvinyl, polystyrene and polyurethane in Pipes, construction industry (Pipes, window frames, insulation, adhesives, sealants), Polylactic acid, Polycaprolactone, Silicone, Polyethylene glycol in medical and pharmaceutical industries (Biodegradable sutures, drug delivery systems, prosthetics, medical tubing), Acrylic polymers in pain industries, Nylon, polyester in textile industries, Polyimides, Epoxy resins, Polyethylene, Polyvinylidene fluoride in electrical industries, etc.

1.5 Polymer Processing

Polymer processing is defined as an "engineering activity concerned with operations carried out on polymeric materials or systems to increase their utility". The set of techniques is used to transform raw polymer materials into finished products with specific shapes, structures, and properties. Detailed methods are discussed (Table 1.4) and illustrated (Fig. 1.18) [41–44].

Table 1.4 Polymer processing techniques

Processing techniques	Process	Type of products	Polymers used
Compression molding	Raw material in the form of polymer granules, molding powder, or the preform These are placed in the preheated cavity of the mold, and close the mold with the desired amount of compression. Heat and pressure ensure a good flow of the resin or the polymer, filling all the intricate parts of the mold cavity. The mold has to be cooled well below the glass transition temperature (Tg) or the heat distortion temperature (HDT) of the polymer before the mold is opened (for thermoplastics) and for thermosets, the product must be ejected at hot condition Poor heat transmission, and slow process	Car panels, fenders, engine components, and interior components Kitchen tools, utensils, appliances, bowls, cups, and plates, and mirror handles Medical syringe stoppers, respirator masks, and rubber clothing	Thermoplastic, thermosets, rubbers, and composites
Transfer molding	The polymer is melted in a separate chamber, called the "plasticizing pot" or the "transfer pot". It is then forced into the mold cavity through narrow, heated channels by applying ram pressure. The mold is kept at a high temperature to cause rapid curing of the resin system Molding cycles are usually shorter because of high temperatures and pressures. Molding of large and sophisticated parts is possible. Uniform curing with minimum shrinkage. Products with high density and mechanical properties are possible	Integrated circuits, plugs, connectors, pins, coils, and studs	Thermoplastic, thermosets, rubbers, and composites

(continued)

Table 1.4 (continued)

Processing techniques	Process	Type of products	Polymers used
Injection molding	Molten thermoplastic is injected into a steel mold cavity under high pressure, a barrel of suitable length pushes the molten polymer to move along its length through successive zones of different temperatures under the screw/ram pressure. The rotation of the screw pumps the molten polymer forward followed by the backward movement of the screw to accumulate enough material to fill the mold. Then it moves forward axially along with the polymer melt like an injection plunger to fill the mold under high injection pressure Reaction injection molding (RIM): Injection of low-viscosity monomers or low polymers into the mold. Suitable for the production of polyurethane by rapid injection of a diol and a diisocyanate into an impingement chamber where they react with each other and fill the mold at a relatively low pressure. Polymerization occurs inside the mold under the influence of catalyst, heat, and pressure. It allows the production of solid molded objects from liquid monomers and low-viscous oligomers	Plastic bottles, bottle caps, legos and toys, computers and other consumer electronics, kitchenware, plastic furniture, flower pots, toilet seats, surgical devices, jewel boxes Molded closures, containers, components, and drinkware Syringes, surgical instruments, and diagnostic devices, etc.	Thermoplastic, thermosets, rubbers, and composites Most commonly used polymers are: acrylonitrile butadiene styrene (ABS), polycarbonate (PC), nylon, polyethylene (PE), polypropylene (PP), etc.
Blow molding	Steps involved: 1. Forming of a molten "parison" usually by extrusion 2. Compressed air is used to blow the molten plastic into the configuration of the mold 3. Plastic mass quickly solidifies coming in contact with the cold mold	Hollow articles such as bottles	The most commonly used polymers are polyethylene (PE) and polypropylene (PP)

(continued)

Table 1.4 (continued)

Processing techniques	Process	Type of products	Polymers used
Extrusion	The polymer melt is propelled by a rotating screw along the extruder and a metal die at the other end shapes the melt into desired form The screw is divided into several zones. In the feed zone, the polymer granules or masterbatch are poured from the hopper and pushed forward into the plasticizing zone where the granules melt under the influence of shearing action and external heat. The molten mass then passes through a compression zone to the metering zone. The metering zone helps maintain a uniform flow rate at the die end and pressurizes the molten mass out of the die through the screen pack and breaker plate assembly. The design of types of extruders varies according to the polymer used. The design variables are the number of screws, the length-to-diameter (L/D) ratio, the screw geometry including its pitch, flight, and their variations along its length, and the die. The shape of the product depends on the shape of the die Various coated or laminated products can also be prepared by extrusion	Pipes, tubing, films, and coatings	Polyethylene (PE), polypropylene (PP), polyvinyl chloride (PVC), acrylonitrile butadiene styrene (ABS), and polyamide
Calendaring	For fast and continuous formation of sheet or film The polymer passes through the nips between the heated rolls to give a product of uniform thickness. The first nip controls the feed rate while the second and the third control the thickness of the product	Rubber sheets, tires, rubber flooring, paper sheets, plastic sheets, adhesive tapes	PVC, Polyethylene (PE), polypropylene (PP), acrylonitrile-butadiene-styrene (ABS), and thermosets
Rotomolding	Use to make large, hollow objects, even of complex shapes The polymer in powdered form is placed inside the heated, rotating mold. The polymer melts and coats the inside of the mold. When the polymer melts completely and sticks to the interior surface of the mold uniformly, it is cooled and solidified. Thus, it replicates the shape and texture of the mold	Water storage tanks, kayaks, road barriers, playground equipment, industrial containers, automotive parts, bins, toys, sports equipment, baby cribs, display mannequins, and fuel cells	Crystalline polymers such as nylon or "Terylene" or PET polyester

(continued)

Table 1.4 (continued)

Processing techniques	Process			Type of products	Polymers used
For fibers Spinning	**Melt spinning**: Simplest, requires no solvents or non-solvents. No need for solvent evaporation, precipitation, or coagulation and drying Specifically, for polyamide, polyester, and polyolefin (polypropylene) fibers	**Dry Spinning**: Polymer solution is spun in a heated vertical chamber and the solvent is removed by evaporation Specifically, for cellulose acetate	**Wet Spinning**: Solvent is used for spinning and leached out by non-solvent. Fibers are coagulated as soon as the spun solution impinges into the coagulating bath (non-solvent medium). Chemical reaction may be involved in the coagulating bath to convert the derivative into final polymer Specifically, for viscose rayon, poly (vinyl chloride) and its copolymers and polyacrylonitrile (acrylic fibers)	Yarn, filaments, fibers	Notable examples are polyamides, polyesters, polyolefins, polyurethanes, polyvinyl chloride (PVC), and polybenzimidazole
Rubber Internal mixtures	Two horizontal rotors with protrusions are placed in a casing Feed door and hopper at the top and a discharge door at the bottom. The rubber is mixed between the two rotors and between each rotor and the body of the casing. The mastication takes place over a wide area. So, better mixing, and takes less time				Natural rubber (NR), styrene-butadiene rubber (SBR), nitrile rubber (NBR), butyl rubber (IIR), ethylene propylene diene monomer (EPDM), and polychloroprene (neoprene)

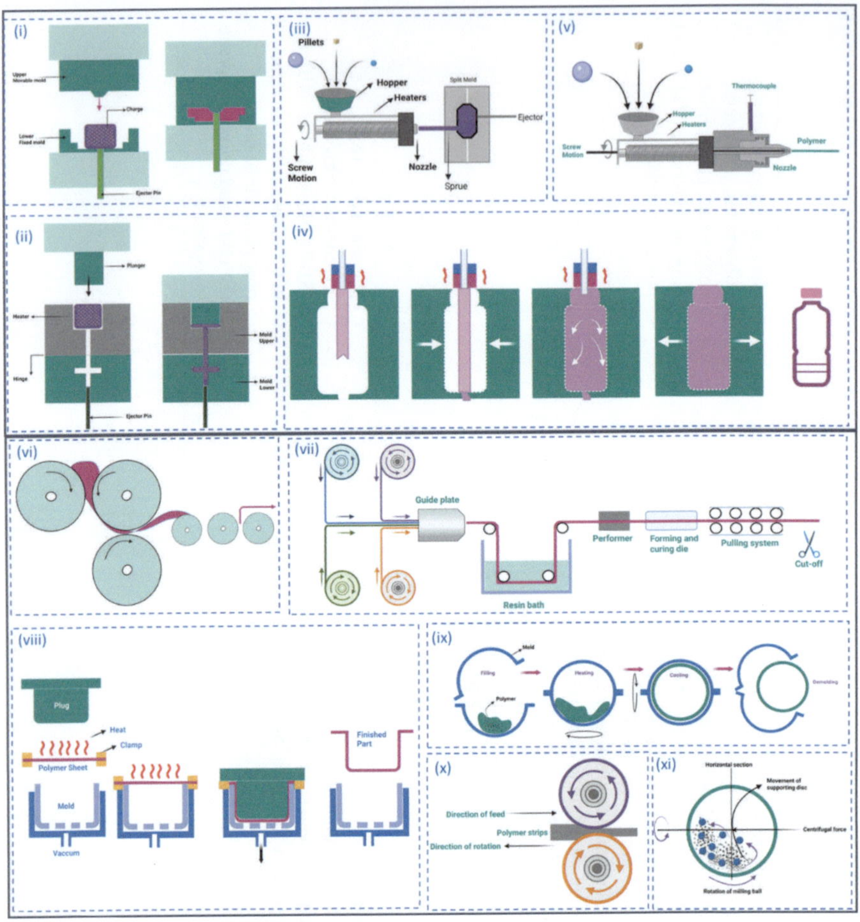

Fig. 1.18 Polymer processing techniques **i** compression molding **ii** transfer molding **iii** injection molding **iv** blow molding **v** extrusion **vi** calendaring **vii** pultrusion **viii** thermoforming **ix** rotomolding **x** two-roll mill **xi** ball mill

1.6 Concept of polymer characterization

Polymer characterization involves the analytical study of polymers to determine their chemical composition, molecular structure, morphology, and physical properties. This process is essential for understanding the **behavior** and performance of polymers, optimizing manufacturing processes, ensuring quality control, and developing new materials.

- Key aspects of polymer characterization include examining the morphology (such as crystalline or amorphous phases), **analyzing** chemical composition, assessing molecular weight and structure, and evaluating physical properties like thermal stability, mechanical strength, and solubility.
- Common techniques used in polymer characterization include spectroscopy (e.g., FTIR, NMR), chromatography (e.g., GPC), microscopy (e.g., SEM, TEM), and thermal analysis (e.g., DSC, TGA).

1.7 Conclusions

This chapter offers a foundational understanding of polymers by tracing their historical development, elucidating core concepts, and exploring their diverse classifications. It outlines key polymerization techniques and highlights various processing methods essential to industrial applications. The chapter concludes with a conceptual overview of polymer characterization, setting the stage for deeper exploration of analytical techniques in subsequent chapters.

Acknowledgements We have used bioRender.com (student licence) for making illustration.

Conflict of Interest Authors declare no competing conflict of interest.

References

1. Griskey RG (1995) Polymer basics, structural characteristics, characterization. In: Griskey RG (ed) Polymer process engineering. Springer, Netherlands, Dordrecht, pp 1–47
2. Suresh B, Subhranjali R, Jeebanjyoti M et al (2023) Polymer chemistry: an overview
3. Batra S (2025) Polymer chemistry essentials. Educohack Press
4. Causin V, Causin V (2015) Polymers: an overview. Polymers on the crime scene: forensic analysis of polymeric trace evidence, pp 9–52
5. Saldívar-Guerra E, Vivaldo-Lima E (2013) Introduction to polymers and polymer types. Handbook of polymer synthesis, characterization, and processing, pp 1–14
6. Sperling LH (2015) Introduction to physical polymer science. John Wiley & Sons
7. Obi BE (2018) Polymer chemistry and synthesis. Polymeric foams structure-property-performance, pp 17–40. https://doi.org/10.1016/B978-1-4557-7755-6.00002-1
8. Wu L, Rondon B, Dym S et al (2023) Regulating cationic polymerization: From structural control to life cycle management. Prog Polym Sci 145:101736. https://doi.org/10.1016/J.PROGPOLYMSCI.2023.101736
9. Umoren SA, Solomon MM, Saji VS (2022) Basic polymer concepts I. Polym Mater Corros Inhib 3–48. https://doi.org/10.1016/B978-0-12-823854-7.00007-2
10. Fukukawa K, Ueda M (2012) Sequence control in one-step polycondensation. Polym Sci Compre Ref 1–10(1–10):71–93. https://doi.org/10.1016/B978-0-444-53349-4.00134-5
11. Freger V, Ramon GZ (2021) Polyamide desalination membranes: formation, structure, and properties. Prog Polym Sci 122:101451. https://doi.org/10.1016/J.PROGPOLYMSCI.2021.101451

12. Wagh SJ, Dhumal SS, Suresh AK (2009) An experimental study of polyurea membrane formation by interfacial polycondensation. J Memb Sci 328:246–256. https://doi.org/10.1016/J.MEMSCI.2008.12.018
13. Heintz AM, Duffy DJ, Hsu SL et al (2003) Effects of reaction temperature on the formation of polyurethane prepolymer structures. Macromolecules 36:2695–2704. https://doi.org/10.1021/ma021559h
14. ZIEGLER-NATTA CATALYSTS
15. Abdel-Alim AH, Hamielec AE (1972) Bulk polymerization of vinyl chloride. J Appl Polym Sci 16:783–799. https://doi.org/10.1002/app.1972.070160321
16. Balke ST, Hamielec AE (1973) Bulk polymerization of methyl methacrylate. J Appl Polym Sci 17:905–949. https://doi.org/10.1002/app.1973.070170320
17. Huskić M, Žigon M (2007) PMMA/MMT nanocomposites prepared by one-step in situ intercalative solution polymerization. Eur Polym J 43:4891–4897. https://doi.org/10.1016/J.EURPOLYMJ.2007.09.009
18. Brooks B (2010) Suspension polymerization processes. Chem Eng Technol 33:1737–1744
19. Lovell PA, Schork FJ (2020) Fundamentals of emulsion polymerization. Biomacromol 21:4396–4441
20. Wang Q, Fu S, Yu T (1994) Emulsion polymerization. Prog Polym Sci 19:703–753. https://doi.org/10.1016/0079-6700(94)90031-0
21. Samakande A, Sanderson RD, Hartmann PC (2007) Synthesis and characterization of novel quaternary ammonium RAFT agents. Synth Commun 37:3861–3872. https://doi.org/10.1080/00397910701568751
22. Hasannia M, Aliabadi A, Abnous K et al (2022) Synthesis of block copolymers used in polymersome fabrication: application in drug delivery. J Control Release 341:95–117. https://doi.org/10.1016/J.JCONREL.2021.11.010
23. Nuyken O, Pask SD (2013) Ring-opening polymerization—an introductory review. Polymers 5:361–403. https://doi.org/10.3390/POLYM5020361
24. Fang X, Simone CD, Vaccaro E et al (2002) Ring-opening polymerization of ϵ-caprolactam and ϵ-caprolactone via microwave irradiation. J Polym Sci A Polym Chem 40:2264–2275. https://doi.org/10.1002/POLA.10306
25. Penczek S, Pretula J, Slomkowski S (2021) Ring-opening polymerization. Chem Teach Int 3:33–57. https://doi.org/10.1515/cti-2020-0028
26. Nuyken O, Pask S (2013) Ring-opening polymerization—an introductory review. Polymers (Basel) 5:361–403. https://doi.org/10.3390/polym5020361
27. Zhang T (2023) Visible light-mediated polymerization for protein-polymer conjugation: a biocompatible tool for therapeutic applications
28. Gibson VC, Marshall EL (2003) Metal complexes as catalysts for polymerization reactions. In: Comprehensive coordination chemistry II. Elsevier, pp 1–74
29. Tsarevsky NV, Matyjaszewski K (2007) Green:atom transfer radical polymerization: From process design to preparation of well-defined environmentally friendly polymeric materials. Chem Rev 107:2270–2299. https://doi.org/10.1021/CR050947P/ASSET/IMAGES/LARGE/CR050947PH00013.JPEG
30. Matyjaszewski K, Xia J (2001) Atom transfer radical polymerization. Chem Rev 101:2921–2990. https://doi.org/10.1021/cr940534g
31. Krishna Rao KSV, Madhusudana Rao K (2012) A review on radical polymerization used for design and development of biomaterials. New Developments. Nova Science Publishers Inc., Radical Polymerization
32. Kato M, Kamigaito M, Sawamoto M, Higashimura T (1995) Polymerization of methyl methacrylate with the carbon tetrachloride/dichlorotris- (triphenylphosphine)ruthenium(ii)/methylaluminum bis(2,6-di-tert-butylphenoxide) initiating system: possibility of living radical polymerization. Macromolecules 28:1721–1723. https://doi.org/10.1021/MA00109A056
33. Matyjaszewski K, Xia J (2001) Atom transfer radical polymerization. Chem Rev 101:2921–2990. https://doi.org/10.1021/CR940534G/ASSET/IMAGES/LARGE/CR940534GF00014.JPEG

34. Mayadunne RTA, Rizzardo E, Chiefari J et al (1999) Living radical polymerization with reversible addition-fragmentation chain transfer (RAFT polymerization) using dithiocarbamates as chain transfer agents. Macromolecules 32:6977–6980. https://doi.org/10.1021/MA9906837/ASSET/IMAGES/LARGE/MA9906837F00003.JPEG
35. Destarac M, Charmot D, Franck X, Zard SZ (2000) Dithiocarbamates as universal reversible addition-fragmentation chain transfer agents. Macromol Rapid Commun 21:1035–1039. https://doi.org/10.1002/1521-3927(20001001)21:15%3c1035::AID-MARC1035%3e3.0.CO;2-5
36. Mayadunne RTA, Rizzardo E, Chiefari J et al (2000) Living polymers by the use of trithiocarbonates as reversible addition-fragmentation chain transfer (RAFT) agents: ABA triblock copolymers by radical polymerization in two steps. Macromolecules 33:243–245. https://doi.org/10.1021/MA991451A/ASSET/IMAGES/LARGE/MA991451AF00003.JPEG
37. Perrier S (2017) 50th anniversary perspective: RAFT polymerization—a user guide. Macromolecules 50:7433–7447. https://doi.org/10.1021/ACS.MACROMOL.7B00767/ASSET/IMAGES/LARGE/MA-2017-00767N_0008.JPEG
38. Nicolas J, Guillaneuf Y, Lefay C et al (2013) Nitroxide-mediated polymerization. Prog Polym Sci 38:63–235. https://doi.org/10.1016/J.PROGPOLYMSCI.2012.06.002
39. Guillaneuf Y, Gigmes D, Marque SRA et al (2006) Nitroxide-mediated polymerization of methyl methacrylate using an SG1-based alkoxyamine: how the penultimate effect could lead to uncontrolled and unliving polymerization. Macromol Chem Phys 207:1278–1288. https://doi.org/10.1002/MACP.200600125;PAGE:STRING:ARTICLE/CHAPTER
40. Kim MS, Chang H, Zheng L et al (2023) A review of biodegradable plastics: chemistry, applications, properties, and future research needs. Chem Rev 123:9915–9939. https://doi.org/10.1021/ACS.CHEMREV.2C00876/ASSET/IMAGES/MEDIUM/CR2C00876_0010.GIF
41. Mills N, Jenkins M, Kukureka S (2020) Processing. Plastics 67–93. https://doi.org/10.1016/B978-0-08-102499-7.00005-9
42. Zia KM, Akram N, Tabasum S et al (2021) Polymer processing technology. Process Technol Bio Based Polym 29–61. https://doi.org/10.1016/B978-0-323-85772-7.00005-7
43. Cantor KM, Watts P (2011) Plastics processing. In: Applied plastics engineering handbook: processing and materials 195–203. https://doi.org/10.1016/B978-1-4377-3514-7.10012-1
44. Soong DS (2000) Polymer processing. Chem Eng Educ 15:355–371. https://doi.org/10.1016/B978-008043417-9/50021-0

Nikhil Kumar is a senior research fellow the Advanced Technology Development Centre, Indian Institute of Technology Kharagpur. He is specialized in fabrication of smart polymeric materials for diagnosis and therapeutics. Nikhil completed his Bachelor of Pharmacy and Master of Pharmacy in Pharmacology. His masters research work was focused on developing dressings for diabetic foot ulcers. His current research revolves around the design and application of smart polymeric biomaterials for advanced diagnosis and therapeutic interventions for cutaneous squamous cell carcinoma. He has published 12 peer-reviewed research articles. He is actively engaged in research focused on polymer characterization, with particular interest in the structural, thermal, and rheological analysis of advanced polymeric materials. Mr. Kumar has been involved in research projects that explore the application of spectroscopic techniques (FTIR, NMR), thermal analysis (DSC, TGA), to evaluate polymer performance under varied environmental conditions. Mr. Kumar is also developing a novel diagnostic tool that integrates multiple characterization techniques FTIR, Raman spectroscopy, AFM and nanoindentation to probe and differentiate pathological tissue conditions such as fibrosis and hyperkeratosis.

Roumita Hore is currently a research scholar at the Rubber Technology Centre, Indian Institute of Technology, Kharagpur, India. She has been awarded the prestigious Prime Minister Research Fellowship in 2022. She has completed her M.Tech (2021) and B.Tech (2017) in Polymer Science and Technology and B.Sc (2014) in Chemistry (Hons) from Calcutta University. Roumita has served as an executive R&D in CICO Technologies limited for 2 years (2017–2019). Her area of research includes the fabrication of sustainable new-generation rubber composites for the application of passive vibration isolation in electric vehicles. Roumita is experienced in polymer characterization, particularly in thermal and chemical analysis of polymeric systems.

Mr. Soumajyoti Ghosh is a research scholar at the Indian Institute of Technology, Kharagpur. He has completed a B.Sc. (2016) in industrial Chemistry and an M.Sc. (2018) in Materials Science from RKMV, Belur Math, West Bengal and Sardar Patel University, Gujarat, respectively. During M.Sc., he worked on Synthesis of Poly (imide-siloxane) copolymer for high-end applications. He also completed his M.Tech (2020) from Indian Institute of technology, Delhi in Polymer Science and Engineering. During this, he completed his thesis work based on Synthesis of Core-Shell Polymer Nanoparticles. He had an experience in Machino Polymers Ltd. as an R&D Executive, Assistant Manager.

Nikhil Dhore is pursuing a Ph.D. at the Indian Institute of Technology Kharagpur, focusing on the development of sustainable polymers for decorative and protective applications. Nikhil has completed his bachelor's in chemical technology (2018) and a master's specializing in paint technology (2020). He worked as a project associate at CSIR-Indian Institute of Chemical Technology, Hyderabad, where he gained 3+ years of hands-on experience in developing sustainable polymers for coating and foam applications and their characterization on FTIR, TGA, DSC, UTM, DLS, electrochemical workstations, salt spray test, weathering tests, and goniophotometry. He was a part of the technology transfer of the "tin-free and solvent-free UV-curable polyurethane acrylate" to Sai Life Sciences, Hyderabad. His current research on the synthesis of water-based polyurethane, non-isocyanate polyurethane, biobased epoxy amine systems, and hydrogel. More than 6+ peer-reviewed international journal articles in his account till 2024.

Chapter 2
Structure-property-Process Correlation of Polymers

Payel Maity, Nikhil Dhore, and Nikhil Kumar

Abstract The structure-property-process correlation in polymers is a fundamental concept that ties together how the molecular arrangement influences the processing and properties of the end product. In this chapter, we will discuss the correlation and influence of the chemical structure, chain length, molecular weight distribution, and crystallinity of a polymer on its chemical, mechanical, thermal, optical, and electrical properties, and morphological characteristics. The polymer as raw material can be synthesized via different polymerization techniques, which will undergo several processes in the presence of fillers and other additives to manufacture the final product. The method and mechanism of polymerization directly impact the purity, architecture, chain length, and molecular weight distribution of a polymer. This chapter will provide information about polymerization and processing techniques, including polymer additives.

2.1 Introduction

The outstanding adaptability of polymers comes from their unique capacity to be adapted to a broad variety of applications, from common home goods to complex aerospace components [1]. This adaptability originates from the complex interplay between their microscopic structure, macroscopic features, and the techniques used in their synthesis and production. The study of structure-property-process (SPP) correlations in polymers is an important element of polymer science because it reveals how the material's molecular architecture and processing circumstances affect its

P. Maity
School of Nanoscience and Technology, Indian Institute of Technology Kharagpur, Kharagpur, India

N. Dhore
Rubber Technology Centre, Indian Institute of Technology Kharagpur, Kharagpur, India

N. Kumar (✉)
Advanced Technology Development Centre, Indian Institute of Technology Kharagpur, Kharagpur, India
e-mail: nikhilsingh25694@gmail.com

S. Chattopadhyay and N. Kumar (eds.), *Polymer Characterization*,
https://doi.org/10.1007/978-981-95-2609-3_2

overall performance [1]. Understanding the SPP connection is critical not only for creating novel polymeric materials but also for optimizing and upgrading current ones to satisfy particular performance requirements. Polymers' molecular interactions are determined by their structure, which is defined by factors such as molecular weight, chain architecture, and crystallinity. These interactions, in turn, affect the material's physical and mechanical characteristics, including tensile strength, elasticity, thermal stability, and chemical resistance.

The processing processes used, including polymerization, extrusion, molding, and curing, are crucial in determining the final structure of the polymer. Scientists and engineers may manipulate processing conditions to alter material qualities to obtain desired results. This delicate correlation of structure, property, and process leverages the value of a comprehensive grasp of polymer science.

In this chapter, we will look at the complicated web of SPP correlations in polymers, including basic concepts, essential factors, and recent advances in the area. We will demonstrate how a thorough grasp of SPP correlations may lead to the inventive design and optimization of polymeric materials, opening the path for cutting-edge technological developments.

2.2 Structure

2.2.1 Chemical Structure and Molecular Arrangement

Structure refers to the arrangement of atoms or molecules within a material. It includes details like molecular configuration, degree of polymerization, and chemical composition. These aspects determine the fundamental properties of the material, such as mechanical strength, thermal stability, and chemical reactivity. The structure of a polymer is crucial in defining its basic characteristics and how it will perform under different conditions. A polymer's chemical structure relates to the arrangement of atoms inside its molecules, which includes the kind of monomer utilized, their architecture (linear, branching, or cross-linked), and the presence of functional groups. This structure provides the basis for the polymer's characteristics [2, 3]. Linear polymers are typically stronger and more rigid, making them easier to dissolve and process. Whereas branched polymers offer more flexibility, and cross-linked polymers exhibit higher thermal stability due to stronger covalent bonds between chains as shown in Fig. 2.1. Additionally, polymer structure influences phase behaviour, enabling unique crystalline formations, and modulates chemical reactivity. These characteristics allow polymers to be customized for specific applications, such as flexible packaging or durable automotive parts.

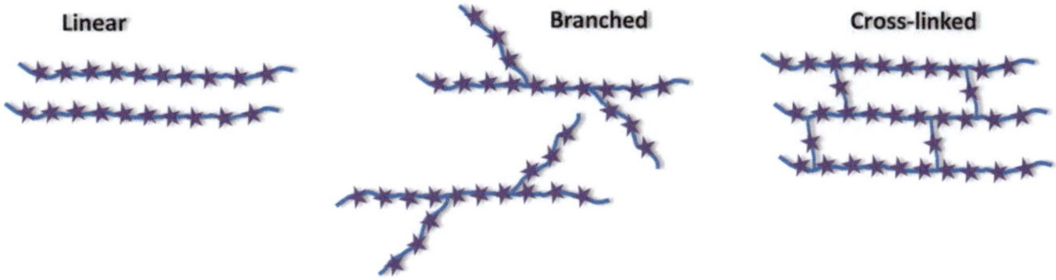

Fig. 2.1 Different arrangements of polymer

2.2.1.1 Monomer Composition

The type of monomers used in a polymer chain determines its fundamental properties.

Hydrocarbon Monomers: These include alkenes like ethylene (C_2H_4) and propylene (C_3H_6). They form polymers like polyethylene (PE) and polypropylene (PP). Polyethylene, made from ethylene monomers, is flexible and tough.

Vinyl Monomers: These include vinyl chloride ($CH_2 = CHCl$) and styrene ($C_6H_5CH = CH_2$). They form polymers like polyvinyl chloride (PVC) and polystyrene (PS). For example, polystyrene, composed of styrene monomers, is rigid and brittle.

Diene Monomers: These include butadiene ($CH_2 = CH–CH = CH_2$) and isoprene ($CH_2 = C(CH_3)–CH = CH_2$). They form elastomers like polybutadiene and polyisoprene (natural rubber) have long range elasticity.

2.2.1.2 Functional Groups

The presence of functional groups (such as hydroxyl, carboxyl, or amine) can significantly alter polymeric properties. These groups can introduce polarity, enabling hydrogen bonding or ionic interactions, which enhance mechanical strength, thermal stability, and chemical resistance.

Polar Groups: Monomers with polar functional groups (like acrylonitrile, $–C\equiv N$) enhance polymer interactions, leading to more substantial materials.

Hydrophilic/Hydrophobic Balance: The balance of hydrophilic (water-attracting) and hydrophobic (water-repelling) groups affects the polymer's solubility and interactions with water.

The polymer solubility parameter, often referred to as the Hildebrand solubility parameter, is a numerical value that indicates the degree of interaction between materials, particularly polymers and solvents. It is derived from the cohesive energy density, which is the energy required to separate molecules from each other. Polymers with similar solubility parameters to a solvent are more likely to dissolve in that solvent. This parameter is useful for predicting the solubility, miscibility, and swelling behaviour of polymers. However, it is most effective for nonpolar and slightly polar systems and may not account for complex interactions like hydrogen bonding.

Hildebrand Solubility Parameter (δ)

$$\delta = \sqrt{\frac{E_c}{V_m}} \tag{2.1}$$

where E_c is the cohesive energy and V_m is the molar volume.

Cohesive Energy Density (CED)

$$CED = \frac{(\Delta H_v - RT)}{V_m} \tag{2.2}$$

where ΔH_v is the heat of vaporization, R is the gas constant, and T is the temperature.

These equations help to determine the solubility and miscibility of polymers in different solvents by comparing their solubility parameters.

2.2.1.3 Copolymers

Random Copolymers: Monomers are randomly distributed along the polymer chain (e.g., styrene-butadiene rubber).

Block Copolymers: Large blocks of each type of monomer are grouped together (e.g., SBS rubber).

Graft Copolymers: Side chains of one type of monomer are grafted onto the backbone of another type (e.g., high-impact polystyrene).

2.2.1.4 Cross-Linking

Cross-linking involves forming covalent bonds between polymer chains, creating a network structure. This significantly enhances mechanical strength, thermal stability, and chemical resistance, making cross-linked polymers like vulcanized rubber and epoxy resins suitable for demanding applications.

Understanding the chemical composition provides insight into how polymers can be tailored for specific applications by adjusting the types and ratios of monomers used, their functional groups, and the polymerization process itself.

2.2.2 *Molecular Weight Distribution (MWD)*

Molecular Weight refers to the mass of a single molecule of a polymer. It is typically measured in daltons (Da) or atomic mass units (amu). Molecular Weight Distribution (MWD) describes the range and distribution of molecular weights present in a polymer sample. It provides insight into the variety of chain lengths in a polymer. Key parameters in MWD are:

Number Average Molecular Weight (Mn): The total weight of all polymer molecules divided by the number of polymer molecules. It provides a measure of the chain length of the polymer molecules.

$$\overline{Mn} = \sum x_i M_i \tag{2.3}$$

where M_i represents the mean molecular weight of the size range *i*, and x_i is the fraction of the total numb of chains within the corresponding size range. This can be measured by using the Gel permeation chromatography (GPC) or size exclusion chromatography (SEC) technique or by using osmometry.

Weight Average Molecular Weight (Mw): Considers the weight of each polymer molecule, providing a measure that emphasizes larger molecules. It is calculated as the sum of the weight of each molecule times its molecular weight, divided by the total weight of all molecules. A weight average molecular weight Mw is based on the weight fraction w_i within the size ranges:

$$\overline{Mw} = \sum w_i M_i \tag{2.4}$$

This can be measured by using the GPC or the light scattering technique.

Both the Mn and the Mw are important, but Mw tends to be more influential on polymer properties. This is because Mw gives more weight to larger molecules, which often have a greater impact on properties like mechanical strength, viscosity, and thermal stability.

Polydispersity Index (Đ): A measure of the broadness of the molecular weight distribution. It is calculated as the ratio of Mw to Mn. A PDI of 1 indicates a uniform polymer with all chains of the same length, while a higher PDI indicates a broader distribution. The PDI plays a crucial role in determining the uniformity and performance of a polymer.

$$Đ = \mathbf{Mw/Mn} \tag{2.5}$$

MWD has a significant impact on the structure-property-process (SPP) correlation of polymers, defining their performance and application across several domains [4, 5]. In terms of structure, MWD controls the order and regularity of polymer chains, which directly affects crystallinity. Polymers having a narrow MWD crystallize more effectively, resulting in greater mechanical strength and chemical resistance, as shown in high-density polyethylene (HDPE). A larger MWD, on the other hand, introduces a variety of link lengths, increasing chain tangling and so improving toughness and impact resistance.

MWD affects polymer characteristics similarly. Mechanical qualities like tensile strength and durability improve when the average molecular weight and MWD increase owing to the existence of long chains that effectively disperse energy. The range of molecular weights also influences thermal qualities, such as glass transition and melting temperatures, which influence thermal stability and window production. A larger MWD improves rheological qualities, which are important for processing since it decreases melt viscosity and increases flow, allowing for procedures like extrusion and injection molding. MWD influences polymer melt behaviour and stability during processing, with a wider MWD providing greater processability and smoother flow under stress due to the consistency of the flow of materials of various sizes, resulting in more efficient manufacturing and higher production rates.

MWD also plays an important part in blending and compounding; various MWDs may be mixed to obtain desired property profiles while balancing mechanical strength and processability. For example, in polymer blends, blending high molecular weight polymers with low molecular weight variations yields materials with appropriate performance properties for some particular applications. Understanding and managing MWD allows polymer scientists to customize polymers to particular demands, maximizing the balance of structural integrity, material characteristics, and processing ease, eventually improving the usefulness and performance of polymer-based products.

2.2.3 Chain Length and Chain Flexibility

Chain length and flexibility significantly influence the SPP correlation in polymers, which affects their performance and utility [6, 7]. The kinetic chain length (v) of the polymer is the number of polymers added during the polymerization to the polymer chain and it can be related to the number average degree of polymerization (DP_n) using Eqs. (2.6) and (2.7).

$$v = DP_n \tag{2.6}$$

$$2v = DP_n \tag{2.7}$$

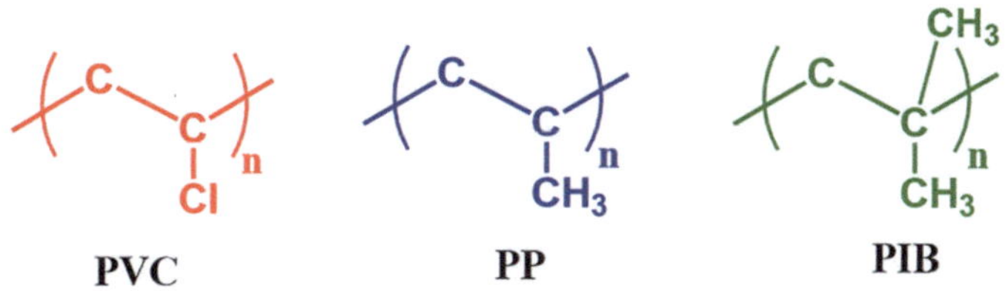

Fig. 2.2 Polymer chains with different side groups

Longer polymer chains with greater molecular weight often improve mechanical qualities such as tensile strength and toughness owing to increased chain entanglement and interaction. These longer chains help to increase deformation resistance and load-carrying capability. Shorter chains, despite having less mechanical strength, may increase processability by decreasing melt viscosity, making the polymer simpler to mold and shape during industrial processes such as extrusion and injection molding.

For example, in polyvinyl chloride (PVC), longer chains make it stronger and more stable when heated. Limited chain movement and strong interactions between chains add to its rigidity. For polypropylene (PP), longer chains mean better strength and heat resistance because they form tight structures. The restricted chain rotation and strong chain interactions help this even more. In polyisobutylene (PIB), longer chains increase the melting point and stability. The reduced chain movement and good interchain interactions boost its overall properties. So, the length of chains, how easily they rotate, and how they interact with each other all play a big role in deciding how these polymers perform (Fig. 2.2).

Polymer chain flexibility is also important since it is governed by variables such as chain conformation and the existence of flexible or stiff segments within the polymer. Flexible chains, which are often the consequence of low glass transition temperatures (T_g), make polymers more bendable and stretchable, which is useful in applications that need elasticity, such as rubber materials. Die swell and surface roughness both influence the elasticity of polymers in distinct ways. Die swell, which occurs when a polymer expands as it exits a die, can alter the molecular orientation and distribution of polymer chains, affecting the material's elastic properties. Significant die swell leads to changes in mechanical behavior due to the relaxation of stretched chains.

Surface roughness impacts elasticity by introducing stress concentrations and micro-defects that can affect how a material interacts with its environment. Smoother surfaces distribute stress more uniformly, maintaining better elastic properties, while rough surfaces can reduce elasticity through localized deformation and stress concentrations. Controlling processing conditions and surface finish is crucial for achieving desired elastic properties in polymers. Rigid chains, which may be formed from high glass transition temperature (T_g) or stiff aromatic rings within the polymer structure, improve qualities such as heat resistance and dimensional stability, making them appropriate for high-temperature environments.

The balance of chain length and flexibility influences the crystallinity of the polymer; flexible chains can inhibit crystallization, resulting in amorphous regions with higher transparency and lower density, whereas rigid chains can promote crystallinity, resulting in materials with greater strength and heat resistance. In processing, the interaction of chain length and flexibility determines the polymer's behaviour under stress and temperature, impacting processing ease and product quality. A polymer with the right chain length and flexibility may strike a good balance between processability, mechanical performance, and thermal stability, making it ideal for applications including packaging, automotive components, and medical devices. Understanding and modifying various features of polymer structure allows scientists and engineers to create materials with specific processibility and optimum performance, revealing the complex relationship between polymer molecular properties and macroscopic behaviours.

2.2.4 *Crystallinity*

Crystallinity refers to the degree to which the polymer chains are ordered and aligned in a regular, repeating pattern. In a crystalline region, polymer chains are packed closely together, leading to higher density and often higher strength and rigidity.

Degree of Crystallinity

In polymers crystalline and amorphous regions can coexist. Disturbance of the arrangement of the long chains (twists, kinks…) results in the local amorphization of the polymer. The degree of crystallinity is a measure of the proportion of crystalline regions within the polymer. It's usually expressed as a percentage.

$$\%\,\text{Crystallinity} = \frac{\rho c(\rho s - \rho a) \times 100}{\rho s(\rho c - \rho a)} \tag{2.8}$$

Where, ρc = density of perfect crystalline polymer, ρa = density of completely amorphous polymer, and ρs = density of analysing partially crystalline polymer.

Crystallinity has a major effect on the SPP correlation of polymers, defining their mechanical, thermal, and barrier characteristics, as well as processability [8, 9]. Crystalline areas in a polymer are highly organized and closely packed, which improves the material's mechanical strength, stiffness, and resistance to deformation under stress. This rise in crystallinity results in increased tensile strength and hardness, making polymers such as polyethylene terephthalate (PET) and polypropylene (PP) excellent for applications that need endurance and structural integrity.

Between crystalline and amorphous polymers there exists a quasi-crystalline polymer (Fig. 2.3). Quasicrystals refer to polymeric materials that exhibit a unique form of ordered structure without periodic repetition. Unlike conventional polymer crystals, which have repeating units arranged in a regular, periodic pattern, quasi-crystalline polymers have an ordered yet non-periodic arrangement of their polymer

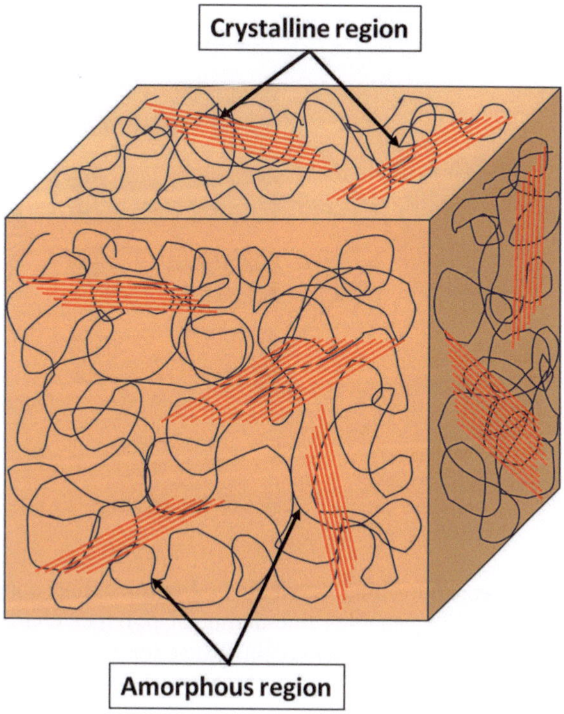

Fig. 2.3 Structure showing amorphous and crystalline portion of the polymer

chains. This unique structure can result in distinctive physical properties, such as unusual mechanical strength, low thermal conductivity, and high resistance to chemical degradation. The discovery of quasi-crystalline polymers opens up possibilities for designing materials with specific properties tailored to various applications. For instance, these polymers could be used in advanced coatings, structural materials, or even in nanotechnology, where their unique properties can be exploited for innovative solutions.

In terms of thermal characteristics, crystallinity increases the melting temperature (Tm) and improves thermal stability, allowing polymers to function at greater temperatures. This is especially significant in applications like packaging and automotive components, where heat resistance is critical. Crystallinity also affects barrier qualities by limiting the permeability of gases and liquids through the polymer matrix. This makes crystalline polymers excellent for food packaging and medical equipment where moisture and gas barriers are needed.

The degree of crystallinity may influence the transparency and optical characteristics of polymers, with greater crystallinity often resulting in more opaque materials owing to light scattering by crystalline areas. Polymers' flow and solidification properties are influenced by their crystallization behaviour during processing. Polymers having a greater degree of crystallinity may have higher viscosity and slower cooling rates, influencing operations such as injection molding, extrusion, and blow

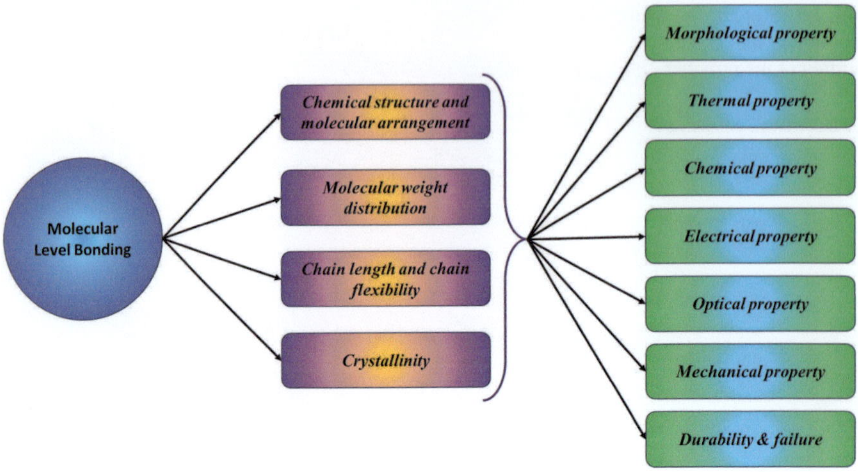

Fig. 2.4 Structure-property relation of the polymer

molding. Controlling crystallinity by processing conditions such as cooling rates and annealing enables fine-tuning of polymer characteristics to match particular application needs. Slow cooling rates, for example, might encourage the formation of larger crystalline areas, which improve mechanical qualities but may complicate processing owing to greater viscosity. Rapid cooling, on the other hand, might result in smaller crystalline areas, which improves processability but may reduce mechanical strength. Understanding and managing crystallinity is, therefore, critical for improving the performance and manufacturability of polymer-based products, demonstrating the complex interaction between polymer structure, characteristics, and processing processes. Polymer scientists and engineers may build materials with specific characteristics for a broad variety of industrial and consumer applications by modifying their crystallinity, illustrating the crucial role of crystallinity in polymer SPP correlation (Fig. 2.4).

2.3 Properties

2.3.1 Chemical Properties

Chemical characteristics have an important role in defining polymers' SPP correlation, impacting their overall performance, reactivity, and applicability for different applications [10]. These qualities are significantly defined by their chemical makeup, which includes the kinds and configurations of monomers, functional groups, and copolymers. Polymers containing polar functional groups, such as hydroxyl (–OH), carboxyl (–COOH), or amine (–NH_2), have greater chemical reactivity, solubility in polar solvents, and enhanced hydrogen bonding, leading to better mechanical strength and thermal stability. Nonpolar polymers, such as polyethylene or polypropylene, are distinguished by their chemical resistance, hydrophobicity, and low surface energy, making them ideal for applications that need moisture and chemical resistance, such as packaging materials and coatings. The presence of particular functional groups also determines the polymer's susceptibility to chemical degradation, oxidation, and hydrolysis, which affects its long-term durability and environmental compatibility. For example, polymers having ester or amide bonds may hydrolyze in the presence of water, reducing their mechanical characteristics and longevity. The chemical structure of the polymer also determines its interaction with fillers, additives, and other components of composite materials, which affects the composite's ultimate qualities and performance. During processing, polymers' chemical characteristics govern their compatibility with various solvents, catalysts, and processing aids, impacting processing ease and product quality. Polymers' chemical reactivity may be used to incorporate cross-linking, grafting, and other chemical alterations, which improve qualities including thermal stability, mechanical strength, and chemical resistance. These alterations may be adjusted to create precise property profiles for particular applications, such as medicinal devices, automotive parts, or electronic components [11, 12]. Polymer chemical resistance is especially relevant in applications that need materials to survive solvents, acids, bases, and other reactive chemicals without considerable degradation.

Understanding and controlling polymers' chemical properties is critical for optimizing their performance, processability, and suitability for various industrial and consumer applications, demonstrating the complex interplay between chemical composition and the broader SPP correlation in polymer science. Polymer scientists may create materials with specific qualities by leveraging chemical properties, assuring their efficacy and dependability in a wide range of applications.

2.3.2 Mechanical Properties

Mechanical qualities are critical in the structure-property-process (SPP) correlation of polymers, determining their suitability for diverse applications by affecting their strength, elasticity, and endurance. These attributes are inextricably tied to the polymer's molecular structure, such as chain length, cross-linking density, and crystallinity. Longer polymer chains and greater molecular weights often improve tensile strength and toughness by increasing chain entanglement and intermolecular forces. Cross-linking, which forms a network structure by connecting polymer chains, considerably improves mechanical strength and elasticity, giving materials like vulcanized rubber higher resilience and deformation resistance. The degree of crystallinity in a polymer influences its mechanical characteristics by forming hard, organized areas that improve stiffness and dimensional stability. Higher crystallinity often leads to greater tensile strength, hardness, and resistance to creep and deformation under load, as shown in materials such as high-density polyethylene (HDPE) and isotactic polypropylene. However, this renders the polymer more brittle and less impact-resistant. Another important consideration is the glass transition temperature (Tg); polymers with higher Tg demonstrate better stiffness and strength at higher temperatures, making them appropriate for high-temperature applications. Cooling rates and shear pressures during molding and extrusion have an influence on mechanical characteristics because they change the microstructure and orientation of the polymer. For example, fast cooling may cause residual stresses and diminish crystallinity, resulting in poor mechanical performance, while regulated cooling can improve crystallinity and strength. Fibers and nanoparticles are examples of additives and fillers that may improve stiffness, reduce brittleness, and reinforce the polymer matrix. The link between mechanical qualities and processing is bidirectional; knowing this interaction enables the adjustment of processing settings to produce the required mechanical properties. Tailoring mechanical characteristics by careful control of molecular architecture, crystallinity, cross-linking, and additives allows for the creation of polymers with particular performance profiles for applications ranging from flexible films to strong structural components. In summary, mechanical properties are critical in the SPP correlation, and improving them requires a thorough knowledge of the molecular and manufacturing parameters that influence the polymer's overall performance and appropriateness for a wide range of applications [12].

2.3.3 Thermal Properties

Thermal characteristics are critical in the structure-property-process (SPP) correlation of polymers, influencing their performance and applicability for a variety of applications [13]. Polymers' thermal behavior is governed by their molecular structure, which includes the kinds of monomers, crystallinity, and chain flexibility. Polymers with high crystallinity, such as polyethylene and polypropylene, have higher melting temperatures (Tm) owing to the ordered arrangement of chains, which takes more energy to disrupt. Amorphous polymers having irregular chain architectures, such as polystyrene and polycarbonate, have lower Tm values and wider glass transition temperature (Tg) ranges. Chain flexibility influences Tg, which is the temperature at which a polymer transforms from a glassy, stiff to a more flexible, rubbery state. Polymers with flexible chains have a lower Tg, which results in higher ductility and impact resistance at lower temperatures. Cooling rates and annealing are two examples of processing variables that impact thermal characteristics by changing the microstructure of the polymer. Slow cooling rates encourage crystallization, which increases Tm and thermal stability, while fast cooling may trap amorphous areas, lowering Tm. Thermal stability, another important attribute, is affected by the presence of stabilizers and additives that prevent thermal deterioration during manufacturing and in end-use applications. The capacity of a polymer to tolerate high temperatures without deterioration is critical in applications such as automobile components and electrical gadgets. The molecular arrangement of the polymer and the presence of conductive fillers also influence thermal conductivity, which is significant for heat dissipation in applications such as thermal management systems. Understanding and manipulating thermal characteristics via molecular structure and processing circumstances allows polymer scientists to create materials with particular thermal behaviours, enhancing their performance for a wide range of industrial and consumer applications [14, 15]. This demonstrates how structure, qualities, and processing all work together to determine polymer thermal properties.

2.3.4 Optical Properties

Polymers' optical characteristics, including as transparency, refractive index, and light scattering, are greatly impacted by their molecular structure, crystallinity, and processing conditions, and therefore play an important role in the structure-property-process (SPP) connection [16]. Higher crystallinity often results in greater opacity owing to light scattering by crystalline areas, while amorphous polymers with disordered chain configurations, such as polycarbonate, are more transparent. The chemical structure, which includes the presence of aromatic rings and conjugated systems, influences light absorption and transmission, as well as color and clarity. Processing factors, such as cooling rates and annealing, influence the size and distribution of crystalline areas, hence affecting transparency and haze. UV stabilisers and pigments, for example, may alter optical characteristics to improve performance in applications such as optical lenses and packaging films. Controlling these parameters allows for the creation of polymers with specific optical characteristics for a wide range of applications, from transparent packaging materials and optical fibres to light-diffusing display screens. This complex interaction of molecular features and processing processes emphasizes the need of knowing and managing the SPP correlation in order to produce desired optical properties in polymer-based materials.

2.3.5 Electrical Properties

Electrical characteristics of polymers, such as conductivity, dielectric constant, and insulating behavior, are inextricably connected to their molecular structure, degree of crystallinity, and processing circumstances, and play an important part in the structure-property-process (SPP) relationship [17]. The inclusion of polar functional groups and conjugated double bonds in the polymer structure increases electrical conductivity, as shown in conductive polymers such as polyaniline and polypyrrole. The degree of crystallinity influences dielectric qualities; greater crystallinity frequently results in lower dielectric loss and better insulating capabilities owing to the restricted mobility of charge carriers in ordered areas. Processing circumstances, such as the addition of dopants and fillers, may drastically affect electrical characteristics by establishing conductive pathways or changing the polymer's molecular structure. For example, the use of conductive fillers such as carbon nanotubes or graphene may significantly improve the electrical conductivity of polymer composites. The combination of molecular structure and processing techniques allows for the customization of electrical characteristics for particular applications such as flexible electronics, capacitors, and insulating materials. Understanding and managing these parameters within the SPP framework enables the development of polymers with optimal electrical performance that are suited for a wide range of technical and industrial applications.

2.3.6 Morphological Characteristics

Morphology refers to the overall form and organization of the material on a larger scale. It encompasses features like crystallinity, phase separation, domain sizes, and the arrangement of different structural elements. Morphology influences properties like optical clarity, toughness, and how the material interacts with its environment. While structure focuses on the microscopic or atomic-level arrangement, morphology describes the macroscopic organization and appearance of the material, both of which are essential in determining its performance and potential applications. Polymer morphological features, such as surface roughness, phase separation, and microstructure, play an important role in the structure-property-process (SPP) relationship, influencing performance and application.10 The molecular structure, such as the length and branching of polymer chains, determines the creation of crystalline and amorphous areas, affecting the overall shape. Higher crystallinity often produces well-defined, ordered microstructures, which improve mechanical strength and thermal stability. Processing parameters, such as cooling rates and shear pressures, have a substantial impact on the final morphology; quick cooling may result in finer microstructures with smaller crystalline areas, while slower cooling encourages the formation of bigger, more defined crystals. Phase separation in block copolymers or polymer blends produces separate morphological domains, which may enhance material qualities such as toughness, impact resistance, and flexibility. Additives and fillers can influence morphology by reacting with the polymer matrix, resulting in unique surface textures and improving attributes like conductivity or barrier performance. Understanding and controlling these morphological properties through precise manipulation of molecular structure and processing techniques allows for the design of polymers with tailored properties, optimizing performance for applications ranging from packaging and coatings to biomedical devices and structural elements. This emphasizes the importance of morphology in the SPP correlation of polymers.

2.3.7 *Durability and Failure*

The structure-property-process (SPP) correlation has a significant impact on polymer durability and failure qualities, which influences their long-term performance and dependability in a variety of applications. The molecular structure, which includes chain length, cross-linking density, and crystallinity, influences the polymer's resistance to physical and chemical degradation. Higher molecular weight and cross-linking improve durability by making the polymer more resistant to stress, wear, and environmental conditions including UV radiation and chemical exposure. Crystalline areas provide reinforcement, resulting in greater tensile strength and resistance to creep and fatigue. Annealing and controlled cooling are two processing procedures that may improve the microstructure of the polymer and increase its endurance. The addition of additives, stabilizers, and fillers may further increase durability by shielding the polymer matrix from oxidative and thermal degradation. The microstructure of the polymer, as well as the presence of faults or impurities introduced during manufacturing, impact failure parameters such as mode and rate of fracture. Understanding and managing these parameters within the SPP framework allows for the development of polymers with specialized durability and failure qualities, assuring their effectiveness and dependability in demanding applications such as automotive components, medical devices, and construction materials. This emphasizes the significance of SPP correlation in enhancing the lifetime and performance of polymer-based goods.

2.4 Process

2.4.1 *Polymerization Techniques*

Polymerization techniques are methods used to synthesize polymers by joining monomers into long chains (Fig. 2.5) [18–20]. Here are the key techniques.

2.4.1.1 Addition (Chain-Growth) Polymerization

Monomers with double or triple bonds (like ethylene or styrene) undergo rapid chain reactions to form polymers. This polymerization method comprises three steps: initiation, propagation (chain growth), and termination (ending the chain growth) [21].

Types

- **Free Radical Polymerization**: Polymerization starts with free radicals (e.g., polystyrene production).
- **Cationic Polymerization**: Polymerization starts with cations (e.g., butyl rubber).

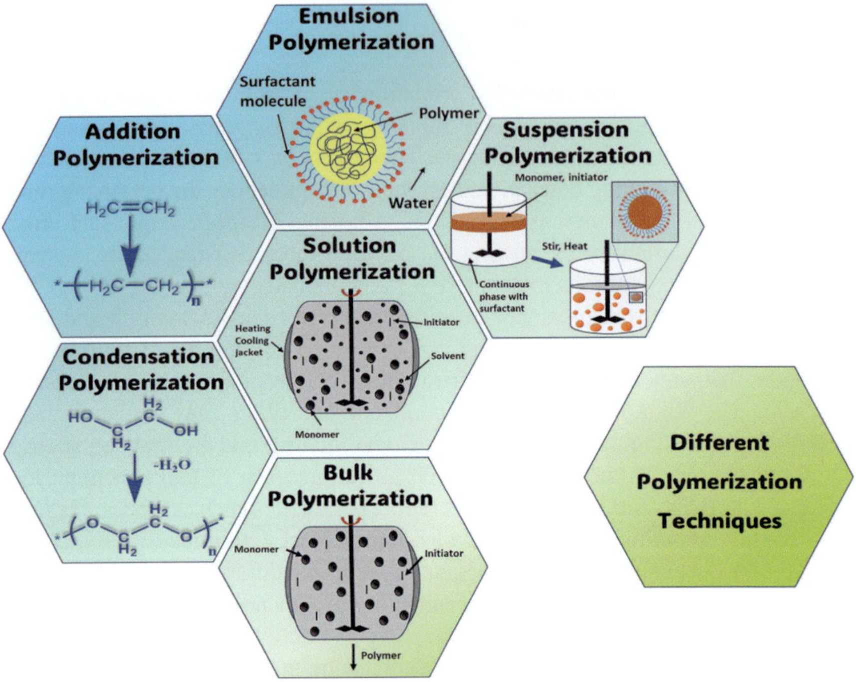

Fig. 2.5 Different polymerization techniques

- **Anionic Polymerization**: Polymerization starts with anions (e.g., polybutadiene).

2.4.1.2 Condensation (Step-Growth) Polymerization

Monomers with two functional groups (like - carboxyl, amine, and hydroxyl) react to form polymers and by-products, typically water or methanol. This technique involves the gradual formation of oligomers, which then combine to form long-chain polymers [22, 23].

Example

Polyesters: Formed from diols and dicarboxylic acids (e.g., polyethylene terephthalate).

Polyamides: Formed from diamines and dicarboxylic acids (e.g., nylon).

2.4.1.3 Emulsion Polymerization

An emulsion polymerization consists of four main components water (solvent, continuous phase), monomer (insoluble or sparingly soluble in water), initiator

(Water soluble), and surfactant or emulsifier (e.g. sodium salt of long chain fatty acid).

This method is used for the preparation of polyvinyl acetate, polychloroprene, butadiene/ acrylonitrile copolymers, etc. [24].

Applications: Used in producing synthetic rubber, adhesives, and coatings.

2.4.1.4 Solution Polymerization

Monomers dissolve in a solvent, which aids in heat dissipation and viscosity control. This technique is suitable for producing polymers with controlled molecular weights and low residual monomers. Examples include acrylics and certain polyurethanes [25].

Applications: Producing high-purity polymers like polyacrylonitrile.

2.4.1.5 Bulk Polymerization

This method involves polymerizing pure monomers without solvents or dispersants, resulting in high-purity polymers. However, heat management is critical due to exothermic reactions [26].

Applications: Used in producing PMMA (polymethyl methacrylate) and polystyrene.

2.4.1.6 Suspension Polymerization

Involves the dispersion of monomer droplets in a continuous aqueous phase, with polymerization initiated by a soluble catalyst [27].

Applications: Used to produce PVC (polyvinyl chloride).

These techniques are foundational for producing a wide array of polymeric materials with tailored properties for diverse applications.

2.4.2 Polymer Processing

Processing of polymers involves a variety of techniques that transform raw polymer materials into finished products with desired shapes, properties, and functionalities [28]. Here's an overview of some common polymer processing methods (Fig. 2.6):

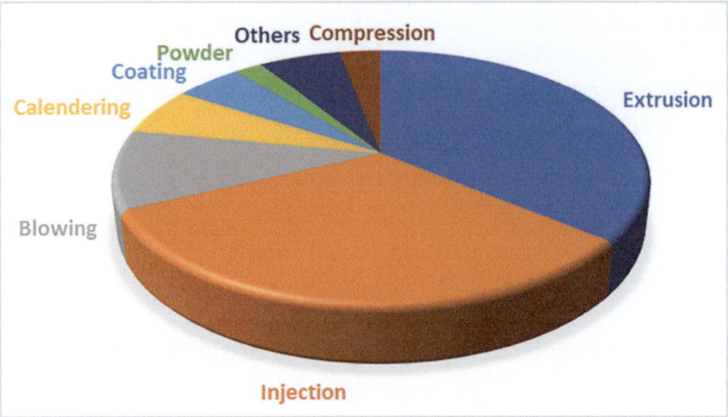

Fig. 2.6 Pie chart showing the distribution of various polymer processing techniques, with injection and extrusion being the most widely used methods

2.4.2.1 Extrusion

Extrusion is one of the most widely used polymer processing techniques. It involves forcing molten polymer through a die to create continuous shapes like pipes, sheets, and films.

Steps

- Polymer pellets are fed into the hopper.
- The material is heated and melted in the barrel.
- A screw pushes the molten polymer through a shaped die.
- The extruded material is cooled and solidified.

Applications

- Producing pipes, tubing, and profiles.
- Manufacturing plastic films and sheets.

2.4.2.2 Injection Molding

Injection molding is used to create complex, three-dimensional shapes with high precision. It's ideal for mass-producing small to medium-sized parts [29, 30].

Steps

- Polymer pellets are fed into the injection molding machine.
- The material is heated and melted.
- The molten polymer is injected into a mold cavity under high pressure.
- The mold is cooled to solidify the part.
- The molded part is ejected from the mold.

Applications

- Manufacturing automotive parts, household items, and medical devices.

2.4.2.3 Blow Molding

Blow molding is used to create hollow objects like bottles and containers. There are three main types: extrusion blow molding, injection blow molding, and stretch blow molding [31, 32].

Steps (Extrusion Blow Molding)

- A tube-like parison is extruded.
- The parison is clamped in a mold and inflated with air.
- The inflated polymer takes the shape of the mold cavity.
- The mold is cooled, and the part is ejected.

Applications

- Producing bottles, containers, and tanks.

2.4.2.4 Compression Molding

Compression molding is used for thermosetting polymers. It involves compressing the polymer into a mold to form the desired shape [33, 34].

Steps

- Polymer material is placed in the mold cavity.
- The mold is heated and closed, applying pressure to shape the material.
- The material cures (sets) and solidifies in the shape of the mold.
- The molded part is ejected.

Applications

- Manufacturing automotive parts, electrical components, and gaskets.

2.4.2.5 Thermoforming

Thermoforming involves heating a plastic sheet until it becomes pliable and then forming it over a mold. There are two main types: vacuum forming and pressure forming [35, 36].

Steps (Vacuum Forming)

- A plastic sheet is clamped and heated.
- The heated sheet is draped over a mold.
- A vacuum is applied to suck the sheet into the mold, forming the desired shape.
- The formed part is cooled and trimmed.

Applications

- Creating packaging, trays, and disposable cups.

2.4.2.6 Rotational Molding

Rotational molding (or rotomolding) is used to create large, hollow parts. It involves rotating a mold to distribute the polymer evenly [37, 38].

Steps

- Polymer powder is placed in a mold.
- The mold is heated and rotated simultaneously around two axes.
- The polymer melts and coats the interior of the mold evenly.
- The mold is cooled while still rotating to solidify the part.
- The molded part is removed.

Applications

- Producing tanks, containers, and playground equipment.

2.4.2.7 3D Printing

3D printing, also known as additive manufacturing, involves creating three-dimensional objects by depositing materials layer by layer based on a digital model. It is widely used for prototyping and producing complex geometries [39, 40].

Steps

1. **Design**: A digital 3D model is created using CAD software.
2. **Slicing**: The model is sliced into thin horizontal layers using slicing software.
3. **Printing**: The printer deposits the material layer by layer, following the sliced model.
4. **Post-Processing**: The printed object may undergo additional steps such as support removal, surface finishing, and curing.

Applications

- Aerospace components, medical/dental devices, automotive parts, consumer products, and custom prosthetics.

2.4.2.8 Calendering

Calendaring is a process used to produce thin, continuous sheets or films from polymers. The process involves passing a heated polymer through a series of rollers, which compress and shape the material into a uniform thickness [41, 42].

Steps

1. **Blending and Fluxing**: The polymer is first blended and fluxed to achieve the desired consistency.
2. **Heating**: The polymer is heated until it becomes pliable.
3. **Rolling**: The heated polymer is passed through a series of rollers, which compress it into a thin sheet.
4. **Cooling**: The sheet is cooled and rolled up for further processing or use.

Applications

- Producing floor tiles, continuous flooring, rainwear, shower curtains, table covers, pressure-sensitive tape, automotive and furniture upholstery, wall coverings, and luminous ceilings.

2.4.2.9 Resin Transfer Molding (RTM)

RTM is a closed-mold process used to manufacture high-performance composite components. It involves injecting resin into a mold cavity containing a fiber reinforcement, resulting in parts with excellent surface quality and complex shapes [43, 44].

Steps

1. **Mold Preparation**: The mold is prepared by applying a gel coat to ensure a high-quality finish.
2. **Preform Placement**: A preform, made of fiber reinforcement (e.g., fiberglass), is placed into the mold cavity.
3. **Mold Closure**: The mold is closed and clamped to create a sealed cavity.
4. **Resin Injection**: Catalyzed resin is injected under pressure into the mold, displacing air through vents and ensuring complete impregnation of the fibers.
5. **Curing**: The resin is cured, typically at elevated temperatures, to harden into a solid composite part.

Applications

Producing automotive parts, creating boat hulls and other marine structures, fabricating prosthetics and other medical devices, producing lightweight, high-strength components for aircraft.

2.4.2.10 Powdering

Powder processing involves using polymer powders to create solid objects through techniques like sintering and compaction. This method is often used for materials that are difficult to shape by melting [45].

Steps

1. **Mixing**: Polymer powder is mixed with a lubricant or binder.

2. **Compacting**: The powder is compressed in a mold to form a "green" compact.
3. **Sintering**: The compact is heated at high temperatures to fuse the powder particles together, forming a solid object.
4. **Cooling**: The sintered object is cooled and may undergo additional finishing processes.

Applications

- Manufacturing ceramics, brittle metals, and high-performance polymers like PTFE (polytetrafluoroethylene).

Each processing method has its advantages and is chosen based on the desired properties, application requirements, and economic considerations. Proper control of processing conditions, such as temperature, pressure, and cooling rates, is essential to achieving optimal performance and quality in the final polymer products.

2.4.3 Polymer Additives

Polymer additives are substances incorporated into polymers to enhance their properties and processing performance. They play a vital role in the structure-property-process (SPP) correlation.

Fillers are crucial in polymer additives due to their extensive impact on various properties and applications. They enhance mechanical strength, thermal stability, and electrical conductivity, while also being cost-effective. For instance, glass fibers increase tensile strength and heat resistance, carbon black boosts UV resistance and conductivity, and calcium carbonate improves rigidity and thermal stability. Fillers reduce shrinkage and warping, improve barrier properties by decreasing permeability to gases and liquids, and influence surface characteristics like roughness and gloss. They also enhance processing efficiency. While other additives like plasticizers, stabilizers, and colorants serve specific purposes, fillers offer comprehensive benefits, addressing multiple property requirements simultaneously. This versatility and ability to tailor polymer performance make fillers more important than other specialized additives in polymer manufacturing (Table 2.1).

Common additives include

By carefully selecting and incorporating these additives, scientists and engineers can tailor polymer properties to meet specific application requirements, optimizing performance and extending the lifespan of polymeric materials.

Table 2.1 Common types of additives used during the processing

Serial number	Type of additives	Function	Examples	Applications	References
1	Plasticizers	Increase the flexibility and reduce the brittleness of polymers	Phthalates, adipates, and citrates	Used in PVC to make flexible products like cables, films, and synthetic leather	[46]
2	Stabilizers	Improve the thermal stability and resistance to degradation caused by heat, UV light, or oxidation	UV stabilizers, antioxidants, and heat stabilizers	Extend the lifespan of outdoor plastics, automotive parts, and packaging materials	[47]
3	Fillers	Improve mechanical properties, reduce cost, and enhance dimensional stability	Talc, calcium carbonate, glass fibers, and carbon black	Used in automotive components, construction materials, and household items	[46]
4	Colourants	Provide colour to polymer products	Pigments (inorganic and organic) and dyes	Used in packaging, toys, automotive parts, and consumer goods	[48]
5	Flame retardants	Reduce the flammability and prevent the spread of fire in polymer products	Halogenated flame retardants, phosphates, and inorganic compounds like aluminium hydroxide	Used in building materials, electronics, textiles, and transportation	[49, 50]
6	Anti-microbial agents	These additives inhibit the growth of bacteria and fungi	zinc oxide, titanium dioxide, chlorhexidine, Essential oils (tea tree oil, eucalyptus oil)	Used in medical devices, personal care products, packaging materials and textiles	[18, 51, 52]

(continued)

Table 2.1 (continued)

Serial number	Type of additives	Function	Examples	Applications	References
7	Anti-oxidants	Prevent degradation caused by oxidation during processing and usage	Hindered phenols, phosphites, and thioesters	Used in packaging, electrical insulation, and automotive parts	[12, 46, 53]
8	Anti-static agents	Reduce static charge buildup on the surface of polymers	Quaternary ammonium compounds and phosphate esters	Used in electronics packaging, films, and textile fibers	[54, 55]
9	Processing aids	Improve processability and reduce defects during manufacturing	Lubricants, release agents, and flow promoters	Used in extrusion, injection molding, and blow molding processes	[56]
10	Impact modifiers	Enhance the toughness and impact resistance of polymers	Elastomeric modifiers, core–shell particles	Used in automotive parts, packaging, and household appliances	[57, 58]
11	Biocides	Inhibit the growth of microorganisms and prevent biodegradation	Fungicides, bactericides, and algicides	Used in medical devices, coatings, and water treatment systems	[59, 60]

2.5 Conclusion

In summary, the intricate web of structure-property-process (SPP) correlations in polymers forms the bedrock of polymer science and technology. By delving into the detailed interplay between chemical structure, molecular arrangement, and processing conditions, we can gain profound insights into how these factors collectively dictate the final properties and performance of polymeric materials. Understanding and manipulating these correlations allow scientists and engineers to tailor polymers for a myriad of applications, ensuring optimal functionality and durability. From enhancing mechanical strength and thermal stability to fine-tuning optical and electrical properties, the precise control of SPP relationships is essential for advancing polymer innovation and meeting the evolving demands of diverse industries. Through this comprehensive exploration, we underscore the significance of a holistic approach to polymer design, paving the way for future advancements and breakthroughs in material science.

Acknowledgements We have used bioRender.com (student license) for making illustration.

Conflict of Interest Authors declare.
No competing conflict of interest.

References

1. Patel GM, Shah V, Bhaliya J, Pathan P, Nikita KM (2022) Polymer-based nanomaterials: an introduction. Smart Polym Nanocomposites Des Synth Funct Prop Appl 27–59. https://doi.org/10.1016/B978-0-323-91611-0.00018-9
2. Meijer HEH, Govaert LE (2005) Mechanical performance of polymer systems: The relation between structure and properties. Prog Polym Sci 30:915–938. https://doi.org/10.1016/j.progpolymsci.2005.06.009
3. Shah V, Bhaliya J, Patel GM, Popaliya M, Mishra A, Patel PR (2023) Biodegradable polymer nanocomposites as electrode materials for electrochemical double-layer capacitors and hybrid supercapacitor applications. In: Biodegradable and Biocompatible Polymer Nanocomposites: Processing, Characterization, and Applications. Elsevier, pp 311–352
4. Nunes RW, Martin JR, Johnson JF (1982) Influence of molecular weight and molecular weight distribution on mechanical properties of polymers. Polym Eng Sci 22:205–228. https://doi.org/10.1002/pen.760220402
5. Gentekos DT, Sifri RJ, Fors BP (2019) Controlling polymer properties through the shape of the molecular-weight distribution. Nat Rev Mater 4:761–774. https://doi.org/10.1038/s41578-019-0138-8
6. Bedford SE, Yu K, Windle AH (1992) Influence of chain flexibility on polymer mesogenicity. J Chem Soc Faraday Trans 88:1765–1773. https://doi.org/10.1039/FT9928801765
7. Nwammuo OP, Maitland GC (1992) Effects of chain flexibility on polymer solution dynamics. J Chem Soc Faraday Trans 88:1803–1818. https://doi.org/10.1039/FT9928801803
8. Qu T, Nan G, Ouyang Y, Bieketuerxun B, Yan X, Qi Y, Zhang Y (2023) Structure-Property Relationship, Glass Transition, and Crystallization Behaviors of Conjugated Polymers. Polymers (Basel) 15:4268. https://doi.org/10.3390/polym15214268
9. Mileva D, Tranchida D, Gahleitner M (2018) Designing polymer crystallinity: An industrial perspective. Polym Cryst 1:. https://doi.org/10.1002/pcr2.10009
10. Guyot A, Bartholin M (1982) Design and properties of polymers as materials for fine chemistry. Prog Polym Sci 8:277–331. https://doi.org/10.1016/0079-6700(82)90002-8
11. Bhaliya JD, Shah VR, Mehta S, Kashyap SS, Borah K, Kutcherlapati SNR (2024) Life cycle assessments, environmental implications, and toxicological evaluations of functionalized polymer nanocomposites. In: Advances in Functionalized Polymer Nanocomposites. Elsevier, pp 1031–1066
12. Shah V, Bhaliya J, Patel GM, Deshmukh K (2022) Advances in polymeric nanocomposites for automotive applications: A review. Polym Adv Technol 33:3023–3048. https://doi.org/10.1002/pat.5771
13. Leszczyńska A, Njuguna J, Pielichowski K, Banerjee JR (2007) Polymer/montmorillonite nanocomposites with improved thermal properties. Part I. Factors influencing thermal stability and mechanisms of thermal stability improvement. Thermochim Acta 453:75–96. https://doi.org/10.1016/j.tca.2006.11.002

14. Bhaliya J, Shah V, Katariya H, Suthar V, Patel G (2023) Fluoropolymer nanocomposites for volatile organic compounds and gas-sensing application. Adv Fluoropolymer Nanocomposites Fabr Process Charact Appl 445–483. https://doi.org/10.1016/B978-0-323-95335-1.00024-4
15. Bhaliya JD, Raju Kutcherlapati SN, Dhore N, Punugupati N, Sunkara KL, Misra S, Joshi SSK (2025) Soybean oil-derived, non-isocyanate polyurethane–TiO2 nanocomposites with enhanced thermal, mechanical, hydrophobic and antimicrobial properties†. RSC Sustain 3:1434–1447. https://doi.org/10.1039/d4su00587b
16. Chelsea Chen X, Green PF (2010) Control of morphology and its effects on the optical properties of polymer nanocomposites. Langmuir 26:3659–3665. https://doi.org/10.1021/la903108u
17. Sichel EK, Gittleman JI, Sheng P (1982) Electrical properties of carbon-polymer composites. J Electron Mater 11:699–747. https://doi.org/10.1007/BF02672392
18. Bhaliya JD, Raju Kutcherlapati SN, Dhore N, Punugupati N, Sunkara KL, Misra S, Joshi SSK (2025) Soybean oil-derived, non-isocyanate polyurethane–TiO2 nanocomposites with enhanced thermal, mechanical, hydrophobic and antimicrobial properties†. RSC Sustain 3:1434–1447. https://doi.org/10.1039/d4su00587b
19. Dhore N, Prasad E, Narayan R, Rao CRK, Palanisamy A (2023) Studies on Biobased Non-Isocyanate Polyurethane Coatings with Potential Corrosion Resistance. Sustain Chem 4:95–109. https://doi.org/10.3390/suschem4010008
20. Joshi SSK, Waghmare JS, Kashyap SS, Bhaliya JD, Sunkara KL, Misra S, Kutcherlapati SNR (2025) Phosphated graphene oxide-enhanced water-based acrylic coatings with improved anti-corrosion and antimicrobial properties. Surf Interf 62. https://doi.org/10.1016/j.surfin.2025.106240
21. Bermeshev MV, Chapala PP (2018) Addition polymerization of functionalized norbornenes as a powerful tool for assembling molecular moieties of new polymers with versatile properties. Prog Polym Sci 84:1–46. https://doi.org/10.1016/j.progpolymsci.2018.06.003
22. Fakirov S (2019) Condensation polymers: their chemical peculiarities offer great opportunities. Prog Polym Sci 89:1–18. https://doi.org/10.1016/j.progpolymsci.2018.09.003
23. Ramakrishnan S (2017) Condensation polymerization. Resonance 22:355–368. https://doi.org/10.1007/s12045-017-0475-0
24. Asua JM (2004) Emulsion polymerization: from fundamental mechanisms to process developments. J Polym Sci Part A Polym Chem 42:1025–1041. https://doi.org/10.1002/pola.11096
25. Ahadi N, Askari S, Fouladitajar A, Akbari I (2022) Facile synthesis of hierarchically structured MIL-53(Al) with superior properties using an environmentally-friendly ultrasonic method for separating lead ions from aqueous solutions. Sci Rep 12:2649. https://doi.org/10.1038/s41598-022-06518-8
26. Balke ST, Hamielec AE (1973) Bulk polymerization of methyl methacrylate. J Appl Polym Sci 17:905–949. https://doi.org/10.1002/app.1973.070170320
27. Brooks B (2010) Suspension polymerization processes. Chem Eng Technol 33:1737–1744. https://doi.org/10.1002/ceat.201000210
28. Vlachopoulos J, Strutt D (2003) Polymer processing. Mater Sci Technol 19:1161–1169. https://doi.org/10.1179/026708303225004738
29. Agrawal AR, Pandelidis IO, Pecht M (1987) Injection-molding process control—a review. Polym Eng Sci 27:1345–1357. https://doi.org/10.1002/pen.760271802
30. Chen Z, Turng LS (2005) A review of current developments in process and quality control for injection molding. Adv Polym Technol 24:165–182. https://doi.org/10.1002/adv.20046
31. Laroche D, Kabanemi KK, Pecora L, Diraddo RW (1999) Integrated numerical modeling of the blow molding process. Polym Eng Sci 39:1223–1233. https://doi.org/10.1002/pen.11509
32. Schmidt FM, Agassant JF, Bellet M (1998) Experimental study and numerical simulation of the injection stretch/blow molding process. Polym Eng Sci 38:1399–1412. https://doi.org/10.1002/pen.10310
33. Jaafar J, Siregar JP, Tezara C, Hamdan MHM, Rihayat T (2019) A review of important considerations in the compression molding process of short natural fiber composites. Int J Adv Manuf Technol 105:3437–3450. https://doi.org/10.1007/s00170-019-04466-8

34. Tatara RA (2023) Compression molding. In: Applied plastics engineering handbook: processing, sustainability, materials, and applications. Elsevier, pp 389–424
35. Throne J (2023) Thermoforming. In: Applied plastics engineering handbook: processing, sustainability, materials, and applications. Elsevier, pp 449–480
36. Truckenmüller R, Giselbrecht S, Rivron N, Gottwald E, Saile V, Van Den Berg A, Wessling M, Van Blitterswijk C (2011) Thermoforming of film-based biomedical microdevices. Adv Mater 23:1311–1329. https://doi.org/10.1002/adma.201003538
37. Gupta N, Ramkumar PL, Sangani V (2020) An approach toward augmenting materials, additives, processability and parameterization in rotational molding: a review. Mater Manuf Process 35:1539–1556. https://doi.org/10.1080/10426914.2020.1779934
38. Rao MA, Throne JL (1972) Principles of rotational molding. Polym Eng Sci 12:237–264. https://doi.org/10.1002/pen.760120402
39. Shahrubudin N, Lee TC, Ramlan R (2019) An overview on 3D printing technology: Technological, materials, and applications. Procedia Manuf 35:1286–1296. https://doi.org/10.1016/j.promfg.2019.06.089
40. Zhou LY, Fu J, He Y (2020) A review of 3D printing technologies for soft polymer materials. Adv Funct Mater 30. https://doi.org/10.1002/adfm.202000187
41. Chong JS (1968) Calendering thermoplastic materials. J Appl Polym Sci 12:191–212. https://doi.org/10.1002/app.1968.070120118
42. Mitsoulis E (2009) Calendering of polymers. In: Advances in Polymer processing: from macro- to nano-scales. Elsevier, pp 312–351
43. Shojaei A, Ghaffarian SR, Karimian SMH (2003) Modeling and simulation approaches in the resin transfer molding process: a review. Polym Compos 24:525–544. https://doi.org/10.1002/pc.10050
44. Sozer EM, Simacek P, Advani SG (2012) Resin transfer molding (RTM) in polymer matrix composites. In: Manufacturing techniques for polymer matrix composites (PMCs). Elsevier, pp 245–309
45. Berens AR, Hopfenberg HB (1978) Diffusion and relaxation in glassy polymer powders: 2. Separation of diffusion and relaxation parameters. Polymer (Guildf) 19:489–496. https://doi.org/10.1016/0032-3861(78)90269-0
46. Battegazzore D, Bocchini S, Alongi J, Frache A (2014) Plasticizers, antioxidants and reinforcement fillers from hazelnut skin and cocoa by-products: extraction and use in PLA and pp. Polym Degrad Stab 108:297–306. https://doi.org/10.1016/j.polymdegradstab.2014.03.003
47. Hodgson JL, Coote ML (2010) Clarifying the mechanism of the denisov cycle: how do hindered amine light stabilizers protect polymer coatings from photo-oxidative degradation? Macromolecules 43:4573–4583. https://doi.org/10.1021/ma100453d
48. Maile FJ, Pfaff G, Reynders P (2005) Effect pigments—past, present and future. Prog Org Coatings 54:150–163. https://doi.org/10.1016/j.porgcoat.2005.07.003
49. Morgan AB, Gilman JW (2013) An overview of flame retardancy of polymeric materials: application, technology, and future directions. Fire Mater 37:259–279. https://doi.org/10.1002/fam.2128
50. De Wit CA (2002) An overview of brominated flame retardants in the environment. Chemosphere 46:583–624. https://doi.org/10.1016/S0045-6535(01)00225-9
51. González-Henríquez CM, Sarabia-Vallejos MA, Hernandez JR (2019) Antimicrobial polymers for additive manufacturing. Int J Mol Sci 20:1210. https://doi.org/10.3390/ijms20051210
52. Francolini I, Piozzi A (2019) Polymeric systems as antimicrobial or antifouling agents. Int J Mol Sci 20:4866. https://doi.org/10.3390/ijms20194866
53. Ingold KU, Pratt DA (2014) Advances in radical-trapping antioxidant chemistry in the 21st century: a kinetics and mechanisms perspective. Chem Rev 114:9022–9046. https://doi.org/10.1021/cr500226n
54. Kosiński S, Rykowska I, Gonsior M, Krzyżanowski P (2022) Ionic liquids as antistatic additives for polymer composites—a review. Polym Test 112:107649. https://doi.org/10.1016/j.polymertesting.2022.107649

55. Yudaev PA, Tamboura B, Chistyakov EM (2023) Antistatic polymer materials. Nanotechnologies Constr 15:139–151. https://doi.org/10.15828/2075-8545-2023-15-2-139-151
56. Wiesinger H, Wang Z, Hellweg S (2021) Deep dive into plastic monomers, additives, and processing aids. Environ Sci Technol 55:9339–9351. https://doi.org/10.1021/acs.est.1c00976
57. Markarian J (2004) Impact modifiers: how to make your compound tougher. Plast Addit Compd 6:46–49. https://doi.org/10.1016/S1464-391X(04)00203-X
58. Greco R (1998) Impact modifiers: (1) mechanisms and applications in thermoplastics, pp 375–385
59. Chen A, Peng H, Blakey I, Whittaker AK (2017) Biocidal polymers: a mechanistic overview. Polym Rev 57:276–310. https://doi.org/10.1080/15583724.2016.1223131
60. Misin VM, Zezin AA, Klimov DI, Sybachin AV, Yaroslavov AA (2021) Biocidal polymer formulations and coatings. Polym Sci - Ser B 63:459–469. https://doi.org/10.1134/S1560090421050079

Chapter 3
Chemical Analysis of Polymers

Roumita Hore, Nikhil Dhore, and Abhijit Bera

Abstract The chapter fundamentally covers the chemical analysis of polymers, predominantly rubber, as example, as per ASTM D297. The chapter provides a basic idea about chemical identification of polymers, which includes analysis of elements, functional groups, and unsaturation present in the polymer backbone. The primary tests include the ignition, pyrolysis, spot test, and boiling with conc. H_2SO_4. The wet test includes the Lassaigne test, the Azo dye test, and the Beilstein test. The concept of Soxhlet extraction is also covered in context with eliminating additives from compounded polymer. The trace element detection is also briefly covered.

Keywords Lassaigne test · Pyrolysis · Azo dye test · Soxhlet extraction · Trace element analysis

3.1 Introduction

Chemical analysis is a crucial method for understanding the chemical composition of the material to analyze the connection between chemical composition and material properties. It is not only a valuable tool for research and development but also an asset for quality control to meet specific technical standards for commercial products. It also deals with the quantification of potentially harmful chemicals present in the compound, ensuring they comply with regulatory limits and standards for consumer safety.

R. Hore (✉) · N. Dhore
Rubber Technology Centre, IIT Kharagpur, Kharagpur, West Bengal, India
e-mail: horeroumita2019@gmail.com

A. Bera
R&D Centre, ALP Overseas Pvt. Ltd., Alwar, Rajasthan, India

S. Chattopadhyay and N. Kumar (eds.), *Polymer Characterization*,
https://doi.org/10.1007/978-981-95-2609-3_3

3.2 Chemical Analysis

3.2.1 Preliminary Tests

The chemical analysis of polymers is primarily accomplished as per ASTM D297.

Ignition Test: The ignition test determines the response of a material to flame. The fire triangle represents the conditions of ignition. It includes heat, oxygen, and fuel. The polymer sample is placed on a nickel spatula during the ignition test and heated gently. The observation consists of the following: smoky or sooty flame, significant odor (sweet or pungent), continuity of flame after removal from fire, and the type of ash remaining at the end of the experiment, Table 3.1. To establish the extent of flame retardancy of the compounded polymer, the ignition test can also be executed using a cone calorimeter or, as per the UL 94 test methodology [1–3].

Table 3.1 Preliminary tests of different rubbers

Rubber	Ignition test	Pyrolytic test	Boiled with conc. 80% H_2SO_4
NR	Smoky flame with a distinctive odor. The flame continues burning on the removal of the flame	Dark color liquid distils out	Decomposes
BR	Smoky flame and continues burning on removal from the flame	–	Decomposes
SBR	Sooty flame and sweet odor of styrene	–	Decomposes
CR	Extinguish on the removal of the flame. Ignition with Cu-wire gives a green flame	Acidic liquid distils, and blue litmus turns red	Decomposes
NBR	Smoky flame	–	Decomposes
IIR	Highly smoky flame	Clear neutral liquid distils	Inert
Silicone rubber	Highly smoky gives white ash	–	Decomposes
EPDM/EPM	Burns with a clear flame (White smoke)	–	Inert
Hypalon	Ignition with Cu wire gives a green flame	Acid liquid distils	Inert
PVC	Extinguish on the removal of flame (with plasticizer is flammable). Ignition with Cu-wire gives a green flame	Acid liquid distils	Inert
EVA	Smoky flame. Continues to burn on the removal of the flame	Acid liquid distils	Inert
Fluoro-rubbers	Burns highly in flame. But extinguishes immediately on the removal of the flame	Acid liquid distils	Inert

The ignition test can also be executed with a copper wire. A small piece of sample is rounded up on a copper wire and held in a flame. Polymers with significant chlorine content respond to this test. The flame turns green for chloroprene rubber (CR) and polyvinyl chloride (PVC).

Pyrolysis: Pyrolysis is a thermochemical decomposition process that causes the degradation of organic materials by heating in the absence of oxygen. During pyrolysis, long-chain polymers are degraded into small molecules, resulting in the production of gases, liquids, and solid residues. The nature of the produced chemicals varies for different polymers [4–6].

The test is carried out by taking ~0.5 g of the sample in a test tube and attaching the side arm. The sample is allowed to decompose by heating with a Bunsen flame. Slowly, the vapors start to appear at the mouth of the side arm. Then the end beneath the surface is immersed in solution I and II, subsequently in the receiver test tubes. The colour change is observed and recorded.

The receiver test tube solutions are further heated in a water bath at 100 °C for 3 min, and any change that occurs is recorded and classified the sample using Table 3.2.

Solution I: 1.0 g of p-dimethylaminobenzaldehyde and 0.01 g of hydroquinone are dissolved in 100 ml of absolute methanol. Then, 5 ml of HCl and 10 ml of ethylene glycol are added. Then the relative density of the solution is adjusted to 0.851 at 25 °C by adding a calculated amount of methanol or ethylene glycol. The reagent is kept in a dark place [7].

Solution II: 2.0 g of sodium citrate ($2Na_3C_6H_5O_7 \cdot 11H_2O$), 0.2 g of citric acid, 0.03 g of bromocresol green, and 0.03 g of Metanil yellow are dissolved in 500 ml of water [7].

Boiled with conc. H_2SO_4: A small piece of polymer sample is heated with conc. H_2SO_4 in a test tube. Unsaturated polymer decomposes, whereas saturated polymer remains inert.

These three tests are considered preliminary tests, and their responses are tabulated in Tables 3.1 and 3.2.

Table 3.2 Respond to solutions I and II after the pyrolysis test

Polymer	Solution I		Solution II
	Initial color	Color after heating	Color
Blank	Pale yellow	Pale yellow	Green
CR	Yellow/Green	Pale yellow/Green	Red
NBR	Orange red	Red	Green
SBR	Yellow green	Green	Green
NR or IR	Brown	Violet blue	Green
IIR	Yellow (Droplet floats)	Pale blue green	Green

Identification by Spot Tests

To perform the test, a strip of each type of spot test paper is wet with the corresponding wetting solution and held in a parallel position about 5 mm above the surface of a heating element pressed against the sample. The color is recorded, and the sample is classified using Table 3.3.

The reagents for spot tests include spot test papers, generally filter papers impregnated with specific solutions, CR-NBR wetting solution, IIR wetting solution, and SBR-NR-IR wetting solution [7].

Preparation of spot test papers

CR-NBR Spot Test Papers: 2.0 g of cupric acetate and 0.25 g of Metanil yellow are dissolved in 500 ml of methanol. The filter papers are impregnated into the solution, dried, and cut into strips.

IIR Spot Test Papers: A blank filter paper is used.

SBR-NR-IR Spot Test Papers: A solution of 3 g of p-dimethylaminobenzaldehyde and 0.05 g of hydroquinone in 100 ml of ethyl ether is prepared. Then, the filter papers are impregnated into the solution, dried, and cut into strips.

These spot test papers should be stored in brown glass bottles in sufficient darkness. Otherwise, they will lose their efficiency if stored in light.

Preparation of wetting solutions

CR-NBR Wetting Solution: 2.5 g of benzidene dihydrochloride is dissolved in a mixture of 500 ml methanol and 500 ml of water. Then, 10 ml of an aqueous solution of hydroquinone (0.1%) is added to the mixture.

IIR Wetting Solution: 5.0 g of yellow mercuric oxide (HgO) is added to a mixture of 15 ml of H_2SO_4 and 80 ml of water. The solution is subjected to continuous heating until the oxide dissolves. Then it is cooled and diluted to 100 ml with water.

SBR-NR-IR Wetting Solution: 30 g of trichloroacetic acid is dissolved in isopropanol and diluted to 100 ml with isopropanol.

All these solutions must be stored in the dark for long-term stability.

Table 3.3 Spot test response of different polymers

Polymer	CR-NBR test	IIR test	SBR-NR-IR test
CR	Red/Violet	Blank/pale brown	Green
NBR	Green	Pale brown	Yellow green
SBR	Blank/pale brown	Brown	Blue green
NR/IR	Blank/pale brown	Brown	Blue
IIR	Blank/pale brown	Yellow	Pale lavender

3.2.2 Elemental Analysis

The polymers contain a limited number of elements with diagnostic significance. Apart from the hydrocarbon backbone, the major elements include nitrogen, sulfur, and halogen. The test for analysing the elements using metallic sodium in a fusion tube is commonly known as the fusion test and broadly known as the Lassaigne test [3, 8].

The polymer sample is taken in a fusion tube and fused with sodium by heating on a Bunsen burner. When the fusion tube turns red hot, it is broken into a mortar pestle and ground uniformly. The finely ground pieces are mixed with distilled water and then filtered. The filtrate is collected for subsequent testing, represented in Table 3.4.

$$Na + C + N \rightarrow NaCN$$

Table 3.4 Elemental analysis of polymers

Experiment	Observation	Inference
Filtrate + freshly prepared $FeSO_4$ solution + 1 drop of conc. H_2SO_4, then 2–3 drop $FeCl_3$ solution is added	Prussian blue color appears as shown in Fig. 3.1	Nitrogen is present (Possible polymers: NBR, PU, etc.)
See Fig. 3.1		
Filtrate + dilute hydrochloride acid (HCl) + lead acetate ($Pb(CH_3COO)_2$)	Black precipitation of lead acetate (PbS) appears as shown in Fig. 3.2	Sulfur is present. (Possible polymers: Hypalon, polysulphide)
Filtrate + a few drops of sodium nitroprusside ($Na_2[Fe(CN)_5NO]$)	Deep violet color appears, Fig. 3.2	
See Fig. 3.2		
Filtrate + HNO_3 (Conc.) + $AgNO_3$	White precipitation appears, Fig. 3.3, which is soluble in ammonium hydroxide (NH_4OH)	Chlorine is present. (PVC, Hypalon, CR, Chloro-butyl)
	A pale-yellow precipitation appears, Fig. 3.3, which is partially soluble in ammonium hydroxide (NH_4OH)	Bromine is present. (Bromo-butyl)
Filtrate is neutralized with dil. HCl. A drop is placed on a dry test paper (produced by immersing in an aqueous solution of Alizarin's)	The red spot will turn yellow	Fluorine is present. (Viton G)
See Fig. 3.3		

$$FeSO_4 + 6NaCN \longrightarrow Na_4[Fe(CN)_6] + Na_2SO_4$$

$$4FeCl_3 + 3Na_4[Fe(CN)_6] \longrightarrow \underset{\text{Prussian blue}}{Fe_4[Fe(CN)_6]_3} + 12NaCl$$

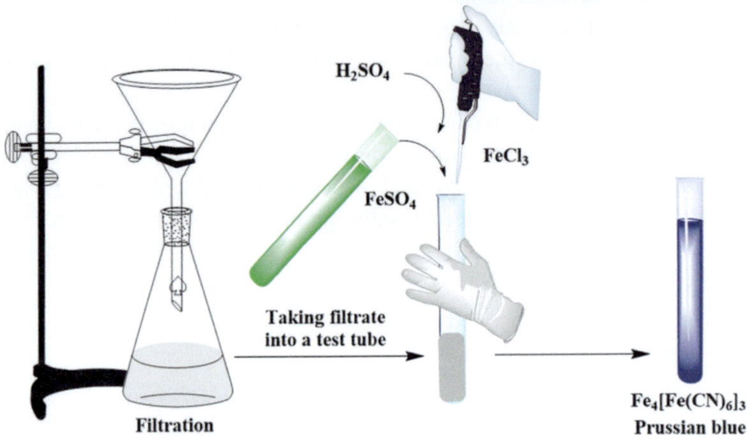

Fig. 3.1 Elemental analysis for nitrogen

$$Pb(CH_3COO)_2 + S^{2-} \longrightarrow \underset{\text{Black}}{PbS\downarrow} + 2CH_3COO^-$$

$$S^{2-} + [Fe(CN)_5NO]^{2-} \longrightarrow \underset{\text{Violet}}{[Fe(CN)_5NOS]^{4-}}$$

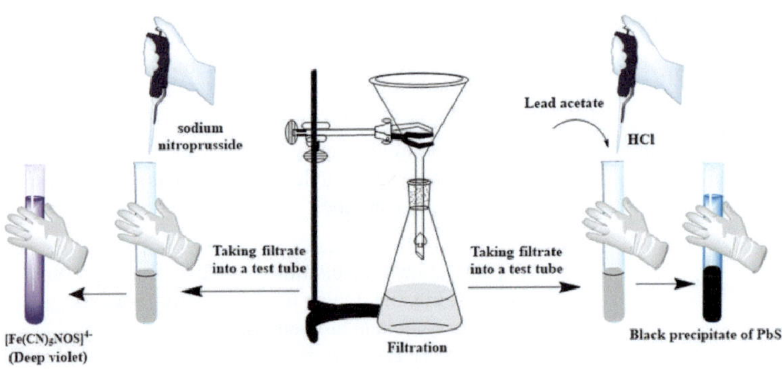

Fig. 3.2 Elemental analysis for sulphur

3.2.3 Functional Group Analysis

The backbone of the polymer is primarily made with hydrocarbon, either with saturation or unsaturation. The major functional group, attached to the polymer backbone,

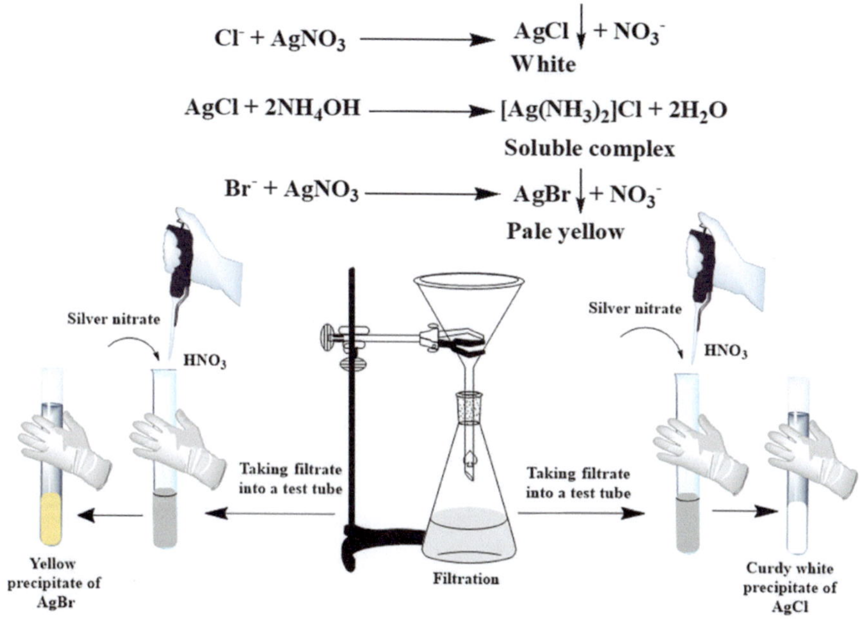

Fig. 3.3 Elemental analysis for halogen (chlorine and bromine)

Table 3.5 Functional group analysis of polymers

Experiment	Observation	Inference
Azo-dye test 1–2 g dried acetone extracted sample is boiled with 20 ml nitric acid (HNO_3) for 1 h. Then the solution is diluted with distilled water and further extracted with ether. The ether extracted sample is further reduced with Zinc (Zn). Sodium nitrite ($NaNO_2$) and HCl are added to the solution, keeping the temperature 0–5 °C, and then β-naphthol is added [3]	A red coloured dye is formed as shown in Fig. 3.4	The **phenyl group** is present
See Fig. 3.4		
Beilstein test A cleaned copper wire is heated, causing the formation of a layer of copper (II) oxide. Then the wire is dipped into the sample and reheated in the flame [3]	A green flame is observed as shown in Fig. 3.5	**Halogen** is present (Cl, Br, I) (Except fluorides)
See Fig. 3.5		

includes aliphatic groups, aromatic rings, and polar groups like halogens or cyanide. Table 3.5 shows the analysis of the phenyl group and halogen present in the polymer.

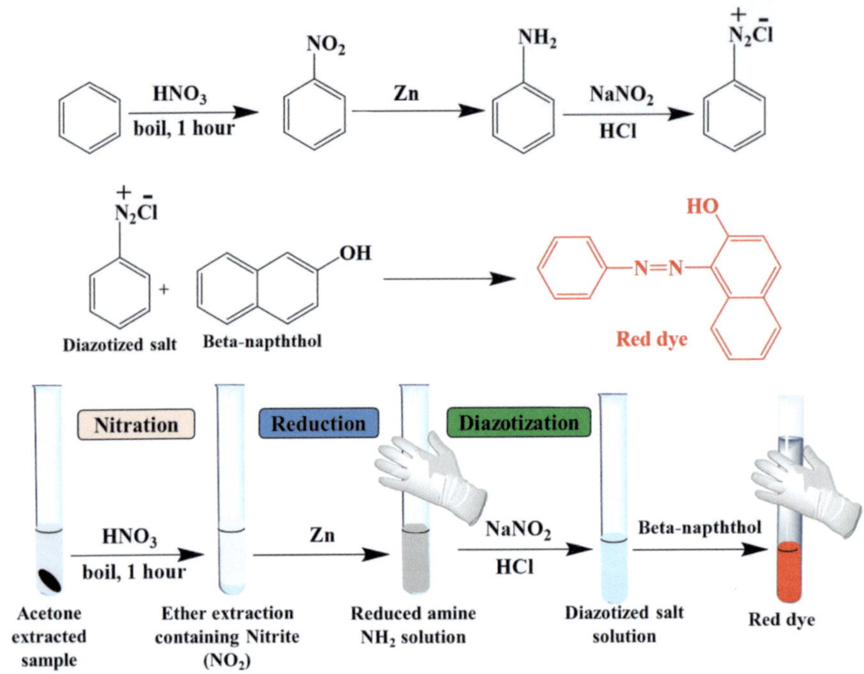

Fig. 3.4 Azo dye test

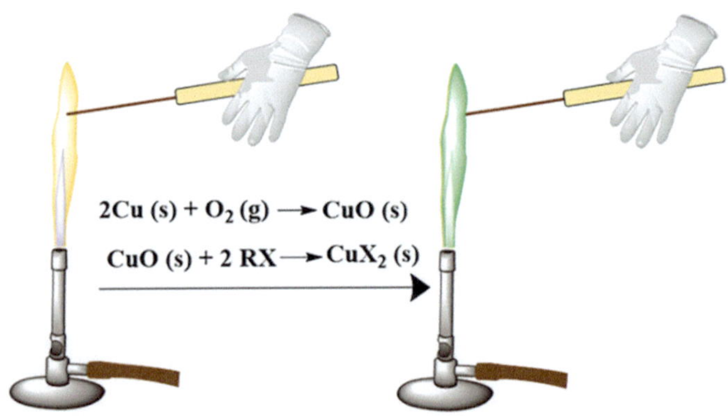

Fig. 3.5 Beilstein test

3.2.4 *Analysis of Unsaturation*

Qualitative analysis of unsaturation: At first, the acetone-extracted rubber is treated with bromine solution (bromine, Br_2, in carbon tetrachloride, CCl_4 solution) in a water bath. CCl_4 acts as a solvent in this reaction, providing a relatively inert medium for the bromination. The color of the Br_2/CCl_4 solution is reddish brown. As the bromination happens, the solution becomes colorless. It ensures the presence of unsaturation in the polymer backbone. In the absence of unsaturation, the bromination will not occur, and the color of the Br_2/CCl_4 solution will remain the same [3].

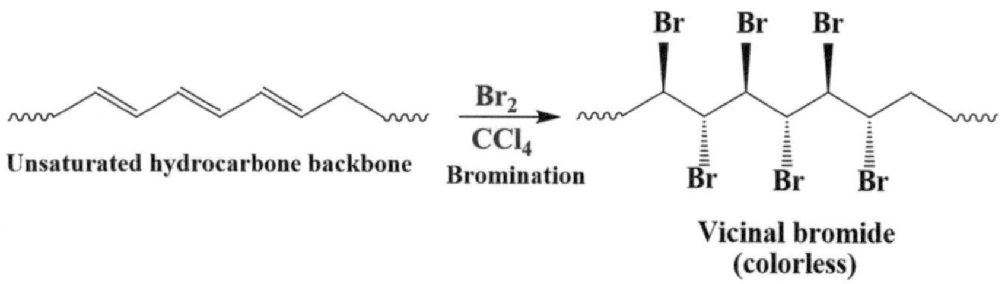

Quantitative analysis of unsaturation: Iodine number analysis

Iodine value measures the total number of double bonds present in a polymer. It is expressed as the grams of iodine that will react with the unsaturation (double bonds) in 100 g of polymer.

Iodine monochloride or monobromide is allowed to react with the sample in the dark. The double bond in the sample will react with iodine and form halides. The quantity of iodine consumed is directly proportional to the number of double bonds present. The amount of unreacted iodine monochloride or monobromide is determined from iodine liberated on the addition of an excess of potassium iodide (KI) by titrating the sample with standard sodium thiosulfate ($Na_2S_2O_3$) using starch as an indicator. The measure of iodine absorbed by the polymer in the reaction gives the degree of unsaturation, Eq. (3.1).

$$\mathbf{Iodinevalue} = \frac{(\mathbf{V}_1 - \mathbf{V}_2)\mathbf{ml} \times \mathbf{M} \times 126.9\,\mathbf{g/mol}}{\mathbf{Wg}} \tag{3.1}$$

V_1 and V_2 are the volumes of thiosulfate used for blank titration and the sample, respectively. M is the concentration of the thiosulfate solution, and Wg is the weight of the solution in gram. The molecular weight of iodine is 126.9 g/mol [9–12].

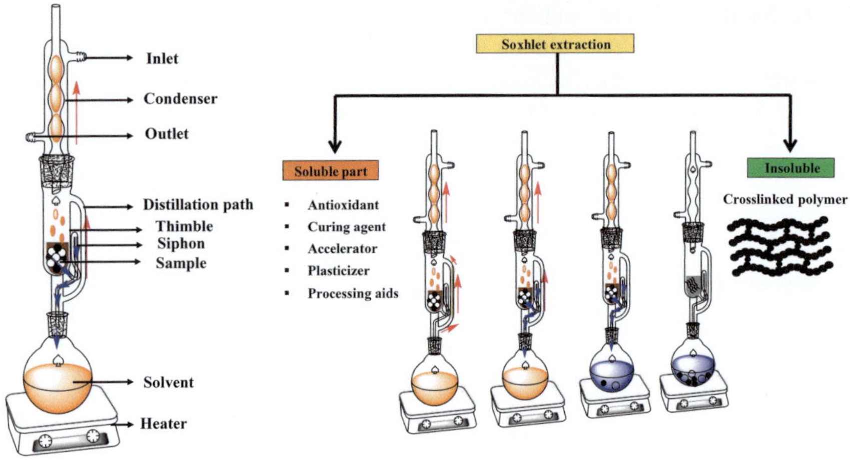

Fig. 3.6 Soxhlet apparatus and Soxhlet extraction

3.2.5 Soxhlet Extraction

Soxhlet extraction is an extraction process to separate low molecular weight compounding ingredients like process oil, plasticizers, antioxidants, and curing agents from compounded polymer. The purpose of Soxhlet extraction is to quantify the concentration of extractable additives and finally obtain the polymer in its pure form. The sample is placed inside the thimble. A suitable solvent is taken in the bottom flask. A reflux condenser is assembled with the Soxhlet extractor attachment. As the heating progresses, the solvent from the bottom flask starts to evaporate and follows a path to the condenser (pathway marked in red, Fig. 3.6). Then, the condensed vapor drips on the sample and dissolves the extractable compounds. Once the extractor chamber fills completely up to a point, the solvent with dissolved extractables siphons back to the bottom flask (pathway marked in blue, Fig. 3.6). This cycle is repeated for several hours. At the end, the extractor chamber remains with the undissolved crosslinked polymer, and the bottom flask contains the extracted additives [13, 14].

3.3 Determination of Trace Elements in Polymers

The presence of trace element determination in polymers is an essential analytical approach, especially for quality control, monitoring of the environment, and understanding the effect of residual catalysts, stabilizers, or contaminants. Few trace elements, such as metal ions (e.g., Fe, Cu, Zn, Ti, Al, Pb), sometimes originate from polymerization catalysts, pigments, stabilizers, or contamination during processing.

The sample should be digested in acid to analyze these elements accurately. Many polymers are chemically inert and primarily organic. They must be broken down to release the inorganic trace elements. This digestion is commonly done using strong acids such as nitric acid (HNO_3), sulfuric acid (H_2SO_4), perchloric acid ($HClO_4$), or aqua regia (a mixture of HCl and HNO_3). In some cases, microwave-assisted digestion is used to enhance the breakdown of the polymer matrix.

Once the polymer has been fully digested and the trace elements are in solution, a variety of chemical methods can be used:

1. **Complexometric Titration** method is widely used for determining metal ions (Ca^{2+}, Mg^{2+}, Cu^{2+}, and Fe^{3+}). EDTA (ethylenediaminetetraacetic acid) is the most common titrant, which forms stable complexes with metal ions. Indicators such as Eriochrome Black T are used to detect the endpoint [15].
2. **Gravimetric Analysis** involves precipitating the trace element as an insoluble compound, filtering, drying, and weighing it. This method is highly accurate but time-consuming and less suitable for ultra-trace levels.
3. **Colorimetric Methods** involve forming a colored complex between the metal ion and a specific reagent. The intensity of the color, measured spectrophotometrically, is proportional to the concentration of the element. For example, 1,10-phenanthroline forms an orange-red complex with Fe^{2+}, while dimethylglyoxime is used for nickel detection [16].
4. **Precipitation and Solvent Extraction** are separation techniques often used before analysis. For instance, a specific metal may be precipitated from solution, filtered, and redissolved for titration or extracted into an organic solvent for further quantification [17].

These chemical methods are generally less sensitive and more labour-intensive than modern instrumental methods (like AAS, ICP-OES, or ICP-MS). However, they still hold value for routine analyses, educational labs, or low-cost environments.

Figure 3.7 shows collective identification techniques used in the rubber industry.

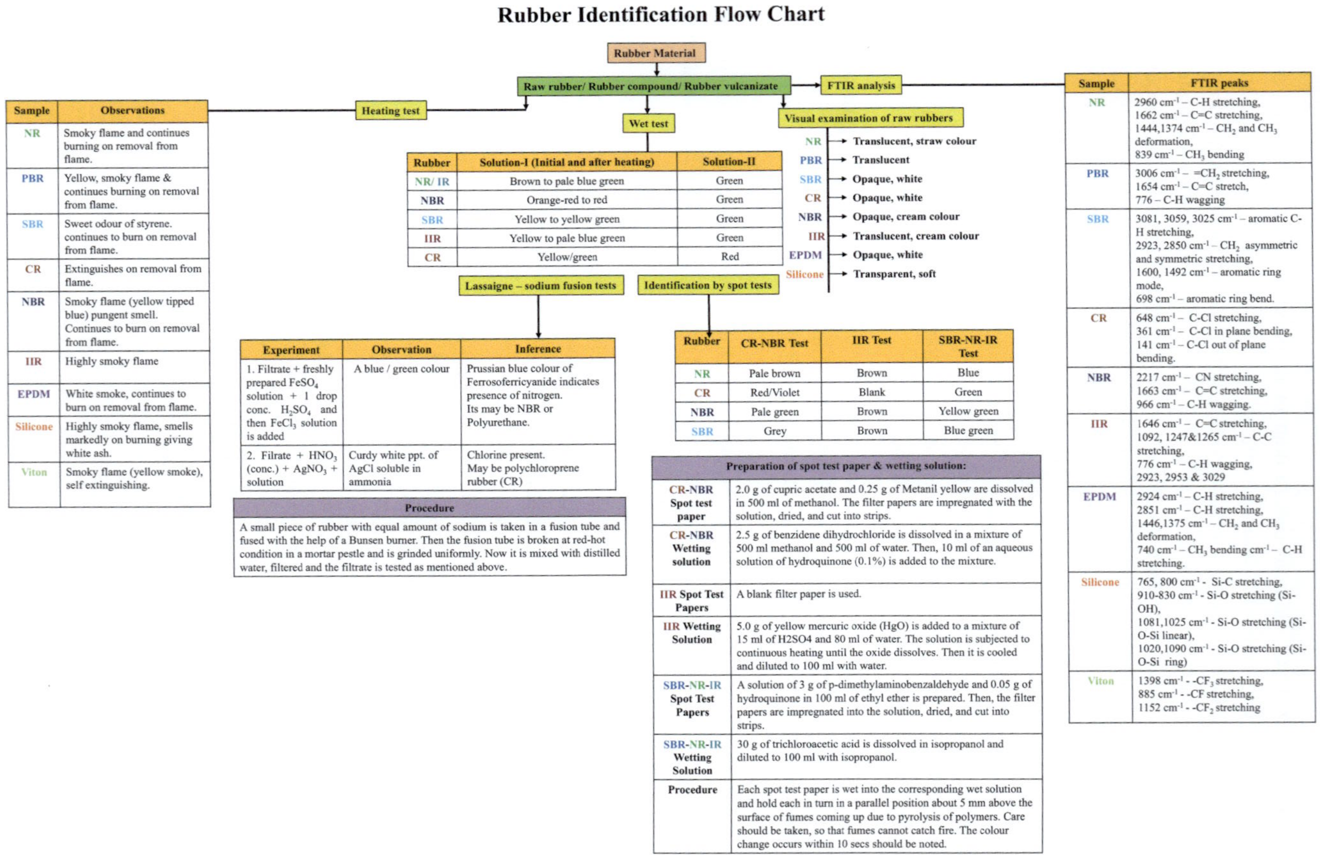

Sample	Observations
NR	Smoky flame and continues burning on removal from flame.
PBR	Yellow, smoky flame & continues burning on removal from flame.
SBR	Sweet odour of styrene. continues to burn on removal from flame.
CR	Extinguishes on removal from flame.
NBR	Smoky flame (yellow tipped blue) pungent smell. Continues to burn on removal from flame.
IIR	Highly smoky flame
EPDM	White smoke, continues to burn on removal from flame.
Silicone	Highly smoky flame, smells markedly on burning giving white ash.
Viton	Smoky flame (yellow smoke), self extinguishing.

Rubber	Solution-I (Initial and after heating)	Solution-II
NR/ IR	Brown to pale blue green	Green
NBR	Orange-red to red	Green
SBR	Yellow to yellow green	Green
IIR	Yellow to pale blue green	Green
CR	Yellow/green	Red

Experiment	Observation	Inference
1. Filtrate + freshly prepared $FeSO_4$ solution + 1 drop conc. H_2SO_4 and then $FeCl_3$ solution is added	A blue / green colour	Prussian blue colour of Ferrosoferricyanide indicates presence of nitrogen. Its may be NBR or Polyurethane.
2. Filrate + HNO_3 (conc.) + $AgNO_3$ + solution	Curdy white ppt. of AgCl soluble in ammonia	Chlorine present. May be polychloroprene rubber (CR)

Procedure
A small piece of rubber with equal amount of sodium is taken in a fusion tube and fused with the help of a Bunsen burner. Then the fusion tube is broken at red-hot condition in a mortar pestle and is grinded uniformly. Now it is mixed with distilled water, filtered and the filtrate is tested as mentioned above.

Rubber	CR-NBR Test	IIR Test	SBR-NR-IR Test
NR	Pale brown	Brown	Blue
CR	Red/Violet	Blank	Green
NBR	Pale green	Brown	Yellow green
SBR	Grey	Brown	Blue green

Preparation of spot test paper & wetting solution:	
CR-NBR Spot test paper	2.0 g of cupric acetate and 0.25 g of Metanil yellow are dissolved in 500 ml of methanol. The filter papers are impregnated with the solution, dried, and cut into strips.
CR-NBR Wetting solution	2.5 g of benzidene dihydrochloride is dissolved in a mixture of 500 ml methanol and 500 ml of water. Then, 10 ml of an aqueous solution of hydroquinone (0.1%) is added to the mixture.
IIR Spot Test Papers	A blank filter paper is used.
IIR Wetting Solution	5.0 g of yellow mercuric oxide (HgO) is added to a mixture of 15 ml of H2SO4 and 80 ml of water. The solution is subjected to continuous heating until the oxide dissolves. Then it is cooled and diluted to 100 ml with water.
SBR-NR-IR Spot Test Papers	A solution of 3 g of p-dimethylaminobenzaldehyde and 0.05 g of hydroquinone in 100 ml of ethyl ether is prepared. Then, the filter papers are impregnated into the solution, dried, and cut into strips.
SBR-NR-IR Wetting Solution	30 g of trichloroacetic acid is dissolved in isopropanol and diluted to 100 ml with isopropanol.
Procedure	Each spot test paper is wet into the corresponding wet solution and hold each in turn in a parallel position about 5 mm above the surface of fumes coming up due to pyrolysis of polymers. Care should be taken, so that fumes cannot catch fire. The colour change occurs within 10 secs should be noted.

Sample	FTIR peaks
NR	2960 cm^{-1} – C-H stretching, 1662 cm^{-1} – C=C stretching, 1444,1374 cm^{-1} – CH_2 and CH_3 deformation, 839 cm^{-1} – CH_3 bending
PBR	3006 cm^{-1} – =CH_2 stretching, 1654 cm^{-1} – C=C stretch, 776 – C-H wagging
SBR	3081, 3059, 3025 cm^{-1} – aromatic C-H stretching, 2923, 2850 cm^{-1} – CH_2 asymmetric and symmetric stretching, 1600, 1492 cm^{-1} – aromatic ring mode, 698 cm^{-1} – aromatic ring bend.
CR	648 cm^{-1} – C-Cl stretching, 361 cm^{-1} – C-Cl in plane bending, 141 cm^{-1} – C-Cl out of plane bending.
NBR	2217 cm^{-1} – CN stretching, 1663 cm^{-1} – C=C stretching, 966 cm^{-1} – C-H wagging.
IIR	1646 cm^{-1} – C=C stretching, 1092, 1247&1265 cm^{-1} – C-C stretching, 776 cm^{-1} – C-H wagging, 2923, 2953 & 3029
EPDM	2924 cm^{-1} – C-H stretching, 2851 cm^{-1} – C-H stretching, 1446,1375 cm^{-1} – CH_2 and CH_3 deformation, 740 cm^{-1} – CH_3 bending cm^{-1} – C-H stretching.
Silicone	765, 800 cm^{-1} - Si-C stretching, 910-830 cm^{-1} - Si-O stretching (Si-OH), 1081,1025 cm^{-1} - Si-O stretching (Si-O-Si linear), 1020,1090 cm^{-1} - Si-O stretching (Si-O-Si ring)
Viton	1398 cm^{-1} - -CF_3 stretching, 885 cm^{-1} - -CF stretching, 1152 cm^{-1} - -CF_2 stretching

Fig. 3.7 Identification of various rubbers by chemical and spectroscopic (FTIR) analysis

Acknowledgements We have used ChemDraw Professional 17.1 (institute access) and bioRender.com (student licence) for making illustration.

Conflict of Interest The Authors declare no competing conflict of interest.

References

1. Babrauskas Vytenis (2014) Ignition Handbook: Principles and applications to fire safety engineering, fire investigation, risk management and forensic science. Fire Science Publishers, A division of Fire Science and Technology Inc.
2. Roenner N, Rein G (2019) Convective ignition of polymers: new apparatus and application to a thermoplastic polymer. Proc Combust Inst 37:4193–4200. https://doi.org/10.1016/j.proci.2018.05.180
3. Vogel AI, Furniss BS (1996) Vogel's textbook of practical organic chemistry. Longman Sci Tech
4. Encinar JM, González JF (2008) Pyrolysis of synthetic polymers and plastic wastes. Kinetic study. Fuel Process Technol 89:678–686. https://doi.org/10.1016/J.FUPROC.2007.12.011
5. Saxena S (2025) Pyrolysis and beyond: sustainable valorization of plastic waste. Appl Energy Combust Sci 21:100311. https://doi.org/10.1016/J.JAECS.2024.100311
6. Jahirul MI, Rasul MG, Chowdhury AA, Ashwath N (2012) Biofuels production through biomass pyrolysis—a technological review. Energies (Basel) 5:4952–5001. https://doi.org/10.3390/EN5124952
7. ASTM D297 (2021) Test methods for rubber products chemical analysis
8. Bowen HJM (n.d.) Identification of polymers in university class experiments
9. Hiscox DJ (1948) Determination of iodine numbers. Anal Chem 20(7):679–680. https://doi.org/10.1021/ac60019a030
10. Rehner J (n.d.) Unsaturation of polyisobutylenes and butyl rubber. Reactions with nitrosobenzene, thiocyanogen, iodine monochloride, and ozone
11. Kemp Aind AR, Peters H (n.d.) Unsaturation of butadiene and related polymers as determined by iodine chloride addition
12. Lee TS, Kolthoff IM, Mairs MA (1948) Determination of unsaturation of synthetic and natural rubbers by means of iodine monochloride. J Polym Sci 3:66–84. https://doi.org/10.1002/pol.1948.120030109
13. Vandenburg Harold J, Clifford Anthony A, Bartle Keith D, Carroll John, Newton Ian, Garden Louise M, Dean John R, Costley Claire T (1997) Critical review: analytical extraction of additives from polymers. Analyst 122:101R–115R. https://doi.org/10.1039/A704052K
14. Möller J, Strömberg E, Karlsson S (2008) Comparison of extraction methods for sampling of low molecular compounds in polymers degraded during recycling. Eur Polym J 44:1583–1593. https://doi.org/10.1016/J.EURPOLYMJ.2008.03.027
15. Kozak J, Townshend A (2019) Titrimetry|overview. Encycl Anal Sci 2019:111–20. https://doi.org/10.1016/B978-0-12-409547-2.14419-1.
16. Jayalakshmi S, Suresh Kumar P, Govindan D, Mariappan M, Ameen F, Veerappan A (2024) Selective colorimetric detection of trace chromium (III) based on silver nanoparticles aggregation via a domino-like reaction. Opt Mater (Amst) 154:115679. https://doi.org/10.1016/J.OPTMAT.2024.115679
17. Aiken G, Leenheer J (1993) Isolation and chemical characterization of dissolved and colloidal organic matter. Chem Ecol 8:135–151. https://doi.org/10.1080/02757549308035305

Chapter 4
Physical Analysis of Polymers

Nikhil Dhore, Nikhil Kumar, and Aparajita Pal

Abstract This chapter provides an in-depth exploration of the fundamental and analytical aspects related to the physical analysis of polymers. It begins with the molecular weight distribution conception and explores into various experimental techniques used for the molecular weight determination of polymers, including ebulliometry, viscometry, osmometry, end-group analysis, light scattering, and ultracentrifugation. The chapter then transitions to the concept of electrical conductivity in polymers, discussing both its theoretical foundation and practical measurement techniques for electrical and thermal conductivity. The latter half of the chapter focuses on the crystallinity of polymers, elaborating on their basic principles, structural implications through unit cells and Miller indices, and influence on material properties. Models such as the fringed-micelle and folded-chain models are described, along with the thermodynamics and kinetics of crystallization. The chapter concludes with a detailed discussion on the analysis of polymer crystallinity using X-ray diffraction (XRD), covering X-ray generation, the mechanism of XRD pattern formation, and analysis of polymer samples. Overall, this chapter offers a comprehensive understanding of key physicochemical properties critical to polymer science and technology.

Keywords Physical analysis · Crystallinity · Electrical properties · Viscometry

4.1 Introduction

Polymers have quietly transformed the modern world. Their usage spans from everyday items like water bottles and food packaging to high-end applications in aerospace, biomedical devices, and flexible electronics [1–4]. Physical attributes of

N. Dhore · A. Pal
Rubber Technology Centre, Indian Institute of Technology Kharagpur, Kharagpur, India

N. Kumar (✉)
Advanced Technology Development Centre, Indian Institute of Technology Kharagpur, Kharagpur, India
e-mail: nikhilsingh25694@gmail.com

S. Chattopadhyay and N. Kumar (eds.), *Polymer Characterization*,
https://doi.org/10.1007/978-981-95-2609-3_4

polymers are required to define their performance in real-world applications. Physical analysis of polymers refers to the measurement of the physical properties of polymer materials that determine their behaviour under various conditions without altering their chemical structure. Mainly, three foundational physical attributes include molecular weight, electrical conductivity, and degree of crystallinity [5–7].

This chapter is divided into three parts: First, how molecular weight and its distribution affect polymer properties, followed by the classical and modern techniques used to determine it, such as ebulliometry, osmometry, light scattering, and ultracentrifugation [8]. Second, we shift focus to electrical conductivity of polymers, a critical parameter in the development of smart and functional materials using methods like the four-point probe method, ASTM D257, electrochemical impedance, rheoelectric measurements, transient hot wire method, steady state, and modified transient phase source. Third, is on understanding the concept of crystallinity using techniques like X-ray diffraction (XRD) [9].

4.2 Concept of Molecular Weight (MW) Distribution

Molecular weight is a fundamental concept in polymer science that directly influences a polymer's physical, mechanical, thermal, and processing properties. Unlike small molecules, polymers have long chains of repeating units (monomers), and these chains often vary in length. As a result, a polymer sample doesn't have a single molecular weight, but a wide molecular weights distribution [10]. The detailed explanation is given in Chaps. 2 and 3.

4.3 Determination of Molecular Weight of Polymer

4.3.1 Ebulliometry

It is a method to measure the number-average molecular weight of polymers by finding the change between the boiling point of a solution containing the polymer and the pure solvent. This method is based on the colligative property. The boiling point elevation is directly proportional to the solute concentration [11, 12].

The basic part of the ebulliometer is the Washburn pump, boiler, RTD probe, condenser, glass shield, side tubing, and heating element. A solution of the polymer is to be prepared in a suitable solvent [13]. The boiling point of a solution is measured in an ebulliometer and compared with the boiling point of a pure solvent. The boiling point difference is related to the MW of the polymer. An ebulliometric constant (K6) is to be used during the calculation, which is specific to the solvent, the polymer, and its concentration. The requirement for the accurate measurement is thermal stability, temperature and concentration equilibrium, and temperature sensing.

1 mol of the polymer sample added to 1 L of water will raise the boiling point by 0.52 °C. If any 500 g of polymer sample is dissolved in 1 L of water, the boiling point of the new solution rises by 0.76 °C.

$$\text{moles} = \frac{\text{mole}}{0.52} \times 0.76°\text{C} = 1.46\,\text{mole (of substance)} \tag{4.1}$$

To find MW of the substance

$$\text{MW(Daltons)} = \frac{500}{1.46\,\text{mole}} \times \frac{\text{mole} \times \text{MW(Daltons)}}{g} = 342\,\text{Daltons} \tag{4.2}$$

This MW correspondence to the MW of sucrose and could be confirmed with further tests [14].

This technique is relatively simple in comparison with other methods, and it is less expensive than the other equipment. It is sensitive to the impurities in the polymer and solvent which leads to the error in the measurements. The elevation is also typically small, so it's best suited for the polymers with lower molecular weights. The precision of this method is limited at high molecular weight [15].

4.3.2 Viscometry

Viscometry, or fluid viscosity measurement, is used to determine the MW of the polymers. Several numbers of viscometers are used, including falling ball viscometers, rotational viscometers, capillary viscometers (Fig. 4.1) [16, 17]. The basic principle of these methods is that the viscosity of the polymer solution is directly related to its MW [11]. This method determines the viscosity-average molecular weight (Mv) of the polymer.

The Mark-Houwink equation is used to relate the viscosity to the average MW. A graph of log [η] versus log M is plotted, and the nature of the graph is a straight line. The value of constant K and a can be obtained from a graph where the ordinate intercept is logK and the slope is a [18].

$$[\eta] = KM^a \tag{4.3}$$

$$\log[\eta] = \log K + a \log M \tag{4.4}$$

The intrinsic viscosity can be measured using viscometers. Assuming a polymer solution or pure solvent is passing via a capillary tube. Let us link the Poiseuille equation at pressure head (P) with the absolute viscosity of a solution of volume passing via the capillary having radius r and length l to their needed flow time [19].

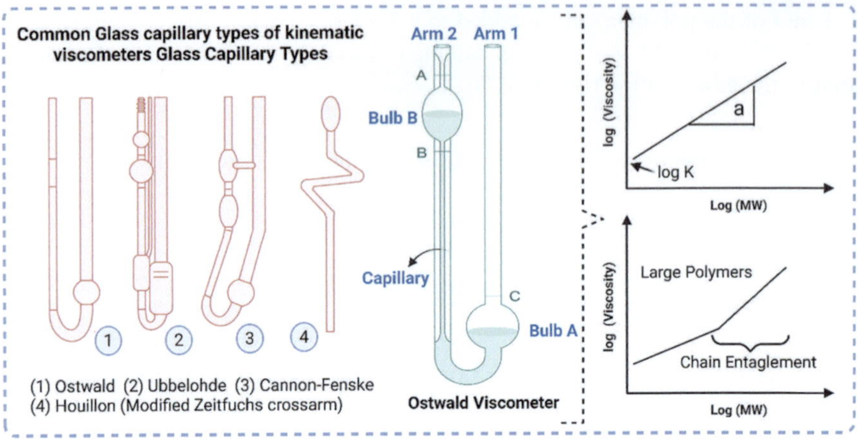

Fig. 4.1 Viscometers used for the determination of molecular weight

$$\eta = \frac{3.14\,Pr^4 t}{8Vl} \tag{4.5}$$

Another equation is Flory-Fox equation which relates the viscosity to the average MW. This equation is given as below [20]:

$$[\eta] = \frac{\Phi(r)^{(2)^{\frac{3}{2}}}}{M} \tag{4.6}$$

where, Φ = Flory constant, R = end-to-end distance, and $(r)^{(2)^{\frac{3}{2}}}$ = Hydrodynamic volume.

Other terms related to the viscometry Table 4.1 [21].

Table 4.1 Terms related to the viscosity

Terms	Definition	Formulae
Relative viscosity (η_r)	Ratio of the viscosity of solution to that of pure solvent	$\eta_r = \frac{\eta}{\eta_o}$ (4.7)
Specific viscosity (η_{sp})	Increment in viscosity due to the presence of solute in solvent	$\eta_{sp} = \frac{\eta}{\eta_o} - 1$ (4.8)
Reduced viscosity (η_{red})	Specific viscosity per unit concentration	$\eta_{\text{red}} = \frac{\eta_{sp}}{\text{c}}$ (4.9)
Inherent viscosity (η_{inh})	Natural logarithm of relative viscosity per unit concentration	$\eta_{\text{inh}} = \frac{\ln \eta_r}{c}$ (4.10)
Intrinsic viscosity $[\eta]$	Limit of reduced or inherent viscosity as concentration approaches zero	$[\eta] = \lim_{c \to 0}\left(\frac{\eta_r}{c}\right)$ (4.11) or $[\eta] = \lim_{c \to 0}\left[\frac{\eta_{sp}}{c}\right]$ (4.12)

4.3.3 Osmometry

Osmometry is absolute method to determine the M_n. In this method, a semipermeable membrane is used to separate the solvent from the polymer solution. This membrane is tightly held in the middle the two chambers as shown in Fig. 4.2. A valve seals the chamber, and a transducer is mounted on a stainless-steel diaphragm, helps continuously measures the internal pressure [22].

It's especially used when M_n is between 20,000 and 30,000 g/mol and up to around 500,000 g/mol. The osmotic pressure gets too low to precisely record data for Mn more than 500,000 g/mol [23]. Possible leakage in the membrane and issues with maintaining symmetry are the drawbacks. However, it has the advantage of not needing calibration and provides an absolute value of M_n. Since osmotic pressure depends on the number of particles in a solution (a colligative property), this technique can compare the osmotic pressure of pure solvents with that of polymer solutions [24].

The capillary rise in a dilute polymer solution is measured without restriction in an automated membrane osmometer, and the osmotic pressure is calculated using a modified form of the Van't Hoff equation [25]:

$$\pi = \frac{cRT}{M_n} + BC^2 \tag{4.13}$$

where, π = Osmotic pressure, C = Polymer concentration, R = Gas constant, T = Absolute temperature, M_n = Number average molecular weight, B = second virial coefficient, accounting for polymer–solvent interactions, and C^2 = square of the concentration (reflects non-ideal behaviour).

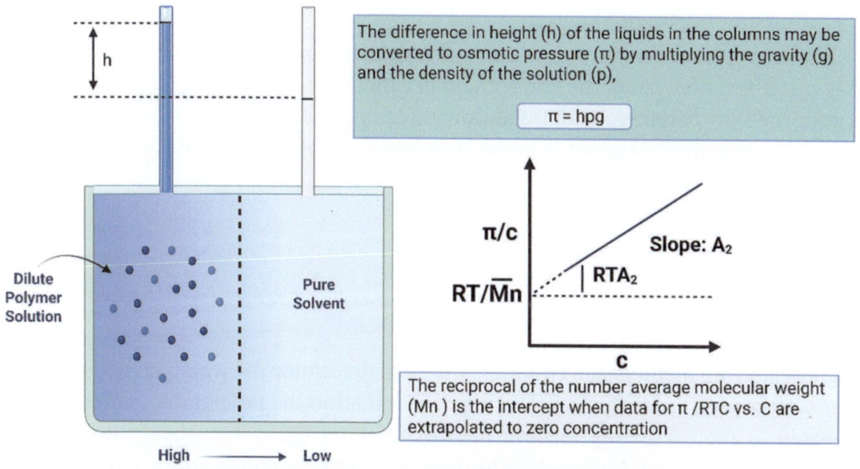

Fig. 4.2 Osmometry technique for molecular weight determination of polymers

4.3.4 End-Group Analysis

End-group analysis measures the number-average molecular weight (M_n) of a polymer by analyzing the end reactive functional groups. This analysis purely relies on quantitatively determining the concentration of these end groups and then calculating the molecular weight based on the relationship between the end group concentration and the degree of polymerization (DP) [26]. This analysis is particularly useful for polymers with relatively low MW, and it utilizes techniques like titration, NMR, or infrared spectroscopy to measure end-group concentrations.

Firstly, determine the functional groups present at the ends of the polymer chains, such as hydroxyl, carboxyl, amine, or ester groups. Choose a method suitable for quantifying the identified end groups.

Titration: For titratable groups like hydroxyl and carboxyl groups, titration with a standard acid or base can be used.

NMR Spectroscopy: ^{1}H NMR can be used to measure the integral of the end-group signal, which is then compared to the integral of the repeating unit.

Infrared Spectroscopy: IR spectroscopy can be used to measure the absorbance of specific end-group vibrations, providing information about end-group concentration.

By using the chosen method, determine the end group concentration in the polymer sample. For titration methods, M_n can be calculated using the following formula:

$$M_n = \frac{\text{Functionality}}{\text{Functional group equivalent}} \tag{4.14}$$

For NMR or IR methods, the average degree of polymerization (DP) or number of repeating units (X_n) is calculated and multiplied by the molecular weight of the repeating unit to obtain M_n.

End-group analysis provides an absolute determination of M_n, meaning it doesn't rely on calibration standards like some other methods. Usually, this method is more appropriate for polymers with rather low molecular weights (under 20,000) because the end group concentration becomes too low to be accurately measured at higher molecular weights [27].

4.3.5 Light Scattering

Light scattering methods are commonly used to determine the weight average molecular weight (M_w) of polymers. During determination the polarizable molecules are exposed to the oscillating electric field of incident light, they scatter the light as shown in Fig. 4.3. As light passes through a polymer solution, some energy is lost due to absorption, conversion into heat, and scattering of light. The amount of light scattered depends on several factors such as the concentration, molecular size, and

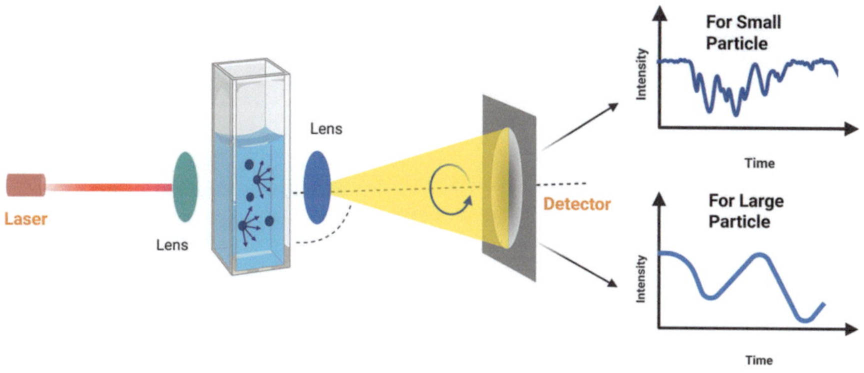

Fig. 4.3 Light scattering method for molecular weight determination of polymers

polarizability of the polymer. The polarizability, which relates to the ease of displacement of the electron cloud of a molecule, is directly linked to the molecular weight of the polymer [28].

$$\frac{I}{I_O} = e^{-\tau l} \tag{4.15}$$

where I = intensity of transmitted light, I_o = intensity of the incident light, τ = attenuation or turbidity coefficient, and l = path length through the medium.

Beer's law for light absorption helps one to explain light scattering measurement [29].

The intensity of scattered light or turbidity (τ), is directly proportional to the square of the difference between the refractive index of the polymer solution (n) and the pure solvent (n_o), proportional to the molecular weight of the polymer (M), and inversely proportional to the fourth power of the wavelength of the employed light (λ) [30].

$$\tau \propto \frac{M(n - n_o)}{\lambda^4} \tag{4.16}$$

where M = molecular weight of the polymer, n = refractive index of the polymer solution, n = refractive index of the solvent, and λ = wavelength of incident light.

4.3.6 *Ultracentrifugation*

Ultracentrifugation measures the molecular weight of molecules, particularly proteins and other macromolecules, by analyzing sedimentation behavior in a high-speed centrifugal field. Ultracentrifugation uses centrifugal force to separate the

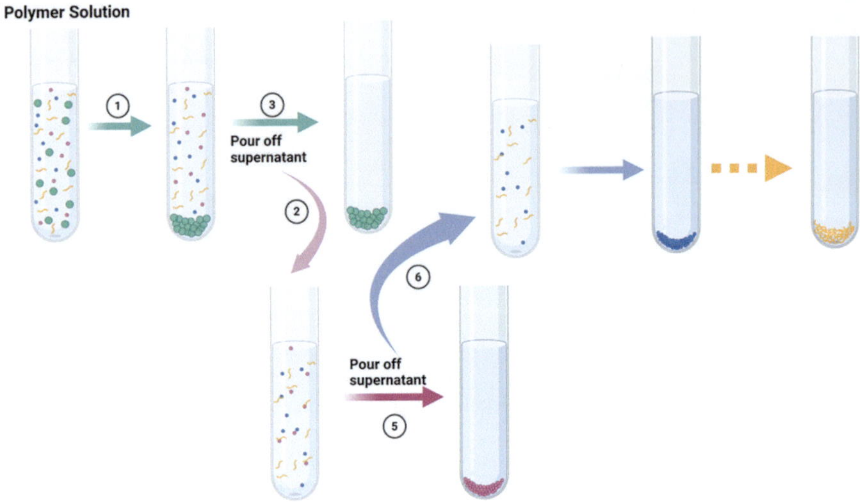

Fig. 4.4 Ultracentrifugation method for molecular weight determination of polymers

different components of a solution based on their density and size (Fig. 4.4) [31]. Two main methods are used for molecular weight determination in ultracentrifugation:

1. Sedimentation Velocity: The Particle's sedimentation in the solution is tracked, which helps to determine the size and shape.
2. Sedimentation Equilibrium: Polymer chains in solution distribute according to their molecular weight and reach an equilibrium concentration that varies based on their weight [32].

A sample is placed in a centrifuge cell and rotated at high centrifugal force in a rotor. Present particles move through the solution, and their movement is monitored with the help of optical detectors [33]. By analysing the rate of sedimentation of polymer particles and equilibrium distribution, the molecular weight, along with other properties, can be measured.

This technique has many different uses for characterizing various macromolecules, including biological moieties such as proteins, nucleic acids, and viruses. It is also considered the most accurate method for determining molecular weight, along with some accurate data on the size, shape, and interactions of macromolecules.

4.4 Concept of Electrical Conductivity of Polymers

Historically, polymers were recognized as insulators. In 1970, Heeger, Shirakawa, and MacDiarmid produced the first intrinsic conductive polymer, leading to the Nobel Prize in 2000 [34]. The polymer was developed by exposing polyacetylene to dopant chemicals, which included oxidizing or reducing agents, as well as electron donors or acceptors. The majority of polymers are insulators because [35]:

- **Molecular Structure**: Polymers are made of long chains of repeating units (monomers) bonded via covalent bonds.
- **Electronic Configuration**: In conventional polymers (like polyethylene, PVC, polystyrene), the electrons are localized in σ-bonds (single covalent bonds).
- **Lack of Free Electrons**: There are no delocalized or free electrons to move under an electric field, which is essential for the conduction of electricity.

The inadequate conductivity of polymers is elucidated by band theory. This theory proposes that the energy levels available to electrons are arranged into permissible bands, with certain energy levels identified as forbidden, referred to as the band gap [36], as illustrated in Fig. 4.5. This idea originates from the use of Schrödinger's Equation in a periodic field of a crystalline substance. The lowest bands are designated as valence bands and exhibit electrical passivity. The uppermost bands involved in electrical conduction are referred to as conduction bands. Conductors possess a partially full valence band, facilitating the excitation of an electron to a higher energy level, as exemplified in Fig. 4.5a. Other conductors, exemplified by divalent metals as shown in Fig. 4.5b, may exhibit an overlap between the vacant conductive band and a completely full valence band (Fig. 4.5d). Semiconductor possesses a comparatively smaller band gap energy than insulators. On the other hand, insulators possess a larger band gap.

Applied voltage causes the ordered movement of charges within the polymer chain to generate an electric current. Throughout this process, the positive charges migrate in the direction toward the electric field, while electrons (from negative charges) traverse in the opposite direction [37]. In most polymeric materials, this movement of electrons is the primary form of electrical conduction. Conductive polymers are distinct from conventional insulating polymers due to their conjugated backbone, which consists of alternating single and double bonds. To stabilize the introduced charges, the polymer undergoes resonance, a phenomenon where double bonds shift along the backbone, enabling movement of the delocalized π-electrons

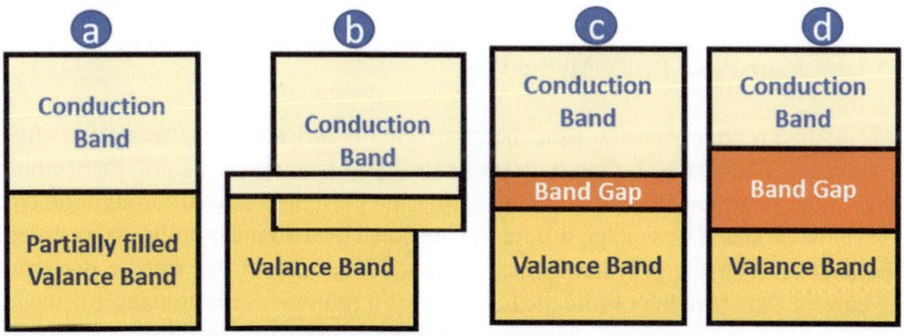

Fig. 4.5 Conduction and valence band concept

leads to stabilization of charges across multiple atoms. This enables delocalization of π-electrons across the chain, allowing charge mobility under an electric field. The phenomenon of double bond migration is referred to as resonance [38].

The process of doping allows additional charge formation (either electrons or holes) into the polymer chain, which helps in a significant increase in conductivity. The process of doping is as follows:

(i) Creates defects or deformations in the polymer chain.
(ii) Leads to the formation of quasi-particles, which are responsible for charge transport:

- Polarons: electron + lattice deformation pairs
- Bipolarons: two charges (same sign) + deformation
- Solitons: observed in specific polymers like polyacetylene.

The electrical conductivity of conductive polymers is strongly influenced by temperature. As the temperature rises, the polymer chains tend to move further apart due to thermal expansion, which can affect how easily electrons can move through the material. However, higher temperatures allow electrons to gain more energy (Boltzmann distribution). This increase in thermal energy makes it easier for electrons to jump from the valence band to the conduction band, enhancing the material's conductivity. Among conductive polymers, polyacetylene, polyaniline, and polypyrrole are the most extensively studied and widely used [39].

Methods that are used to measure the electrical conductivity of polymers:

- **Electrical**: Four Point Probe Method, ASTM D257, electrochemical Impedance, and Rheoelectric measurements
- **Thermal**: Transient hot wire method, steady state, and modified transient phase source.

4.5 Measurement of Conductive Properties of Polymers

4.5.1 Electrical Conductivity Measurement Techniques

4.5.1.1 Four-Point Probe Method

This method is widely used for the measurement of electrical properties. It uses four probes placed in a straight line with the exam equal spacing on the polymer sample which is to be tested (Fig. 4.6). Then, a constant current is passed through the two outermost probes. The voltage differential between the two inner probes is recorded. The resistance of the polymer is measured with the help of the voltage drop and the current. The benefits of this method are that it minimizes the contact resistance, thus providing more accurate measurement, and is ideal for the thin films without damaging them. For the above benefits, this method is preferred over the two-probe method [40].

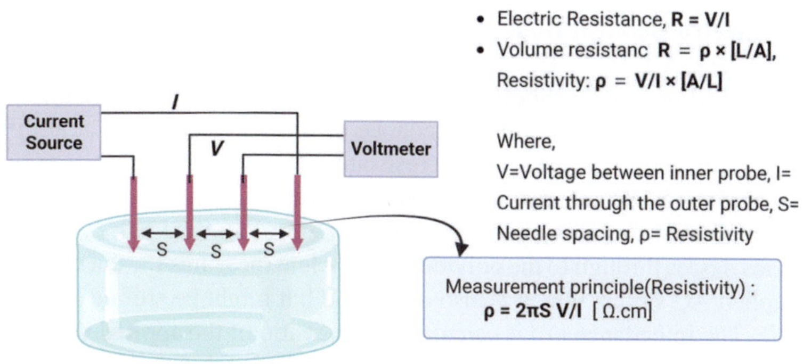

Fig. 4.6 Four-point probe method

4.5.1.2 ASTM D257—Standard Test Methods for DC Resistance or Conductance of Insulating Materials

This method was developed by ASTM International for the volume resistivity and surface resistivity measurement of electrically insulating materials like polymers (figure–). A clean, dry lat sample (rectangular or circular) is needed for the measurement. Moisture can affect the readings. The electrodes should be guarded properly on the sample to minimize the leakage current. For the surface resistivity electrodes are placed on the same surface and for the volume resistivity electrodes are placed on opposite surfaces. The parameters like temperature and relative humidity must be controlled at around 23 ± 2 °C and 50 ± 5% RH. Then a DC voltage (100–500 V, sometimes 1000 V for highly resistive materials) is applied across the electrodes. The resulting current is measured after a stabilization period. Resistivity is calculated using formulas based on electrode geometry and measured resistance. This test assesses polymers used in anti-static or dielectric applications [41].

Volume Resistivity can be expressed as follows:

$$\rho_v = R \times \frac{A}{t} \tag{4.17}$$

where R = measured resistance (Ω), A = electrode area (cm^2), and t = sample thickness (cm)

Surface resistivity can be expressed as follows:

$$\rho_s = R \times k \tag{4.18}$$

where k = geometric constant based on electrode configuration (unitless).

4.5.1.3 Electrochemical Impedance Spectroscopy (EIS)

Electrochemical Impedance Spectroscopy (EIS) can be used to measure the conductivity of polymers by probing the impedance of the polymer material at different frequencies. Impedance is the effective resistance of an electrical circuit or component to alternating current, resulting from the cumulative effects of ohmic resistance and reactance. A minor alternating current (AC) sinusoidal pulse of fluctuating frequencies passes through to the polymer. The system measures the current response to the applied AC signal, which is also a sinusoid but might be shifted in phase and amplitude. The impedance (Z) is calculated as the ratio of the applied voltage to the resulting current. The impedance is a complex number with a real part (resistance) and an imaginary part (reactance). The resistance component of the impedance, particularly the series resistance (R_S), is extracted from the impedance spectrum. The conductivity (σ) can be calculated using the following formula [42, 43]:

$$\sigma = \frac{1}{R_s \times \left(\frac{\text{thickness}}{\text{area}}\right)} \tag{4.19}$$

The data from this study is often presented as a Nyquist plot, which reveals information about the resistance and other interfacial properties. A Nyquist plot is a valuable tool for visualizing and analyzing the complex impedance of a system. It plots the imaginary part of the impedance against the real part, revealing key information about the electrochemical processes occurring at different frequencies. The shape of the Nyquist plot, particularly the presence and characteristics of semicircles, provides insights into the behaviour of various components like resistances, capacitances, and Warburg impedances [44]. EIS is non-destructive technique. EIS provides more than just conductivity; it can also reveal information about charge transfer, diffusion, and other interfacial processes occurring within the polymer over a wide frequency range [45].

4.5.1.4 Rheo-Electric Measurements

Rheo-electric measurements are used to investigate the impact of processing history and thermo-mechanical factors on the electrical conductivity of molten polymers or reactive resins containing conductive fillers. This technique combines rheological (flow) and electrical conductivity measurements to understand how these materials behave under stress and temperature. It's particularly useful for studying the behaviour of polymer composites with conductive fillers, where the flow and electrical properties are intimately linked. Rheo-electric measurements can reveal how shear flow affects the arrangement and structure of conductive fillers within the polymer matrix, impacting conductivity, and also predict the material performance [46].

4.5.2 *Thermal Conductivity Measurement Techniques*

4.5.2.1 Transient Hot Wire Method

The transient hot wire method measures thermal conductivity by observing the temperature increase of a thin wire submerged in the substance being tested. A thin, vertical, resistive wire (usually platinum or nichrome) is placed in the polymer fluid, resin, or material whose thermal conductivity needs to be determined (Fig. 4.7) [47]. A step voltage is applied to the wire, causing it to heat up quickly (Joule effect), and the wire effectively acts as a heat source. The temperature of the wire is continuously monitored using its resistance, which changes with temperature. The temperature rise rate correlates with the thermal conductivity of the surrounding substance. The thermal conductivity of the material can be determined by analyzing the time-dependent temperature variations during the test [48].

$$\Delta T = \frac{q}{4\pi k}\ln\left(\frac{4\alpha t}{r^2}\right) \tag{4.20}$$

where $\Delta T(t)$ = temperature rise of the wire at time t, q = heat per unit length per unit time (W/m), k = thermal conductivity of the sample (W/m·K), α = thermal diffusivity of the sample (m^2/s), r = radius of the wire (m), and t = time (s).

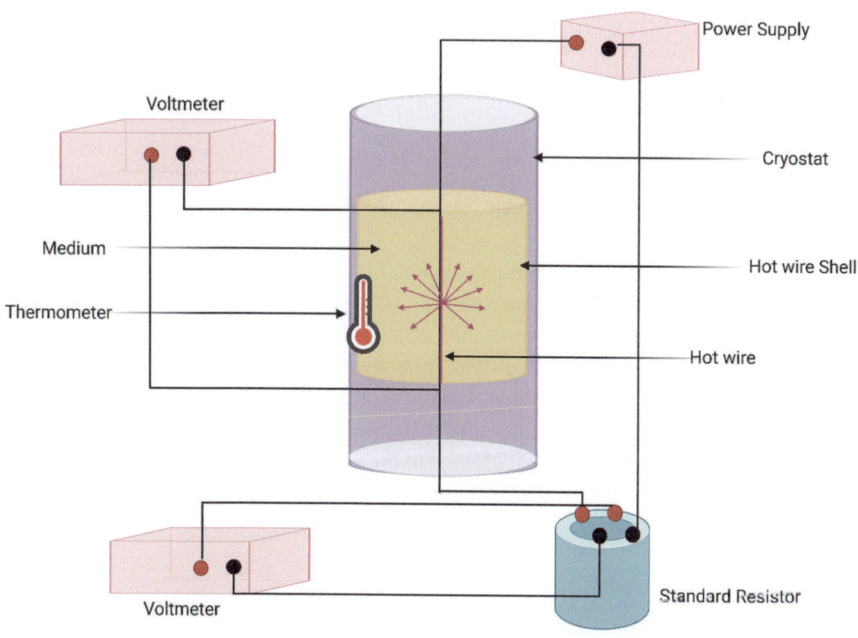

Fig. 4.7 Transient hot wire method

In many practical implementations, this is simplified by measuring ΔT over a short time and calculating k from the slope of the ΔT versus ln(t) plot:

$$k = \frac{q}{4\pi\left(\frac{d\Delta T}{d\ln t}\right)} \tag{4.21}$$

This method is fast and accurate for fluids and some solid polymers. It also requires sample preparation efforts and a highly sensitive method.

4.5.2.2 Steady-State Method

The steady-state method determines the thermal conductivity by applying a constant heat flow through a polymer sample and recording the resulting temperature gradient across it. This is a simple technique but requires careful insulation to minimize heat loss. A known heat source is applied to one face of the polymer sample, and the other face is kept at a cooler, controlled temperature (Fig. 4.8). Once thermal equilibrium is reached, i.e., steady-state, temperature sensors record ΔT [49]. The thermal conductivity of the material is determined using Fourier's equation of heat conduction, based on heat input to the system and geometric characteristics [50]:

$$k = \frac{Q \times L}{A \times \Delta T} \tag{4.22}$$

where k = thermal conductivity (W/m·K), Q = heat flow rate (W)—often provided by a heater, L = thickness of the sample (m), A = cross-sectional area through which heat flows (m^2), and ΔT = temperature difference across the sample (K).

4.5.2.3 Modified Transient Plane Source (MTPS)

The Modified Transient Plane Source (MTPS) method is utilized to assess thermal conductivity, thermal effusivity, and other associated thermal properties of materials. It's known for its speed, precision, and ease of use, which makes it popular for quality control and various applications. It uses a sensor with a heating element and a thermometer that is fabricated on a thin and highly thermal-conductivity substrate. A controlled heat pulse is sent to the sample by the heating element. The thermometer measures the resulting temperature change in the sample, and the temperature change data is noted, which helps to determine the thermal conductivity and effusivity of the material [51].

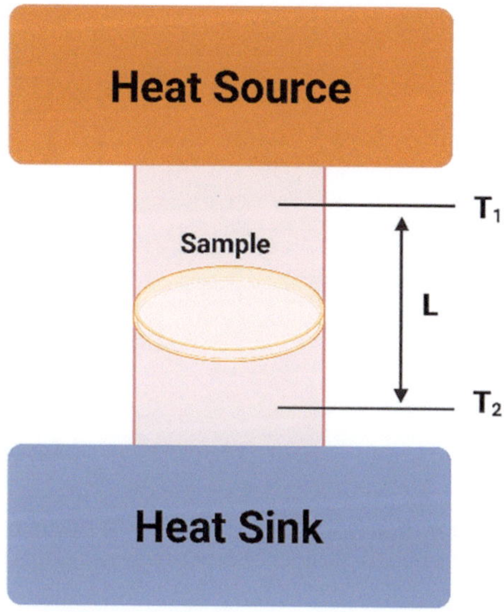

Fig. 4.8 Steady-state method

4.6 Crystallinity of Polymers: Basic Concept

4.6.1 General Concept of Crystallinity

Crystallinity describes the extent of molecular organization inside a material. In a crystalline material, the atoms are arranged in a regular and repetitive pattern that extends three-dimensionally. This repeating three-dimensional arrangement of atoms is called a crystal lattice (Fig. 4.9a). The positions of the crystal lattice where atoms are located are defined as lattice points. Depending on the arrangement of atoms and symmetry of the structure, the crystal lattice can be of different types. Some of the common types include simple cubic, body-centred cubic, hexagonal close-packed, face-centred cubic, etc. (Fig. 4.9b).

In contrast to crystalline solids, if the atoms are positioned in a more random and irregular manner, the material is referred to as amorphous.

4.6.2 Unit Cells and Miller Indices

Crystalline materials are defined by a systematic arrangement of atoms that is consistently replicated throughout the structure. The repeating units are referred to as unit cells, and a crystal lattice comprises the systematic spatial arrangement of these unit cells. Parallel atomic planes crossing with unit cells are utilized for defining the directions and distances within crystals. Often it is called Miller indices to define the orientation of the planes (100, 001, 010, 200, 111, etc.) in a crystal lattice (Fig. 4.9c). The

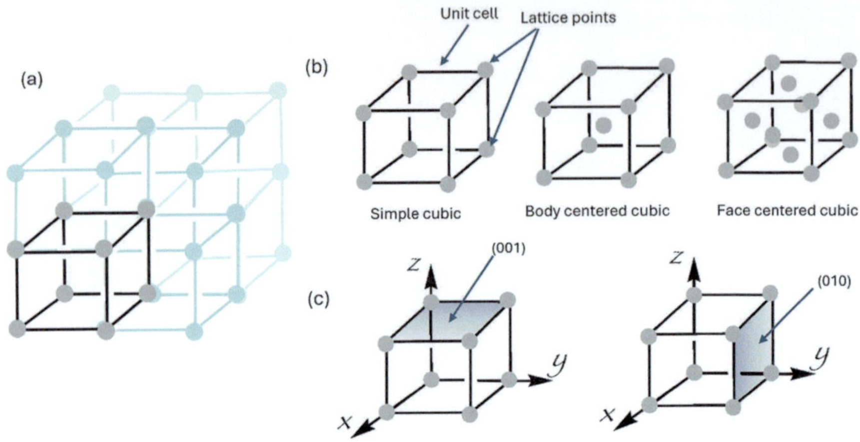

Fig. 4.9 Schematic diagram of **a** crystal lattice, **b** different types of unit cells, **c** Miller indices of different lattice planes

Miller indices are typically written as (hkl), where h, k, and l correspond to the intercepts of the plane with the *x*, *y*, and *z* axes, respectively. There are fourteen different types of configurations available in crystals, depending upon the atomic arrangements, which are called Bravais lattices. These fourteen types are classified by seven basic lattice systems, e.g., cubic, triclinic, monoclinic, hexagonal, orthorhombic, tetragonal, and rhombohedral.

4.6.3 Influence of Crystallinity on Various Properties of Materials

Crystallinity plays a crucial role in determining various properties of materials, including mechanical strength, density, optical characteristics, melting point, rigidity, electrical conductivity, chemical and thermal stability (Table 4.2).

4.6.4 Crystallinity in Polymers

Crystallinity in polymers differs from the concept of crystal lattice observed in small molecules. In polymers, the large polymer chains fold in a way that forms short-range ordered domains on a certain length scale. Most polymers are semi-crystalline, meaning these ordered domains are embedded within a larger amorphous phase. The size scale of these domains is distinct from that of atomic materials [52, 53]. The orderly stacking of folded polymer chains is called lamellae. Lamellae typically

Table 4.2 Key differences between crystalline and amorphous materials

Property	Crystalline materials	Amorphous materials
Mechanical strength	More rigid and less prone to deform when subjected to stress	More flexible and often unable to sustain under mechanical forces due to poor intermolecular interactions
Density	Higher, the molecules are tightly packed due to an ordered structure	Low density due to the presence of more voids between the atoms
Optical property	Generally translucent or opaque as the incident light is scattered in different directions from the various atomic planes present in the material	Transparent as the light can pass through without scattering due to the absence of any specific atomic planes
Melting point	Show a narrow and distinct melting point (T_m) as the phase transition (solid to liquid) occurs	T_m does not exist. However, amorphous materials gradually start to soften from the glassy state over a range of temperatures. The temperature at which an amorphous material transitions from a glassy state to a rubbery state is called the glass transition temperature (T_g)

radiate outward from a central nucleus [3, 4], forming three-dimensional (3D) spherical entities, referred to as spherulites. The size of these spherulites is governed by the nucleation density: a higher number of nuclei yields smaller spherulites, whereas a lower nucleation density results in larger, more brittle structures. Hence, nucleating agents are often used to enhance the nucleation rate to reduce spherulite size.

4.6.5 *Fringed-Micelle Model*

The fringed micelle model was initially formulated to explain the morphology of partially crystalline synthetic polymers that crystallize from an undercooled melt or concentrated solution, resulting in a metastable network with a relatively low degree of crystallinity. Typically, such polymers possess tiny crystalline areas measuring approximately 100 Å in size [54]. Within these regions, segments of polymer chains are regularly aligned in parallel and tightly packed into a crystalline lattice. However, the polymer chains are much longer than the dimensions of the individual crystallites, extending from one crystallite through the intervening amorphous regions and continuing to another crystallite, and so on (Fig. 4.10a). This arrangement creates a structure in which both crystalline and amorphous phases coexist, contributing to the material's unique properties.

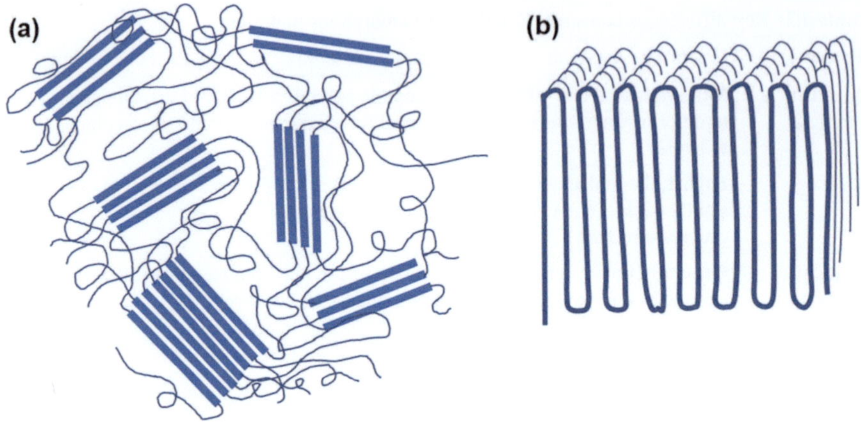

Fig. 4.10 **a** Fringed-micelle model, **b** folded-chain model

4.6.6 Folded-Chain Model

The Hoffman-Lauritzen theory of polymer crystal formation originated from an investigation made in 1957 [55], indicates that typically single crystals of polyethylene can be cultivated from dilute solutions. This theory is fundamentally based on the folded-chain model (Fig. 4.10b), which has been extensively validated for single polymer crystals. According to the model, polymer chains fold back on themselves, forming folded-chain lamellae. The interlamellar amorphous regions serve to bind these lamellae together within the bulk polymer structure.

4.6.7 Thermodynamics of Crystallization

The crystallization phenomenon in polymer is driven either by thermodynamic or kinetic factors. The Gibbs free energy (G) of a system is expressed as, $G = H - TS$, where H, T, and S are enthalpy, temperature, and entropy of the system, respectively [56]. A system attains equilibrium when G reaches a minimum value. Polymer melt consists of randomly coiled macromolecular chains exhibiting a high entropy relative to extended chain conformations, leading to a low G value. Crystallization occurs when the system is cooled down below the melting temperature (Tm) of the polymer. The generation of conformally ordered polymer chains during crystallization leads to a significant decline in the entropy (ΔSm) of the system [57, 58]. Additionally, H decreases as crystallization is an exothermic process. If the negative change in enthalpy, the latent heat (ΔHm) surpasses the value of TmΔSm, ΔG become negative and crystallization will be thermodynamically favoured. However, in thermodynamics, the processes are quasi-static in nature. When a polymer melt is cooled down rapidly, crystallization will be kinetically driven.

Crystallization from dilute polymer solution attains a higher degree of perfection compared to its melt-crystallized counterparts [59–61]. In a dilute solution, polymer chains exist as isolated coils, free from strong entanglements that could hinder the crystallization process. The absence of strong chain entanglements allows greater chain mobility, facilitating orderly arrangement during crystallization.

4.6.8 Kinetics of Crystallization

The kinetics of crystallization can be modelled under various conditions, including scenarios with no ongoing nucleation (i.e., pre-existing nuclei), constant nucleation rates, growth at linear velocity, diffusion-controlled growth, or anisotropic growth geometries such as needles or plates. The Johnson–Mehl–Avrami (JMA) model provides a theoretical framework for describing isothermal phase transformations, specifically the conversion of a parent phase (α) into a product phase (β) through nucleation, subsequent growth, and the impingement of growing regions. This model operates under three primary assumptions: (a) the transformation occurs within an effectively infinite volume, (b) nucleation events are spatially random, and (c) the growth of the new phase is isotropic [62]. The Avrami equation, derived from this model, is widely utilized to quantitatively characterize crystallization kinetics.

$$1 - X = \exp(-Kt^n) \tag{4.23}$$

where, t represents crystallization time and X denotes the volume fraction of the product phase β. The parameters K and n are specific to the phase transformation mechanism [63]. The Avrami equation has been developed for idealized geometries—spheres, discs, and line segments—corresponding to three-dimensional, two-dimensional, and one-dimensional isotropic growth at a constant rate, respectively. The exponent n in the Avrami equation is given by $n = m + g$, where m represents the dimensionality of growth ($m = 3$ for volumetric, $m = 2$ for planar, and $m = 1$ for linear growth), and g reflects the nucleation type, with $g = 0$ for instantaneous nucleation and $g = 1$ for sporadic nucleation.

4.6.9 Factors Affecting Crystallinity in Polymers

The rate, ability, and degree of crystallization of a certain polymer are influenced by various factors, e.g., rate of cooling, melt temperature, amount of branching, tacticity, and molar mass of the polymer [64]. The presence of any additives, like nucleating agents, also plays a crucial role in determining the rate of crystallization in polymeric materials. Polymers can crystallize by cooling from the melt, inducing mechanical strain, or gradually evaporating the solvent. When polymers crystallize, the unit cells that describe their crystalline structure can belong to various crystal systems (Fig. 4.11), including monoclinic, simple cubic, triclinic, etc.

The molecular structure and intermolecular forces profoundly influence the extent of crystallinity and crystal structure in polymeric materials. Atactic polymers such as poly(methyl methacrylate) (PMMA) [65, 66], polystyrene (PS) [67], poly(vinyl chloride) (PVC), etc., are fully amorphous in nature. However, syndiotactic PS exhibits semi-crystalline characteristics due to its orderly and regular structure. This arrangement allows polymer chains to pack densely without obstruction from pendant phenyl groups, facilitated by strong π-stacking interactions. High-density polypropylene (HDPE) has an almost perfectly linear structure [68], allowing it to achieve a highly crystalline state, typically ranging from 80 to 85% crystallinity. In contrast, low-density polyethylene (LDPE) contains numerous long-chain branches and short-chain (ethyl and butyl) [69], which prevent tight molecular packing, limiting its crystallinity to approximately 55% [70, 54]. The crystal structures of frequent polymers are given below in Table 4.3.

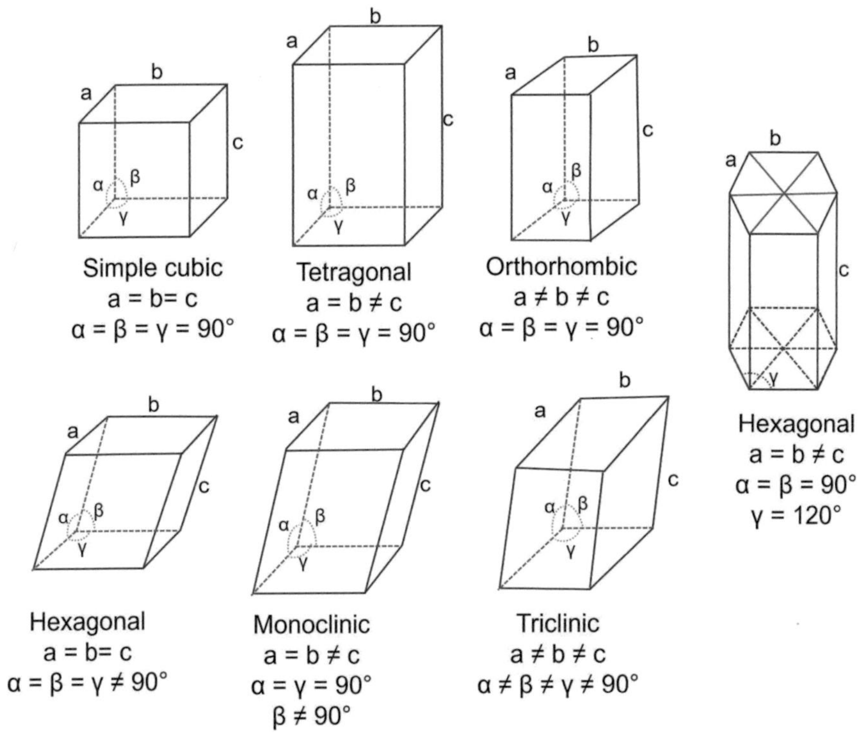

Fig. 4.11 Different types of crystal lattices present in the polymer system

Table 4.3 Crystal structures of polymers

Polymer repeat unit	Crystal system	No. of units
Polyethylene	Orthorhombic	4
Polypropylene (isotactic)	Monoclinic	12
Polypropylene (syndiotactic)	Orthorhombic	8
Polystyrene (isotactic)	Trigonal	18
Polytetrafluoroethylene	Triclinic	13
Poly(vinyl chloride) (syndiotactic)	Orthorhombic	4
Poly(vinyl alcohol) (atactic)	Monoclinic	2
Poly(vinyl fluoride) (atactic)	Orthorhombic	2
1,4-Polyisoprene (cis)	Orthorhombic	8
1,4-Polyisoprene (trans)	Orthorhombic	4
Poly(ethylene terephthalate)	Triclinic	1
Nylon 6, α	Monoclinic	8
Nylon 6,6, α	Triclinic	1

4.7 Measurement of Crystallinity of Polymers: XRD

XRD is called a non-destructive method [71], used to determine the structure of crystal, phase composition and crystallinity degree of semi-crystalline polymers.

4.7.1 Generation of X-Ray

X-ray generates when a high-energy electron beam is bombarded to an atom. As high-energy electrons collide with an atom, it can eject a core electron from the innermost shell, creating a vacant place. An electron from a higher energy level quickly transcends to fill the vacancy, releasing excess energy in the form surplus energy expressed as electromagnetic radiation, specifically X-rays. The released X-rays can be categorized as Kα or Kβ based on the nature of the transition. These characteristic X-ray emissions have specific wavelengths unique to each element, determined by the energy difference between the shells (Fig. 4.12). Additionally, for obtaining a monochromatic X-ray source (Cu Kα), a filter (nickel) is applied to block the unwanted radiation.

4.7.2 Mechanism of XRD Pattern Formation

As a beam of monochromatic X-ray is directed towards a crystalline material, XRD patterns are generated due to scattering from the periodic crystalline planes of the

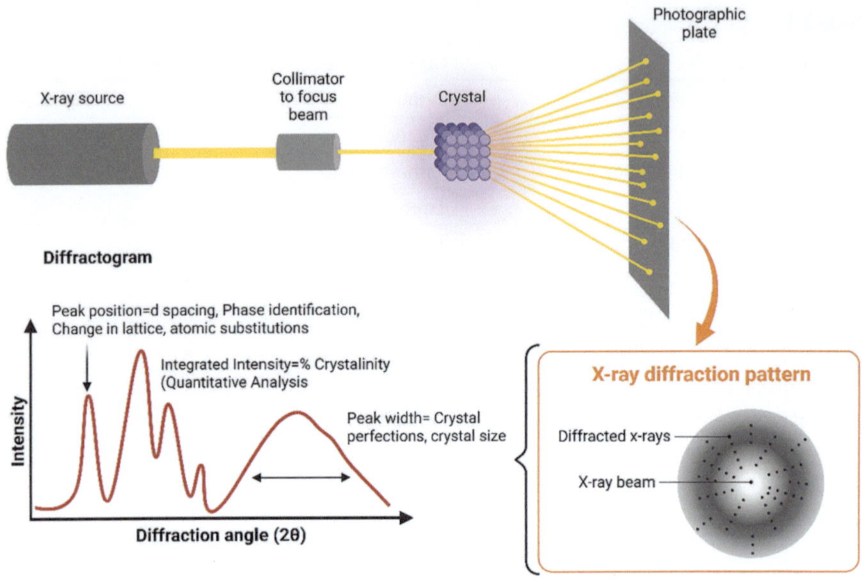

Fig. 4.12 Generation of X-ray diffractogram

substrate. These scattered waves can interfere in two ways: constructive interference or destructive interference. In the regions of destructive interference, no peaks are observed while constructive interference follows Bragg's law (Fig. 4.13).

According to the law, constructive interference occurs when the path difference (d) between the scattered X-rays equals an integer (n) multiple of the wavelength (λ). Thus, diffraction happens when the reflection from the crystalline planes satisfies this condition for constructive interference.

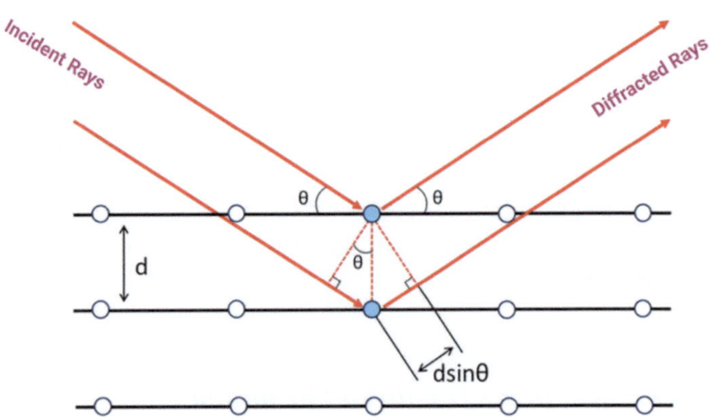

Fig. 4.13 X-ray diffraction

$$n\lambda = 2d\,\mathrm{Sin}\theta \tag{4.24}$$

where, θ is the angle of incidence.

The X-ray detector records the intensity of diffracted beams at different 2θ angles, generating an XRD pattern. This pattern comprises a sequence of peaks matching to the various planes of crystallography (hkl) of the substance being studied. Each crystalline material has a unique diffraction pattern, serving as a "fingerprint" for phase identification and crystallinity assessment.

4.7.3 Study of Polymer Samples from XRD

XRD gives observations of the interior structures at a dimension scale ranging from 0.1 to 100 nm. Polymers can be analyzed as powders, thin films, or fibers. When an X-ray beam incident upon a polymer sample, a fraction of the X-ray is scattered coherently by the regularly spaced atoms in the crystallites. The coherent scattering from a single atom in a crystal interacts with that from neighboring atoms, leading to the formation of a diffraction pattern, which is used for the structural analysis of the matter. XRD patterns are always plotted against 2θ, which corresponds to the total angle between the incident and scattered X-ray beams.

Polymers generally crystallize in multiple polymorphic [72] forms depending on processing conditions (e.g., temperature, pressure, the rate of cooling during crystallization, etc.). Each polymorph has a unique crystalline structure with different densities, symmetries, and packing arrangements of polymer chains. Through XRD, these polymorphs can be identified based on the d-spacing values, intensity, and position of the diffraction peaks.

The diffraction pattern from a semi-crystalline material comprises of two distinct phases, sharp crystalline peaks and a broad amorphous halo. The width of the diffraction peaks helps to calculate the crystallite size (D) and can be estimated using the Scherrer equation [73].

$$D = \frac{K\lambda}{\beta\ \cos\theta} \tag{4.25}$$

where K is described as the Scherrer constant, a function of the shape of the crystallites. λ denotes the wavelength of the X-ray. β represents the whole width at half maximum of the associated peak. If the breadth increases, the crystallite size diminishes, signifying a more flawed crystal structure.

If there is significant strain in the polymer crystallites (e.g., due to defects or deformation), the peak broadening will be influenced by both size and strain. In such cases, more advanced methods like Williamson Hall analysis are used to separate the contributions of size and strain.

Degree of Crystallinity (Xc) in a polymer sample can be analysed using the following equation [74].

$$Xc = \frac{Ac}{(Aa + Ac)} \tag{4.26}$$

where Aa and Ac the integrated areas under the broad amorphous haloc and crystalline peaks respectively, are estimated from the XRD pattern.

In pure polyethylene (PE), the crystalline regions produce sharp diffraction peaks because the polymer chains are well-ordered and form a regular, repeating lattice. When polyethylene (PE) is mixed with an amorphous polymer like polyisobutene (PIB) or polybutadiene (PBD), several changes can occur in the crystalline structure, e.g., the occurrence of various defects can lead to the broadening of the peaks in diffraction. The broader peaks in diffractions indicate reduction in the average size of the crystals attributed to the inhomogeneity and structural disruption of PE by PIB or PBD. Miscibility of the two polymers leads to a single homogeneous phase where the individual polymer chains interact with each other at the molecular level, hindering the nucleation and growth of the crystalline domains. Additives such as plasticizers, nanofillers, or nucleating agents also affect the crystallization behaviour and structural matrices. Figure 4.14 is the description of the crystallographic structure of a protein sample.

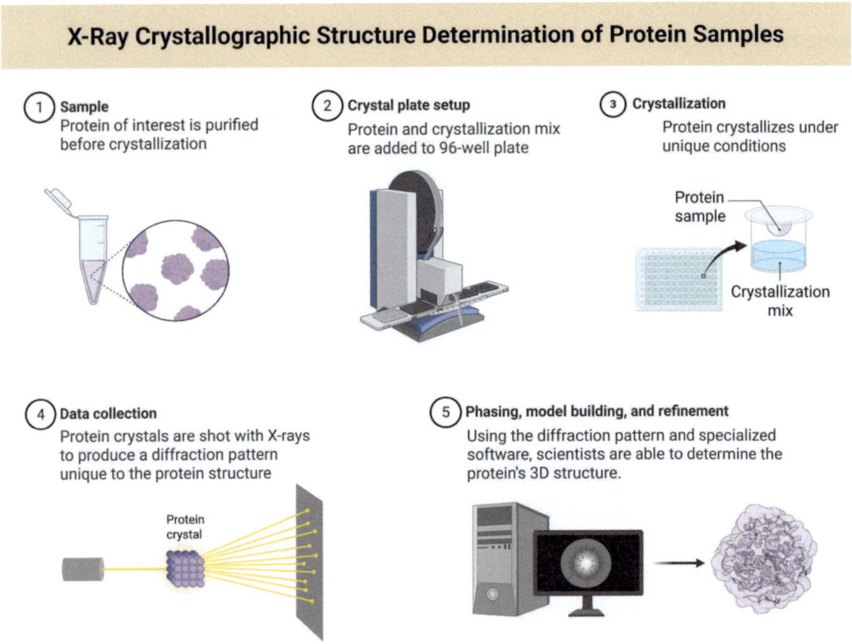

Fig. 4.14 Steps involved in the determination of the X-ray crystallographic structure of the protein sample

4.8 Conclusions

In this chapter, we have explained the key physicochemical parameters that define the structure–property relationship of polymers. A detailed discussion on molecular weight distribution and its determination through various classical and modern techniques highlighted the importance of molecular size in influencing polymer behaviour and performance. The section on electrical conductivity of polymers emphasized the growing significance of conductive polymers and outlined the principles and methodologies used to evaluate their electrical and thermal conductive properties. Furthermore, the concept of crystallinity in polymers was elaborated through theoretical models, thermodynamic and kinetic perspectives, and its effect on material characteristics. Finally, the application of X-ray diffraction (XRD) in assessing polymer crystallinity was presented as a vital analytical approach in polymer science. Altogether, the topics covered in this chapter serve as a foundation for understanding and engineering the properties of polymeric materials.

Acknowledgements We used the Biorender student license (BioRender.com) for the illustrations in this article.

Competing Interest The authors declare no competing interests.

References

1. Khan S, Iqbal A (2023) Organic polymers revolution: applications and formation strategies, and future perspectives. J Polym Sci Eng 6:3125
2. Rosato DV (2011) Plastics end use applications. Springer Science & Business Media
3. Peters EN (2005) Plastics: Information and properties of polymeric materials. Mech Eng Handb: Mater Mech Design 1:335–379
4. Gahlawat G (2019) Polyhydroxyalkanoates biopolymers: production strategies. Springer
5. Campbell D, Pethrick RA, White JR (2017) Polymer characterization: physical techniques. CRC Press
6. Mandelkern L (1985) The relation between structure and properties of crystalline polymers. Polym J 17:337–350
7. Ravve A, Ravve A (2012) Physical properties and physical chemistry of polymers. Principles of polymer chemistry, pp 17–67
8. Billmeyer Jr FW (1965) Characterization of molecular weight distributions in high polymers. In: Journal of Polymer science part c: polymer symposia. Wiley Online Library, pp 161–178
9. Uzun İ (2023) Methods of determining the degree of crystallinity of polymers with X-ray diffraction: a review. J Polym Res 30:394
10. Nunes RW, Martin JR, Johnson JF (1982) Influence of molecular weight and molecular weight distribution on mechanical properties of polymers. Polym Eng Sci 22:205–228
11. Umoren SA, Solomon MM (2016) Polymer characterization: polymer molecular weight determination. Polymer science: research advances, practical applications and educational aspects, pp 412–419
12. Johnston SM (1906) VIII—on the elevation of the boiling points of aqueous solutions of electrolytes. Earth Environ Sci Trans R Soc Edinb 45:193–240
13. Blackmore WR (1960) Design problems in ebulliometry of polymers. Rev Sci Instrum 31:317–321

14. Molecular Weight (MW) Determination by ebulliometry. http://www.monashscientific.com.au/EbulliometerMW.htm
15. Iland P (2000) Techniques for chemical analysis and quality monitoring during winemaking. Patrick Iland Wine Promotions
16. Anderson AM, Bruno BA, Smith LS (2014) Viscosity measurement. AA 2:809
17. Karam HJ (1963) Viscometers and their use. Ind Eng Chem 55:38–43
18. Holdcroft S (1991) Determination of molecular weights and Mark-Houwink constants for soluble electronically conducting polymers. J Polym Sci B Polym Phys 29:1585–1588
19. Sutera SP, Skalak R (1993) The history of Poiseuille's law. Annu Rev Fluid Mech 25:1–20
20. Chiang R (1965) Intrinsic viscosity-molecular weight relationship for fractions of linear polyethylene1. J Phys Chem 69:1645–1653
21. Cragg LH (1946) The terminology of intrinsic viscosity and related functions. Rubber Chem Technol 19:1092–1098
22. Larkins MC, Thombare A (2023) Osmometer
23. Podzimek S (2011) Light scattering, size exclusion chromatography and asymmetric flow field flow fractionation: powerful tools for the characterization of polymers, proteins and nanoparticles. John Wiley & Sons
24. Akhter M, Alam MM (2023) Colligative properties. In: Physical pharmacy and instrumental methods of analysis. Springer, pp 21–44
25. Lima EC, Gomes AA, Tran HN (2020) Comparison of the nonlinear and linear forms of the van't Hoff equation for calculation of adsorption thermodynamic parameters (Δ S° andΔ H°). J Mol Liq 311:113315
26. Garmon RG (1975) End group determinations. Ph E Slade Polym Mol Weights Part II 31
27. Schröder E, Müller G, Arndt K-F (2022) Polymer characterization. Walter de Gruyter GmbH & Co KG
28. Debye P (1947) Molecular-weight determination by light scattering. J Phys Chem 51:18–32
29. Bier M (1957) [5] Light-scattering measurements
30. Schröer W, Köser J, Kuhnen F (2007) Light–scattering in turbid fluids: the single-scattering intensity. J Mol Liq 134:40–48
31. Ohlendieck K, Harding SE (2018) Centrifugation and ultracentrifugation. Wilson Walker's Principles Tech Biochem Mol Biol 1:424–453
32. Cole JL, Lary JW, Moody TP, Laue TM (2008) Analytical ultracentrifugation: sedimentation velocity and sedimentation equilibrium. Methods Cell Biol 84:143–179
33. Cvjetkovic A, Lötvall J, Lässer C (2014) The influence of rotor type and centrifugation time on the yield and purity of extracellular vesicles. J Extracell Vesicles 3:23111
34. Scott JC (2010) History of conductive polymers. Nanostructured conductive polymers, pp 1–17
35. Kar P (2013) Doping in conjugated polymers. John Wiley & Sons
36. Taherian R (2019) The theory of electrical conductivity. Electr Conduct Polym-Based Compos Exp Modelling Appl 1–18
37. Jaiswal M, Menon R (2006) Polymer electronic materials: a review of charge transport. Polym Int 55:1371–1384
38. Sobczyk L, Grabowski SJ, Krygowski TM (2005) Interrelation between H-bond and Pi-electron delocalization. Chem Rev 105:3513–3560
39. Mironov VS, Kim JK, Park M et al (2007) Comparison of electrical conductivity data obtained by four-electrode and four-point probe methods for graphite-based polymer composites. Polym Test 26:547–555
40. Ghorbani MM, Taherian R (2018) 12–methods of measuring electrical properties of material. Electr Conduct Polym-Based Compos: Exp Modelling Appl 365
41. Inzelt G, Láng GG (2010) Electrochemical impedance spectroscopy (EIS) for polymer characterization. Electropolymerization: Concepts, Mater Appl 51–76
42. Lvovich VF (2012) Impedance spectroscopy: applications to electrochemical and dielectric phenomena. John Wiley & Sons
43. Kaur K, Goyal S, Bhattacharjee B, Kumar M (2017) Electrochemical impedance spectroscopy to study the carbonation behavior of concrete treated with corrosion inhibitors. J Adv Concr Technol 15:738–748

44. Laschuk NO, Easton EB, Zenkina OV (2021) Reducing the resistance for the use of electrochemical impedance spectroscopy analysis in materials chemistry. RSC Adv 11:27925–27936
45. Prasad PE, Dhore N, Palanisamy A, Rao CRK (2024) Anti-corrosion studies on cardanol epoxy coatings cured with redox-active aromatic trimer and tetramer oligoanilines. Prog Org Coat 197:108802
46. Liu Q, Richards JJ (2023) Rheo-electric measurements of carbon black suspensions containing polyvinylidene difluoride in *N*-methyl-2-pyrrolidone. J Rheol (N Y N Y) 67:647–659
47. Healy JJ, De Groot JJ, Kestin J (1976) The theory of the transient hot-wire method for measuring thermal conductivity. Physica B+ c 82:392–408
48. Bouguerra A, Aït-Mokhtar A, Amiri O, Diop MB (2001) Measurement of thermal conductivity, thermal diffusivity and heat capacity of highly porous building materials using transient plane source technique. Int Commun Heat Mass Transfer 28:1065–1078
49. Reiter M, Hartman H (1971) A new steady-state method for determining thermal conductivity. J Geophys Res 76:7047–7051
50. Narasimhan TN (1999) Fourier's heat conduction equation: history, influence, and connections. Rev Geophys 37:151–172
51. Harris A, Kazachenko S, Bateman R et al (2014) Measuring the thermal conductivity of heat transfer fluids via the modified transient plane source (MTPS). J Therm Anal Calorim 116:1309–1314
52. Hammond C (2015) The basics of crystallography and diffraction. Basics Crystallogr Diffract. https://doi.org/10.1093/ACPROF:OSO/9780198738671.001.0001
53. Holder CF, Schaak RE (2019) Tutorial on powder X-ray diffraction for characterizing nanoscale materials. ACS Nano 13:7359–7365. https://doi.org/10.1021/ACSNANO.9B05157/ASSET/IMAGES/MEDIUM/NN-2019-051577_0009.GIF
54. Young RJ, Lovell PA (2011) Introduction to polymers, 2rd edn, pp 1–653. https://doi.org/10.1201/9781439894156/INTRODUCTION-POLYMERS-ROBERT-YOUNG-PETER-LOVELL/RIGHTS-AND-PERMISSIONS
55. Cheng SZD, Lotz B (2005) Enthalpic and entropic origins of nucleation barriers during polymer crystallization: the Hoffman-Lauritzen theory and beyond. In: Polymer. Elsevier BV, pp 8662–8681
56. Zhu L, Liu L, Varlas S et al (2023) Understanding the seeded heteroepitaxial growth of crystallizable polymers: the role of crystallization thermodynamics. ACS Nano 17:24141–24153. https://doi.org/10.1021/ACSNANO.3C09130/ASSET/IMAGES/LARGE/NN3C09130_0008.JPEG
57. Strobl G (2006) Crystallization and melting of bulk polymers: New observations, conclusions and a thermodynamic scheme. Prog Polym Sci 31:398–442. https://doi.org/10.1016/J.PROGPOLYMSCI.2006.01.001
58. Strobl G (2005) A thermodynamic multiphase scheme treating polymer crystallization and melting. Eur Phys J E 18:295–309. https://doi.org/10.1140/EPJE/E2005-00032-Y
59. McHugh AJ, Burghardt WR, Holland DA (1986) The kinetics and morphology of polyethylene solution crystallization. Polymer (Guildf) 27:1585–1594. https://doi.org/10.1016/0032-3861(86)90109-6
60. Riande E, Fatou JMG (1976) Crystallization of dilute polyethylene solutions: influence of molecular weight. Polymer (Guildf) 17:99–104. https://doi.org/10.1016/0032-3861(76)90075-6
61. Epp J (2016) X-ray diffraction (XRD) techniques for materials characterization. Materials characterization using nondestructive evaluation (NDE) methods, pp 81–124. https://doi.org/10.1016/B978-0-08-100040-3.00004-3
62. Ponížil P, Procházka J (2006) Polymer crystallization and Avrami equation
63. Wang RY, Zou SF, Jiang BY et al (2017) A generalized Avrami equation for crystallization kinetics of polymers with concomitant double crystallization processes. Cryst Growth Des 17:5908–5917. https://doi.org/10.1021/acs.cgd.7b01016
64. Milner ST (2011) Polymer crystal-melt interfaces and nucleation in polyethylene. Soft Matter 7:2909–2917. https://doi.org/10.1039/C0SM00070A

65. Fox TG, Garrett BS, Goode WE et al (1958) Crystalline polymers of methyl methacrylate. J Am Chem Soc 80:1768–1769. https://doi.org/10.1021/JA01540A068
66. de Boer A, van Ekenstein GORA, Challa G (1975) Crystallization of isotactic poly(methyl methacrylate) from the melt. Polymer (Guildf) 16:930–932. https://doi.org/10.1016/0032-3861(75)90216-5
67. Ouellette RJ, Rawn JD (2018) Synthetic polymers. Organ Chem 861–887. https://doi.org/10.1016/B978-0-12-812838-1.50027-X
68. Maxfield J, Mandelkern L (1977) Crystallinity, supermolecular structure, and thermodynamic properties of linear polyethylene fractions. Macromolecules 10:1141–1153. https://doi.org/10.1021/MA60059A046
69. Mandelkern L, Maxfield J (1979) Morphology and properties of low-density (branched) polyethylene. J Polym Sci Part A-2, Polym Phys 17:1913–1927. https://doi.org/10.1002/POL.1979.180171108
70. Paajanen A, Vaari J, Verho T (2019) Crystallization of cross-linked polyethylene by molecular dynamics simulation. Polymer (Guildf) 171:80–86. https://doi.org/10.1016/J.POLYMER.2019.03.040
71. Garcia-Granda S, Montejo-Bernardo JM (2004) X-ray diffraction—powder. Encycl Anal Sci 389–397. https://doi.org/10.1016/B0-12-369397-7/00671-3
72. Lu G, Li G, Yang X (2008) Morphology and crystalline transition of poly(3-butylthiophene) associated with its polymorphic modifications. Macromolecules 41:2062–2070. https://doi.org/10.1021/MA7026512
73. Asai S, Shimada Y, Tominaga Y, Sumita M (2005) Characterization of higher-order structure of poly(ethylene-2,6-naphthalate) treated with supercritical carbon dioxide. Macromolecules 38:6544–6550. https://doi.org/10.1021/MA050373Q/ASSET/IMAGES/LARGE/MA050373QF00012.JPEG
74. Doumeng M, Makhlouf L, Berthet F et al (2021) A comparative study of the crystallinity of polyetheretherketone by using density, DSC, XRD, and Raman spectroscopy techniques. Polym Test 93:106878. https://doi.org/10.1016/J.POLYMERTESTING.2020.106878

Nikhil Dhore is pursuing a Ph.D. at the Indian Institute of Technology Kharagpur, focusing on the development of sustainable polymers for decorative and protective applications. Nikhil has completed his bachelor's in chemical technology and a master's specializing in paint technology and has hands-on experience of 3+ years in the coating field research. His current research on the synthesis of water-based polyurethane polymer, non-isocyanate polyurethane, and biobased epoxy amine systems. More than 6+ peer-reviewed international journal articles in his account.

Nikhil Kumar is a senior research fellow the Advanced Technology Development Centre, Indian Institute of Technology Kharagpur. He is specialized in fabrication of smart polymeric materials for diagnosis and therapeutics. Nikhil completed his Bachelor of Pharmacy and Master of Pharmacy in Pharmacology. His masters research work was focused on developing dressings for diabetic foot ulcers. His current research revolves around the design and application of smart polymeric biomaterials for advanced diagnosis and therapeutic interventions for cutaneous squamous cell carcinoma. He has published 12 peer-reviewed research articles. He is actively engaged in research focused on polymer characterization, with particular interest in the structural, thermal, and rheological analysis of advanced polymeric materials. Mr. Kumar has been involved in research projects that explore the application of spectroscopic techniques (FTIR, NMR), thermal analysis (DSC, TGA), to evaluate polymer performance under varied environmental conditions. Mr. Kumar is also developing a novel diagnostic tool that integrates multiple characterization techniques FTIR, Raman spectroscopy, AFM and nanoindentation to probe and differentiate pathological tissue conditions such as fibrosis and hyperkeratosis.

Aparajita Pal is pursuing Ph.D. in Rubber Technology Centre IIT Kharagpur. Her area of work is hydrogel based flexible electronics. She has completed her Masters' from IIT Delhi, in Fibre Science and Technology, where she has extensive worked on wet spinning technique.

Chapter 5
Mechanical Analysis of Polymer

Abhay Kumar, Nikhil Kumar, and Manas Jain

5.1 Introduction

Polymers, as the Greek words *poly* (many) and *meros* (parts), are materials composed of long molecular chains of monomers (repeating structural units). The chains interact with each other through diverse arrangements—linear, branched, crosslinked, or networked, as shown in Fig. 5.1a. The mechanical behavior of polymers is closely related to the arrangement and orientation of their chains [1]. These orientations are divided into two parts, i.e., amorphous (Randomly oriented long chains imparting elasticity like rubber) and crystalline (Orderly pattern chains imparting stiffness like plastic) [2]. Based on the orientation, polymers encompass various materials that differ in mechanical properties, including rubbers, plastics, and fibers. The mechanical behavior of rubber/elastomers is quite complex, especially rubber, which consists of long chains that follow crosslinking through vulcanization to enhance strength, as shown in Fig. 5.1b [3], while entanglements contribute to their strength, elasticity, and toughness [4]. Plastics are one kind of synthetic polymer whose mechanical responses depend on composition and structure. Plastics, such as polyethylene or polypropylene, are flexible, sustaining larger deformation before breaking. Plastics like polystyrene or polycarbonate have higher tensile strength and are likely to be brittle under stress. Fibers, whether natural or synthetic, have mechanical behaviors that primarily depend on their polymer composition and the

A. Kumar (✉)
Apollo Tyres Ltd., Chennai, India
e-mail: abhay24491@gmail.com

N. Kumar
Advanced Technology Development Centre, Indian Institute of Technology Kharagpur, Kharagpur, India

M. Jain
Department of Aerospace Engineering, Indian Institute of Technology Kharagpur, Kharagpur, India

S. Chattopadhyay and N. Kumar (eds.), *Polymer Characterization*,
https://doi.org/10.1007/978-981-95-2609-3_5

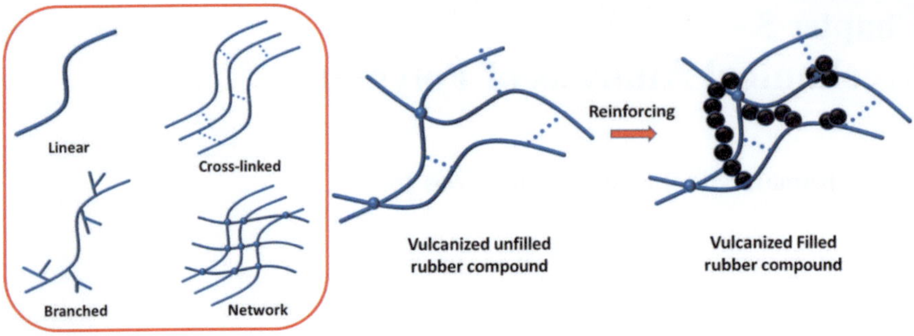

Fig. 5.1 **a** Polymer chain arrangement, **b** unfilled rubber crosslinking, and **c** filled rubber crosslinking

way they are processed. Natural fibers tend to be more flexible and absorb moisture, affecting their strength and stiffness, such as wool or cotton. Synthetic fibers, such as nylon or Kevlar, are made to have high tensile strength and elasticity and are used in applications where durability and resistance to wear are essential. In general, the mechanical response depends on compositions, orientation, crosslinking, and entanglements, which provide inherent polymer properties. These properties include modulus, stiffness, toughness, resilience, and hysteresis, and the ability to withstand impact, resistance to indentation (hardness), and recovery after compressive stress (compression set).

Compared to metals exhibiting plastic deformation [5], polymers exhibit complex non-linear viscoelastic behavior [6–8], making their mechanical characterization highly complex. The complex mechanical behavior is influenced by crosslinking and majorly enhanced by the incorporation of fillers such as carbon black, silica, graphene, and fibers [9], as shown in Fig. 5.1c. The fillers enhance mechanical strength and stiffness and improve durability [4]. Therefore, understanding the mechanical properties of polymers is essential for modifying them for specific applications in the automotive, biomedical, aerospace, and industrial sectors.

The mechanical behavior of polymers is characterized by two methods, i.e., static and dynamic testing methods [2, 6, 10]. Static testing includes modulus, stiffness, toughness, resilience, and hardness measurements to assess the polymer's ability to withstand constant or gradually applied loads. These measurements provide essential data on material deformation under tensile, compressive, or shear loads. Dynamic testing delves into time-dependent behaviors such as creep, stress relaxation, and flex fatigue, vital for materials applications subjected to cyclic or fluctuating loads [11]. Dynamic tests provide additional insights into the polymeric behavior for assessing long-term performance and durability under operational conditions. Combining both testing methodologies ensures a broad understanding of polymers that can respond under real-world loading conditions.

Additionally, the failure mechanism is quite intriguing and crucial for determining the life of the industrial product. The polymer failure mechanisms are classified into brittle and ductile failure, which are influenced by factors such as molecular structure, temperature, and strain rate [12]. Brittle failure, characterized by rapid fracture with minimal deformation, is often observed in glassy polymers or thermosets. Ductile failure occurs in materials with high elongation, where yielding, necking, and significant deformation lead to the formation of the precursor (formation of micro-cracks), ultimately culminating in failure [13]. Adding fillers and plasticizers can alter the failure mode, making failure prediction and prevention a key aspect of material engineering [14–16]. Furthermore, abrasion resistance and heat build-up are critical for evaluating wear performance and thermal stability in applications like tires, conveyor belts, and industrial seals. Therefore, failure modes are crucial to understanding and optimizing material properties, preventing premature failure, and enhancing longevity.

This chapter delves into the complete mechanical analysis of polymers, beginning with the deformation modes and progressing through static and dynamic testing procedures. The properties, including modulus, stiffness, toughness, resilience, hysteresis, impact resistance, and hardness, are discussed in detail, followed by an in-depth depiction of failure mechanisms, abrasion resistance, and thermal effects. Finally, the section precedes an in-depth analysis of stress–strain curves of different polymers, distinguishing between rubbers, plastics, and fibers while examining the influence of fillers. The complete chapter aims to deliver a broad understanding of mechanical properties, defining polymeric materials' functionality and reliability in engineering products.

5.2 Modes of Deformation

The polymers can be tested in various deformation modes to understand complex three-dimensional mechanical responses, primarily classified as tensile, compressive, shear, and bulk (or hydrostatic) deformation modes [1, 17, 18]. Tensile and compressive deformation occur when the material is deformed in one direction. Tensile deformation is led by stretching under pulling force, causing elongation of less than or higher than 100% strain [8]. In contrast, compressive deformation reduces height under pushing force without friction, causing compression that is always less than 100% strain [6]. Shear deformation occurs when the load is applied parallel to each other but in opposite directions, leading to internal distortion. In application, most rubber products experience shear deformation, and this mode is the primary factor for failure. Bulk or hydrostatic deformation occurs when constant pressure is applied from all directions, resulting in a reduction in volume without directional strain. As shown in Fig. 5.2, these deformation modes show a polymer's mechanical response, affecting its strength, flexibility, and overall durability.

The polymer deformation behavior is significantly influenced by its molecular orientation and the presence of fillers. In randomly arranged polymers, deformation tends to be uniform, exhibiting isotropic behavior. In contrast, oriented polymers, such as drawn fibers or stretched films, exhibit anisotropic behavior (direction-based

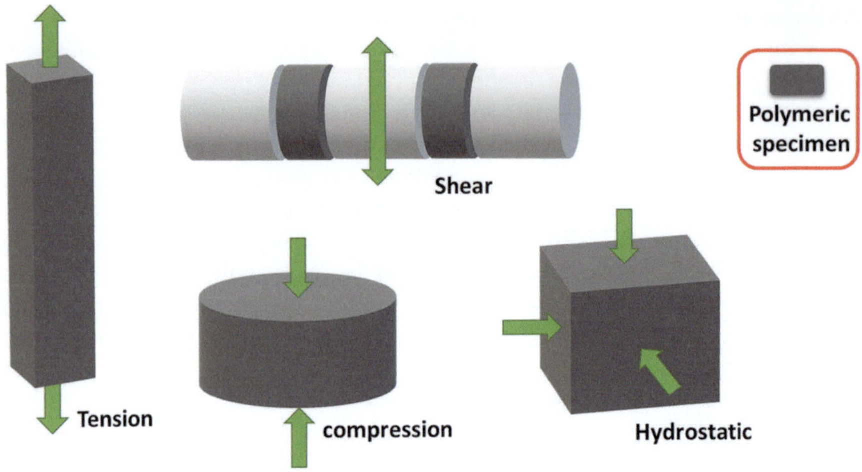

Fig. 5.2 Different deformation modes for a polymeric specimen

behavior) [19]. The rubber filled with filler such as carbon nanotubes, graphene flakes, or glass fiber with an aspect ratio (width to height ratio) of more than one also displays the directional property, which means it behaves differently through different deformation modes. Moreover, homogenously dispersed fillers in a rubber matrix show isotropic behavior; conversely, they form a higher aspect ratio if not distributed homogeneously, featuring direction-based properties. Fillers, for example, carbon black and silica in a rubber matrix, enhance stiffness and strength by restricting chain mobility, ultimately improving load-bearing capacity under tensile and compressive deformation [6, 10]. Graphene and fiber reinforcements, such as glass or aramid fibers, significantly enhance tensile strength, shear resistance, and impart various advanced properties, making them suitable for composite applications [20–23]. In contrast, soft fillers or plasticizers may enhance ductility, reducing stiffness and facilitating longer deformation. Hence, the combination of polymer structure, filler type, and molecular orientation shows whether a material will behave as rigid, flexible, or highly stretchable under different deformation modes.

Additionally, time is a critical factor influencing deformation modes and mechanical response, particularly under constant versus time-dependent loading conditions, due to the contribution of recovery stress. This time dependency necessitates two principal testing approaches: static testing, which involves applying a sustained or gradually increasing load to assess the material's mechanical behavior, and dynamic testing, which employs cyclic or fluctuating loads to evaluate time-dependent properties and long-term performance.

5.3 Static Testing

Static testing of polymer involves evaluating material behavior such as modulus, stiffness, toughness, resilience, hysteresis, impact resistance, hardness, and compression set on constantly applied loads. Rubber undergoes large deformations with minimal force, showing highly elastic behavior followed by a time-dependent response. In contrast, plastics display elastic behavior, with a well-defined yield point followed by ductile deformation. Furthermore, crosslinking, physical entanglements, and filler content influence mechanical behavior during static testing, making it more complex [1]. The mechanical responses during static tests at constant load are discussed in detail.

5.3.1 Modulus

Modulus (E) is determined by the slope or the ratio of stress to strain ($E = \sigma/\epsilon$) of a material's stress–strain curve at a particular deformation. Rubber initially exhibits low modulus and non-linear larger deformation due to its hyperelastic (time-independent) behavior (). This behavior depends on crosslinking, physical entanglements, and filler content. In rubber terminology, three moduli are taken into account, namely 100% modulus (modulus at 100% strain), 200% modulus (modulus at 200% strain), and 300% modulus (modulus at 300% strain) for getting the desired properties [24], in contrast to the engineering materials where Young's modulus is considered as the singular choice of modulus. Majorly, the modulus is the crucial criterion for application in tires, mounts, isolators, seals, gaskets, etc. Plastics show a moderate modulus, which varies significantly between thermoplastics (low modulus, ductile) and thermosets (high modulus, brittle). Fiber modulus varies depending on composition and orientation, as high-modulus fibers exhibit greater stiffness, meaning they deform less under the same load than low-modulus fibers. High-modulus fibers are crucial in applications requiring thermal stability, such as aerospace or high-performance composites.

There are different types of modulus depending on deformation types, such as Young's modulus (tensile), shear modulus, and bulk modulus. These moduli in the elastic region are key parameters that are interlinked with a relation as mentioned in the equations [25]:

$$\sigma = \boldsymbol{E}\epsilon; \tau = \boldsymbol{G}\gamma; \boldsymbol{E} = 3\boldsymbol{K}(1 - 2\boldsymbol{\mu}); \boldsymbol{E} = 2\boldsymbol{G}(1 + \boldsymbol{\mu})$$

where σ and ϵ are tensile stress and strain; τ and γ are shear stress and strain; $\boldsymbol{E}, \boldsymbol{G}, \boldsymbol{K}, \boldsymbol{\mu}$ are Young's modulus, Shear modulus, Bulk modulus, and Poisson ratio.

5.3.2 Stiffness

Stiffness is a material's resistance to deformation under a specifically applied load. Rubber stiffness directly depends on strain levels, with less stiffness at initial strains, followed by increased stiffness at higher strains due to chain alignment. Reinforcing with fillers like carbon black, silica, etc., significantly alters the chain crosslinking and physical entanglement. Plastics range from flexible (low stiffness) to rigid (high stiffness), depending on the polymeric structure and filler content. Fiber stiffness is influenced by its geometry and particularly cross-sectional area. A stiff fiber will always maintain its shape under loading, which is suitable for structural applications.

5.3.3 Toughness

Toughness is a parameter that defines a material's ability to absorb energy before failure at various loading conditions and is represented by the area under the stress–strain curve. The rubber compound has higher toughness due to its ability to undergo large deformations before breaking. Various plastics show toughness, with thermoplastics being generally tougher than brittle thermosets. Toughness in fibers combines both strength and ductility, allowing them to endure high-impact forces or sudden loading without failure, which is vital for protective gear or impact-resistant composites.

5.3.4 Resilience

Resilience is a material's ability to recover from its original shape after deformation. This property is quantified by the energy stored during elastic deformation. Rubber is a highly resilient polymer, recovering after deformation due to its coiled flexible chains. The resilience is influenced by filler in the filled rubber compound, which makes it resistant to fully returning to its original shape, creating a filler-filler breakage (permanent deformation). Plastics display moderate resilience, such as thermoplastics being more resilient than brittle thermosets. Fibers have high resilience, returning efficiently to their original shape upon stress removal, which is essential in dynamic applications like textiles or actuators.

5.3.5 Hysteresis

Hysteresis is the energy loss during cyclic loading, quantified as a difference between loading and unloading stress–strain curves. Rubber exhibits significant hysteresis at a high strain rate, dissipating energy as heat due to internal friction. The hysteresis depends on crosslinking, entanglement, and filler content. Moreover, it also depends on filler types [6]. Plastics also display hysteresis, but to a lesser extent than rubber if loaded within the yield point. Fibers with low hysteresis are preferred in damping or cyclic loading environments because they dissipate less energy, whereas high-hysteresis fibers are beneficial for vibration absorption.

5.3.6 *Impact Resistance*

Impact resistance measures a material's ability to withstand forces without failure at sudden loading. Rubber has excellent impact resistance due to its flexibility and ability to absorb energy. Plastics such as thermoplastics, like polyethylene, are highly impact resistant, whereas thermosets are inclined to be brittle. Fibers with high-impact resistance dissipate significant energy upon sudden loading, making them suitable for ballistic applications, sports equipment, and automotive components.

5.3.7 *Hardness*

Hardness is a material's resistance to indentation or scratching. Rubber displays low hardness that can be boosted by altering various factors, such as curing agents and fillers. Plastics start from soft (low-hardness thermoplastic) to rigid (high-hardness thermosets). Static hardness tests include Shore durometer (for rubber), Rockwell, and Brinell (for plastics and metals). The hardness of a fiber is significant when assessing wear resistance in composite matrices or textiles. Hard fibers extend the material's service life when exposed to friction or contact wear.

5.3.8 *Compression Set*

A compression set is a permanent deformation of a material upon being compressed for a long duration. Due to its elastic nature, the rubber has a low compression set, meaning it regains most of its shape. However, permanent deformation increases with temperature and stress. The compression set is increased in the filled rubber compound due to the presence of filler. Plastics vary widely, with some plastics showing high residual deformation. This is an important property that is essential for rubber seals. Fibers with a low compression set retain their original dimensions better, which is critical for applications requiring repeated compression and recovery cycles.

Fibers are recognized as high-strength, lightweight material alternatives to conventional rubber. Carbon, glass, and aramid fibers are the most widely used

Table 5.1 Comparison of properties between three industrial fiber materials

Property	Carbon fiber	Glass fiber	Aramid fiber
Modulus	High	Moderate	Moderate
Stiffness	High	Moderate	Good
Toughness	Low	Moderate	High
Resilience	High	Good	Excellent
Hysteresis	Low	Moderate	Low
Impact resistance	Low	Good	Excellent
Hardness	High	High	Moderate
Compression set	Low	Low	Low

fiber types, each offering distinct mechanical properties. The following Table 5.1 shows a comparative overview of these properties, emphasizing their advantages and suitability for engineering and industrial uses.

5.3.9 Testing Techniques

Below is a detailed explanation of various testing techniques used for static mechanical testing of polymers and how they relate to these mechanical properties.

5.3.9.1 Tensile and Compression Testing

Tensile testing involves stretching a polymer specimen at a controlled rate (strain rate) until failure. It is governed by various standards, such as rubber ASTM D412 [26] and plastic and fiber ASTM D885 [27]. The strain rate significantly affects the polymeric responses. Polymers behave more ductile at lower strain rates, neglecting the time effect; they tend to become more brittle and stiff at high strain rates. Different properties such as modulus of elasticity (initial slope of stress–strain curve), stiffness (related to the modulus of linear elastic materials), toughness (area under complete stress–strain curve until break), resilience (area under stress–strain curve in the elastic region), and yield strength (onset of plastic deformation) can be measured. Compression testing is performed in a machine similar to tensile testing. However, polymer samples in compression testing are compressed between two plates. The compression test is used mainly in cushioning or load-bearing applications, where modulus and stiffness are the primary criteria. In polymers, results change significantly from tensile tests due to differing molecular alignment under compression.

5.3.9.2 Flexural (Bending), Hardness, and Quasi-Static Impact Testing

The Flexural test applies a load at one or two points on a beam supported at both ends (3-point or 4-point bending). This is analogous to Young's modulus but in bending. This test is used when materials are subjected to bending in service (e.g., plastic rulers, panels).

Hardness testing measures resistance to localized plastic deformation. The major tests for Hardness determination include Shore Hardness and Rockwell Hardness. This property is related to the material's resistance to surface indentation or penetration.

Quasi-static Impact testing simulates impact at relatively low velocities, often using drop weights or pendulum systems at controlled speeds. This test records the maximum energy absorbed before failure, providing toughness.

There are ASTM standards available for determining test methods to determine these properties [26, 28–30].

5.4 Dynamic Testing

Dynamic testing evaluates the viscoelastic properties of rubber and plastic under cyclic and oscillatory conditions, capturing behaviors like stress relaxation, creep, flex fatigue, and damping [31]. Polymers show complex mechanical responses upon subjecting to various strains, frequencies, and temperatures, where energy is absorbed and dissipated as heat due to internal friction. During dynamic testing, the lag/dissipation is recorded at a higher frequency, creating a time recovery response. The dynamic testing for the polymer is mainly conducted on a dynamic mechanical analysis (DMA) instrument that helps to quantify properties such as storage modulus (elastic response, *E'*), loss modulus (energy dissipation, *E"*), complex modulus ($|E^*| = \sqrt{E\prime + iE^{"}}$) and tan delta (damping characteristics).

The dynamic response for elastic, viscoelastic, and viscous conditions is shown in the schematic representation of Fig. 5.3. The stress calculation for viscoelastic materials is represented by equations:

$$\sigma(\omega) = E^*(\omega).\varepsilon(\omega)$$

$$\sigma(\omega) = [E\prime(\omega) + iE''(\omega)].\varepsilon(\omega)$$

$$\sigma(\omega) = E\prime(\omega)[1 + i\eta(\omega)].\varepsilon(\omega)$$

where $\omega =$ angular frequency in radians per second, input stress $\sigma(t) = \sigma_o.sin(\omega t)$, output strain $\varepsilon(t) = \varepsilon_o.\sin(\omega t - \delta)$, and $\eta(\omega) = \frac{E^{"}(\omega)}{E\prime(\omega)} = tan\delta$.

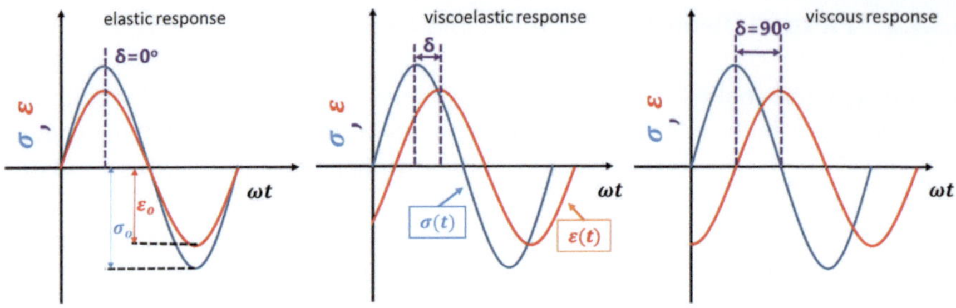

Fig. 5.3 Schematic representation of elastic, viscoelastic, and viscous responses

This dynamic response is crucial for various applications such as vibration isolation, tire performance, and so on, where materials undergo continuous deformation cycles. Dynamic testing thus provides deeper insights into rolling resistance, fatigue resistance, durability, and dynamic thermal effects, which are limited in static tests. These dynamic properties can be tested to understand various properties described in detail, such as creep, stress relaxation, and flex fatigue, whose schematic representation is illustrated in Fig. 5.4.

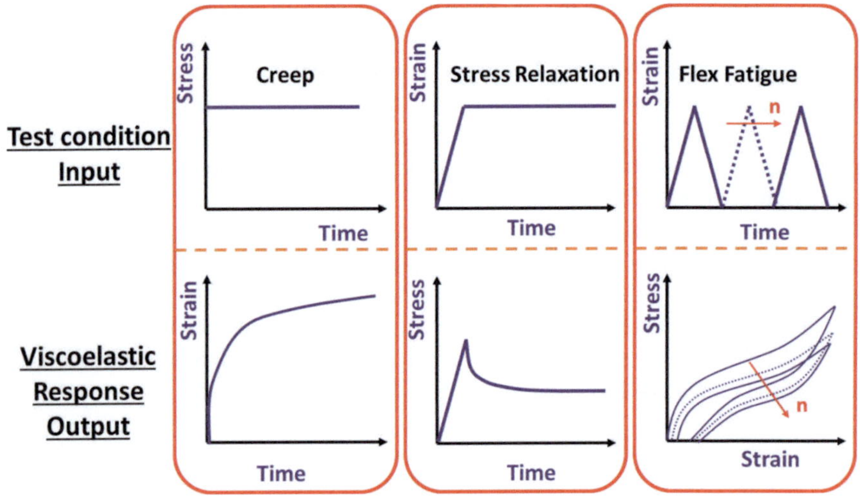

Fig. 5.4 Schematic representation of various dynamic properties test condition input and their responses

5.4.1 Creep

Creep is a time-dependent property that is the deformation of the material under constant load over a definite time until failure, including rubbers, plastics, and biopolymers [32]. During creep at room temperature, plastics show slow creep and irreversible deformation. Plastics exhibit visco-plastic behavior at higher temperatures, where deformation initially recovers elastically but eventually leads to permanent strain over time. Under creep, rubbers show notable elastic deformation; eventually, they fail from molecular slippage. Though it still undergoes slow elongation with time, crosslinking affects instant strength against deformation [33]. Reinforcing with fillers helps limit chain mobility, increasing creep resistance [34]. Though their behavior is affected by moisture absorption, biodegradability, and molecular orientation, biopolymers also exhibit time-dependent deformation. Their mechanical stability is improved by reinforcement with nano-fillers, including graphene or cellulose fibers, qualifying them for sustainable packaging and biomedical implants. Under sustained loads, fibers with low creep resistance slowly elongate, causing dimensional instability and mechanical degradation in long-term use. The fiber type, temperature, and loading influence how much creep occurs. For instance, aramid fibers often display a higher creep resistance than glass fibers, whereas carbon fibers show minimal creep due to their high modulus and crystalline structure. For applications involving extended mechanical loads, such as pressure pipes, seals, and the automotive, medical, and aerospace industries, the creep behavior of polymers is crucial.

5.4.2 Stress Relaxation

Stress relaxation is a time-dependent property where stress decreases over constant strain over time [35]. The stress relaxation happens due to molecular rearrangement and energy dissipation of the material. Polymers exhibit significant stress decay due to polymer chains realignment and relaxation of entanglements. Rubbers exhibit considerable stress relaxation due to their intercoiled flexible polymer chains that show viscoelastic response. Higher crosslinking and filler loading create an interconnected network, lowering the relaxation effect and enhancing long-term performance. Their crystallinity influences plastics' stress relaxation. Stress relaxation is highly dominant in biopolymers such as collagen, gelatin, and polylactic acid (PLA); however, mechanical behavior is highly impacted by hydrogen bonding interactions, moisture content, and degradation rate. Biopolymers are particularly suited for long-term load-bearing applications that can enhance stress retention when reinforced with nano-fillers like graphene or cellulose fibers. Fibers exhibit higher stress relaxation where consistent tension or preload is required in the application, such as tendons in composite structures or tension members in cables. Aramid fibers have good resistance to stress relaxation due to their high tensile strength and molecular alignment, while glass fibers may exhibit greater stress decay under prolonged strain. A stress relaxation behavior study is essential for designing polymer products subjected to load over time without significant performance loss.

5.4.3 Flex Fatigue

Flex fatigue refers to the deterioration of a material's mechanical properties over time when subjected to repeated bending or flexing cycles, which finally leads to cracks and fractures. Due to the presence of flexible disoriented polymer chains, rubber goes on reorientation upon loading. The reorientation leads to energy dissipation during each cycle, which causes the initiation of crack precursors, further leading to crack growth and, eventually, failure. The presence of filler majorly influences the flex fatigue property of rubber due to the presence of small precursors that significantly increase the deterioration of the material [36–38]. Flex fatigue testing is directed by bending a specimen in a cycle until failure to calculate the polymer lifespan under repeated mechanical stress. This property is needed for rubber applications such as tires, conveyor belts, diaphragms, isolators, and seals that help to analyze and modify rubber formulation to enhance durability.

Plastics behave differently from rubber in flex fatigue because of their higher molecular orientation. During cyclic loading, the polymers are less likely to reorient, which can lead to failure and the generation of many precursors [39]. Thermoplastics have fatigue cracks due to polymer chain slippage, but thermosets with stiff crosslinked structures may break more rapidly due to their brittle nature. Fiber-reinforced polymers (FRP), such as carbon-fiber-reinforced epoxy composites, offer superior fatigue resistance, making them ideal for aerospace and automotive applications [40, 41]. The moderate to poor fatigue resistance of biopolymers such as collagen, chitosan, and polylactic acid (PLA) is determined by their molecular structure, rate of disintegration, and degree of hydration [42]. Fatigue-induced weakening caused by protein denaturation and water absorption affects the performance of natural biopolymers, such as collagen-based scaffolds, in tissue engineering applications. However, reinforcing the biopolymers with cellulose nanofibers or bioceramic fillers improves fatigue resistance, making them appropriate for biomedical and biodegradable packaging applications.

The flex fatigue property is also important in textiles, flexible composites, and moving structural elements. Aramid fibers have high toughness and energy absorption capacity, contributing to their superior flex fatigue resistance. Meanwhile, glass fibers are more brittle and prone to breaking when flexed frequently. Although carbon fibers perform well under moderate cyclic conditions, their flex fatigue behavior varies according to matrix and composite architecture. Table 5.2 displays the various fiber characteristics based on creep, stress relaxation, and flex fatigue.

Table 5.2 Different fiber property descriptions based on creep, stress relaxation, and flex fatigue

Property	Carbon fiber	Glass fiber	Aramid fiber
Creep resistance	Excellent—minimal deformation under sustained load due to high stiffness and thermal stability	Moderate—shows some creep at elevated temperatures or prolonged loading	Good—relatively stable under load, though performance can vary by humidity and temperature
Stress relaxation	Very low—retains stress well under constant strain, ideal for tension-critical structures	Moderate to high—loses stress more rapidly under long-term strain	Low—maintains tension effectively, with better performance in dry conditions
Flex fatigue resistance	Good—performs well in composite form, though sensitive to matrix properties and loading conditions	Fair to poor—brittle nature leads to cracking under repeated flexing	Excellent—high durability and energy absorption make it ideal for repeated flexural stress

5.4.4 *Testing Techniques*

Various dynamic testing techniques, i.e., strain amplitude sweep, frequency sweep, and temperature sweep, are described in sections [11, 18].

5.4.4.1 Strain Amplitude Sweep

Strain amplitude sweep testing is crucial for understanding the strength change for a polymeric material over increasing strain levels. This will determine the material's non-linear viscoelastic response. Depending on crosslinking, the filled rubber compound shows different behaviors upon being subjected to a strain amplitude sweep. An unvulcanized filled rubber shows a significant decrease in the modulus due to filler-filler disassociation, also known as the Payne effect [43]. However, vulcanized rubber also exhibits decreases in modulus due to stress-softening related to molecular reorientation, also known as the Mullins effect [44]. At higher strains, stiffness rises for filled rubber compounds due to the alignment of polymer chains. Plastics, especially those with viscoelastic properties, also show strain-dependent behavior but typically exhibit less strain softening than rubbers and more crystal plane slippage, leading to permanent deformation. Strain amplitude sweeps are conducted using dynamic mechanical analysis (DMA), where a material is subjected to oscillatory strain at a fixed frequency. The test helps to evaluate energy dissipation, modulus changes, and fatigue resistance in dynamic applications like gaskets and vibration dampers.

5.4.4.2 Frequency Sweep

Frequency sweep testing characterizes the time-dependent properties of the polymer, which vary with different loading frequencies. This is crucial for understanding viscoelastic responses and important for constitutive models such as Williams-Landel-Ferry (WLF) [45, 46]. Rubber has high viscous behavior at lower frequencies, allowing molecular chains to relax in time. However, it behaves more elastically at higher frequencies due to limited molecular mobility. Moreover, damping depends on vibration frequency, which is crucial for noise, vibration, and harshness (NVH) applications. Plastics also show frequency-dependent behavior, particularly the thermoplastics that show significant time-dependent deformation. Using DMA, frequency sweeps help to determine the glass transition temperature (T_g), damping properties, and stiffness variations for applications like vibration isolators, shoe soles, and damping pads. High-frequency response is crucial in applications where polymers experience rapid cyclic loading, such as automotive bushings and aerospace components.

5.4.4.3 Temperature Sweep

Temperature sweep testing characterizes mechanical properties that vary over temperature. Temperature sweep tests determine transitions of storage modulus of a rubber, such as the T_g (the temperature at which rubber changes from a glassy to a rubbery state) and crosslinking effects [47]. At temperatures below T_g, uncrosslinked rubber acts as a brittle solid, whereas at very high temperatures, it softens and loses elasticity. The temperature range over which it is in a rubbery state to retain flexibility is critical for applications like automotive tires, industrial seals, and aerospace components. During a temperature sweep, plastics have two essential zones of temperature, i.e., T_g and T_m (melting temperature), that help us select between thermoplastics and thermosets for different applications. Polypropylene (PP) and polyethylene (PE) show changes in stiffness around their T_g, which affects their performance in applications such as packaging, automotive interiors, and medical devices. Temperature sweep testing is critical for biopolymers such as polyhydroxyalkanoates (PHA) and polylactic acid (PLA), as they tend to degrade, limiting their use in high-temperature environments thermally. This affects the stability of biodegradable applications. However, advancements in bio-composites and filler modifications improve their temperature resistance for use in sustainable packaging and biomedical applications.

5.5 Brittle and Ductile Failure

Failure mechanisms are critical in understanding material performance for applications where important factors are safety and durability. The materials show two significant types of failure, i.e., brittle and ductile failure [48–50]. These two failures are associated with the yielding (the stress stage at which material begins to deform plastically, i.e., the onset of plasticity) of the material.

5.5.1 *Brittle Failure*

Brittle failure is a sudden and catastrophic failure showing less deformation. This failure does not show yield points and breaks before yielding, showing little or no plastic behavior. Temperature is the key influencing factor that makes a brittle material transition from brittle fracture to ductile fracture as the temperature increases. Rubber is a non-linear viscoelastic material showing no yielding at room temperature. Rubber is elastic and doesn't fail upon twisting as a Brittle material does. However, rubber shows brittle failure below the T_g. The presence of filler in the filled rubber compound alters the T_g. In addition, rubber may show brittle failure for aged or degraded rubber (flexibility loss), for example, cracking in old tires or seals due to ozone exposure or heat aging. Rubber components in dynamic applications must avoid brittleness for reliability.

5.5.2 *Ductile Failure*

Ductile failure is a gradual failure that occurs after yielding. This failure is subjected to yielding and shows plastic deformation (permanent deformation) after the yield point. Temperature plays a significant role in analyzing the transition from ductile material to brittle material as temperature decreases. Plastic and fiber show yielding and ductility. However, plastics and rubber show brittle failure below T_g. Rubber does not show plasticity. However, the filled rubber compound (fillers at high loading level) shows some permanent deformation but no yielding due to filler-filler breakage [6]. Plastics can show necking or tearing in tensile tests.

5.5.3 *Fractography*

Fractography relates to the fracture study of polymeric surfaces, vital in identifying failure modes in polymer components. Using scanning electron microscopy (SEM), fractography discloses surface features such as furrows, striations, dimples, and cleavage facets to provide insight into the failure [51–53]. Furrows show crack propagation direction and are often associated with fatigue failures, where repeated loads lead to progressive crack growth. Due to notches, welds, or material defects in the polymeric material, stress concentration zones often serve as crack initiation points, and their effects are visible when fracture patterns are analyzed. Observing failure patterns, it can be determined whether a failure was due to fatigue, brittle fracture, ductile overload, or environmental factors like corrosion or hydrogen embrittlement.

5.6 Abrasion

Abrasion resistance is the material property that can withstand gradual wear and surface degradation due to friction or mechanical contact [54]. It is critical in applications where continuous scraping, rubbing, or impact occurs. Abrasion resistance in rubber depends on factors like polymer type, filler content, and crosslinking density. High-abrasion-resistant rubbers are widely used in tires, conveyor belts, and industrial rollers. Fillers such as carbon black improve rubber compounds' durability by improving wear resistance [55]. For dynamic applications like off-road tires, abrasion testing ensures durability. The DIN Abrasion Test measures weight loss after abrading against an abrasive surface. The abrasion resistance for plastics varies based on the polymer molecular structure. Ultra-high molecular weight polyethylene (UHMWPE) and polyamide (Nylon 6, 6) have outstanding wear resistance and are used for gears, bearings, and conveyor systems. However, some plastics like polyvinyl chloride (PVC) and polystyrene (PS) have lower abrasion resistance, restricting their applications in low-friction environments. Biodegradable polymers have lower abrasion resistance; however, natural fibers or filler modifications improve wear resistance, making them suitable for sustainable automotive and packaging applications. Rubber abrasions are divided into hysteretic and adhesive abrasions, which act independently or together depending on the rubber compound and contact conditions [56].

5.6.1 Hysteretic Abrasion

Hysteretic abrasion is a type of wear that arises due to hysteresis (internal energy) loss when the rubber deforms repeatedly under dynamic loading. The contact between rubber and the rough surface experiences micro-deformations around surface asperities. Due to continuous cyclic loading, energy is dissipated as heat, creating fatigue and gradual breaking of polymer chains around fillers in the matrix. This type of wear is mainly seen in highly filled rubber compounds such as those used in tires.

5.6.2 Adhesive Abrasion

The adhesive abrasion occurs on the rubber surface that adheres to the abrasive counterpart (like a road or rough test surface) in this mechanism. When the rubber drags on the road surface, the adhesive bonds break, tearing away small fragments of rubber compound. This abrasion is greatly enhanced due to contact pressure, sliding speed, and the chemical reaction between the rubber and the surface.

These two adhesion mechanisms are heavily influenced by factors such as rubber hardness, filler content, crosslink density, and temperature. To avoid or optimize the abrasion resistance, maintained formulations are required to dissipate heat with low hysteresis and less adhesion to the surface that can lead to tearing.

5.7 Heat Build-Up

Heat build-up is a property where heat accumulates in a polymer when subjected to cyclic or repetitive loading, leading to internal friction and a temperature rise [57–61]. The phenomenon is crucial for rubber compounds as they are prepared for dynamic applications and are subjected to repeated deformation. The primary factor for heat build-up is thermal conductivity, which increases with poor thermal conductivity. Heat build-up also leads to aging and cracking over a long period of cyclic loading. This reduces the lifespan of polymeric material due to thermal degradation, such as in tires or vibration isolators. The rubber that shows low hysteresis, like butadiene rubber (BR), is partially incorporated to minimize heat build-up and improve fuel efficiency. Heat-resistant fillers, such as silica, reduce heat generation, making silica-filled rubber compounds more suitable for automotive and aerospace applications. Plastics such as polycarbonate (PC) and acrylonitrile–butadiene–styrene (ABS) are expected to have heat build-up under repeated stress, leading to structural failure. Heat build-up in a biopolymer is a significant challenge, as they degrade at lower temperatures due to unsaturated bonds and impurities. PLA and PHA have limited thermal stability, restricting their use in high-heat applications. However, composite formulations with bio-fillers like lignin and nanocellulose improve heat resistance, making it sustainable for structural applications in packaging, medical implants, and automotive interiors. The Goodrich Flexometer is used to determine heat build-up under cyclic loading.

5.8 Stress–Strain Curves of Different Polymers

The mechanical behavior of materials is primarily linked to the stress–strain curve when subjected to tensile or compressive forces. The shape and characteristics of this curve depend significantly on factors like molecular structure, the extent of crosslinking, and the presence of fillers. Unfilled rubber materials display a unique S-shaped stress–strain curve during loading and undergo significant reversible deformation [1]. During initial strain loading, the polymer chains are coiled (a state of negative entropy), and upon higher strain, the chains start to uncoil. As the strain increases, the polymer chains align more, leading to strain-induced crystallization (SIC), particularly in natural rubber [62]. This alignment stiffens the material, raising the stress levels. Strain is generally categorized into elastic, plastic, and viscoelastic types [18]. Elastic deformation is defined as the material's ability to regain its original state of strain upon load removal. Viscoelastic deformation involves time-dependent strain recovery, encompassing both elastic and viscous behaviors, leading to energy loss and hysteresis. Plastic deformation occurs in plastic and filled rubber compounds, resulting in permanent deformation. In metals and plastics, permanent deformation is accompanied by the achievement of a yield point. However, plastic deformation results in the yield of surface-free plasticity in filled rubber compounds. These deformations determine the complex mechanical response. Cross-play with these properties is crucial for designing materials with optimized performance in automotive, aerospace, and biomedical engineering industries.

Unfilled rubbers, such as natural rubber (NR) and styrene-butadiene rubber (SBR), display low tensile strength. In contrast, filled rubbers with filler, such as carbon black and silica, show increased stiffness, tensile strength, and improved tear resistance. Nano-fillers such as graphene and Nano-clay aid in toughness, making the rubber compound resistant to crack initiation and fatigue failure. Plastics display linear elastic zones at low strains, followed by ductile or brittle failures after reaching the yield point. The yielding is mainly observed in thermoplastics, such as polyethylene (PE), polypropylene (PP), and polycarbonate (PC), where molecular chains undergo plastic deformation before failure. Plastics like polyether ether ketone (PEEK) and acrylonitrile–butadiene–styrene (ABS) display strain hardening, where the stress rises after yielding due to polymeric chain alignment and entanglements. On the contrary, thermosetting plastics, such as epoxy and phenolic resins, lack a yield point, and brittle failure occurs due to their rigid crosslinked structure. Reinforcing plastics with glass, carbon, or aramid fibers increases stiffness and tensile strength for aerospace and automotive applications. Nano-fillers, as described above with plastics, also contribute to enhanced modulus and toughness, allowing advanced high-performance composite materials. Biopolymers, like polyhydroxyalkanoates (PHA), polylactic acid (PLA), and starch-based polymers, have a wide range of stress–strain behaviors that depend on their molecular structure and degree of crystallinity. Many biopolymers behave similarly to plastics, displaying an initial elastic region and then a rapid failure, particularly when unmodified. However, reinforcing natural fibers (cellulose, lignin, hemp) and bio-based plasticizers helps enhance toughness, introducing a yielding phase similar to ductile plastics. Many biopolymers exhibit time-dependent mechanical behavior, where continuous loading leads to creep and stress relaxation. Hybrid composites are invented by reinforcing with nanocellulose, chitin, or lignin nanoparticles to improve ductility and impact resistance. Hybrid composites have many applications, from biodegradable packaging and medical implants to structural composites.

Various factors, such as temperature, strain rate, molecular weight, and filler loading, influence the stress–strain response of polymers. The major influencing factor is temperature, as many polymers exhibit glass transition temperature (T_g), i.e., phase transition zone from rigid to rubbery state. Rubber below its T_g displays brittleness, while plastics develop a ductile nature at higher temperatures. The strain rate significantly influences the material's behavior as rubbers become stiffer. At the same time, loading and decay are observed while unloading under a high strain rate due to viscoelastic effects, whereas plastics exhibit brittle failure at higher strain rates. Reinforcement with fillers plays a significant role, as mechanical properties can be enhanced while playing with the filler content and uniform dispersion. This leads to higher tensile strength due to crosslinking and physical entanglements. These factors are vital in aiding various properties of the polymer and critical for material selection in high-performance industry products.

5.8.1 Constitutive Models for Hyperelastic and Viscoelastic Materials

The stress–strain behavior under various loading conditions is defined using mathematical expressions that form constitutive laws. Polymers such as rubber, plastics, fibers, and biopolymers have complex, non-linear mechanical responses, including elasticity, plasticity, and time-dependent behavior, which require advanced constitutive models. Polymers often exhibit hyperelastic behavior during loading conditions, especially for rubbers that sustain large deformations. The rubber stress–strain relationship can be modeled through strain energy functions. Some simpler models, such as neo-Hookean or Mooney-Rivlin, can predict the stress–strain response for lower strain levels. Advanced constitutive laws such as Ogden, Yeoh, and Arruda-Boyce models [8, 63–65] are required for higher strain levels. At the same time, unloading polymer exhibits time-dependent behaviors at high strain rates. Viscoelastic models define time-dependent responses that can incorporate both elastic and viscous components. The viscoelastic model takes care of the rate and history of the loading and, hence, accurately captures stress relaxation, creep, and hysteresis.

The choice of constitutive model largely depends on the polymer and its application. Hyperelastic material models are used for extensive strain application without unloading/recovering the specimen, as represented through the green line in Fig. 5.5a. However, polymers exhibit time-dependent behavior best described through viscoelastic models like the Maxwell or Kelvin-Voigt models [11]. More advanced models like the Prony series [8], Bergstrom-Boyce [66–68], and PRF [6, 69] models are preferred for non-linear viscoelastic models. The presence of fillers generally adds permanent deformation while unloading the polymer due to filler-filler and filler–polymer breakage, which creates isotropic damage in the compound, as represented by the black dotted line in Fig. 5.5a. The filler-filler breakage is called the Payne Effect [70], and filler–polymer damage is called the Mullins effect [44]. The

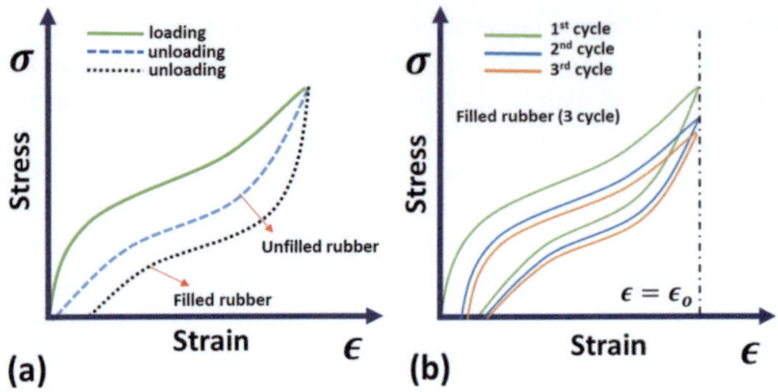

Fig. 5.5 Schematic representation of **a** loading and unloading for unfilled and filled rubber compounds and **b** three cycles of loading and unloading for filled rubber representing stress-softening

damage in filled rubber compounds is a continuous process that depends on strain and cycle number, as shown in Fig. 5.5b. The isotropic damage due to these breakages is modeled through different models, such as the Ogden-Roxburgh [71] and Abhay-Gonzalez [6] models. Abhay-Gonzalez has characterized the filler's isotropic damage through yield-surface-free elastoplastic criterion and, for the first time, quantified the recoverable and unrecoverable stress-softening parameters through their model [6]. The adequate use of various phenomena, such as elastic, viscoelastic, plastic, and stress-softening, and their models can define various complex mechanical behaviors in engineering products.

References

1. Ward IM, Sweeney J (2012) Mechanical properties of solid polymers. Polymer 13
2. Morton M (2013) Rubber technology. Springer Science & Business Media
3. Winspear GG (1958) The Vanderbilt rubber handbook
4. Kumar A et al (2021) Entangled network influenced by carbon black in solution SBR vulcanizates revealed by theory and experiment. Rubber Chem Technol 94:324–338
5. Kumar A et al (2021) Assessment of notch effect based on finite element analysis and digital image correlation technique. Steel Res Int 92:2000376
6. Kumar A, Muhammad SB, Hossain SJ, Chattopadhyay S, Gonzalez M (2023) Mechanical characterization of filled elastomer nanocomposites under finite-deformation cyclic loading using yield-surface-free plasticity. Compos Sci Technol 241
7. Kaliske M, Rothert H (1997) Formulation and implementation of three-dimensional viscoelasticity at small and finite strains. Comput Mech 19:228–239
8. Kumar A et al (2022) Hyper-viscoelastic characterization of highly filled rubber compound: extending approach for geometrical defect analysis. Polym Eng Sci 62:447–460
9. Mermet-Guyennet M et al (2017) Dependence of nonlinear elasticity on filler size in composite polymer systems. Rheol Acta 56:583–589
10. Kumar A et al (2023) Green approaches for functionalized silica-based natural rubber nanocomposites using eco-friendly sodium alginate biopolymers for superior static and dynamic behaviors. ACS Sustain Chem Eng 11:6866–6878

11. Ferry JD (1948) Viscoelastic properties of polymer solutions. J Res Natl Bur Stand 1934(41):53–62
12. Bin Ahmad Z, Ashby MF (1988) Failure-mechanism maps for engineering polymers. J Mater Sci 23:2037–2050
13. Holmes J, Vlandis G, Stachurski Z, Das R, Compston P (2021) Failure behaviour in woven thermoplastic composites subjected to various deformation modes. Compos Part A Appl Sci Manuf 146:106410
14. Bazhenov S (1998) Fillers: their effect on the failure modes of plastics. Plastics additives: an AZ reference, pp 252–259
15. Liew KB, Tan YTF, Peh K-K (2014) Effect of polymer, plasticizer and filler on orally disintegrating film. Drug Dev Ind Pharm 40:110–119
16. Pérez E, Alvarez V, Pérez CJ, Bernal C (2013) A comparative study of the effect of different rigid fillers on the fracture and failure behavior of polypropylene based composites. Compos B Eng 52:72–83
17. Treloar LRG (1944) Stress-strain data for vulcanized rubber under various types of deformation. Rubber Chem Technol 17:813–825
18. Bergstrom JS (2015) Mechanics of solid polymers: theory and computational modeling. William Andrew
19. Nguyen TD, Jones RE, Boyce BL (2007) Modeling the anisotropic finite-deformation viscoelastic behavior of soft fiber-reinforced composites. Int J Solids Struct 44:8366–8389
20. Kim H, Miura Y, MacOsko CW (2010) Graphene/polyurethane nanocomposites for improved gas barrier and electrical conductivity. Chem Mater 22:3441–3450
21. Kotal M, Banerjee SS, Bhowmick AK (2016) Functionalized graphene with polymer as unique strategy in tailoring the properties of bromobutyl rubber nanocomposites. Polymer (United Kingdom) 82:121–132
22. Goswami M et al (2020) Optimization of graphene in carbon black-filled nitrile butadiene rubber: constitutive modeling and verification using finite element analysis. Polym Compos 41:1853–1866
23. Fouchier N, Nadot-Martin C, Conrado E, Bernasconi A, Castagnet S (2019) Fatigue life assessment of a short fibre reinforced thermoplastic at high temperature using a through process modelling in a viscoelastic framework. Int J Fatigue 124:236–244
24. Bhowmick A (2018) Rubber products manufacturing technology. Routledge
25. Dieter G (1961) Mechanical metallurgy—Dieter_George Ellwood.pdf, pp 370–371. Preprint at https://doi.org/10.1093/oxfordhb/9780199238828.013.0020
26. ASTM D412–16 (2021) Standard test methods for vulcanized rubber and thermoplastic elastomers. 16:1–14
27. ASTM D885/D885M (2003) Standard test methods for tire cords, tire cord fabrics, and industrial filament yarns made from manufactured organic-base fibers
28. (2008) Test method for tensile properties of polymer matrix composite materials. Preprint at https://doi.org/10.1520/D3039_D3039M-08
29. (2021) Test method for rubber property durometer hardness. Preprint at https://doi.org/10.1520/D2240-15R21
30. (2025) Test methods for rubber property compression set. Preprint at https://doi.org/10.1520/D0395-18R25
31. Gent AN (1962) Relaxation processes in vulcanized rubber. I. Relation among stress relaxation, creep, recovery, and hysteresis. J Appl Polym Sci 6:433–441
32. Morro A (2017) Modelling of viscoelastic materials and creep behaviour. New Trends Dyn Stab. https://doi.org/10.1007/s11012-016-0585-x
33. Plazek DJ (1966) Effect of crosslink density on the creep behavior of natural rubber vulcanizates. J Polym Sci Part A-2 4:745–763
34. Mostafa A, Abouel-Kasem A, Bayoumi MR, El-Sebaie MG (2009) On the influence of CB loading on the creep and relaxation behavior of SBR and NBR rubber vulcanizates. Mater Des 30:2721–2725

35. Compounds R, Elastomers T, International A (2003) Standard practice for stress relaxation testing of raw rubber, unvulcanized. Current 09:1–16
36. Le Cam JB, Huneau B, Verron E, Gornet L (2004) Mechanism of fatigue crack growth in carbon black filled natural rubber. Macromolecules 37:5011–5017
37. Wang XL, Liao MY, Xu Y, Liu XA (2018) Fatigue crack propagation characteristics of rubbery materials under variable amplitude loading. Results Phys 10:233–240
38. Stoček R, Gehde M, Michael H, Maňas S (2008) Determination of fatigue behaviour and crack propagation of filled vulcanized rubber. Mtf.Stuba.Sk 105–112
39. Tunnicliffe LB, Robertson CG, Mars WV (2023) A microscopy investigation of rubber compound crack precursors and tensile fracture surfaces. Rubber Chem Technol 96:514–529
40. Dew-Hughes D, Way JL (1973) Fatigue of fibre—reinforced plastics: a review. Composites 4:167–173
41. Alam P, Mamalis D, Robert C, Floreani C, Brádaigh CMÓ (2019) The fatigue of carbon fibre reinforced plastics-A review. Compos B Eng 166:555–579
42. McKeen LW (2016) Fatigue and tribological properties of plastics and elastomers. William Andrew
43. Payne AR (1962) The dynamic properties of carbon black loaded natural rubber vulcanizates. Part II J Appl Polym Sci 6:368–372
44. Mullins L (1969) Softening of rubber by deformation. Rubber Chem Technol 42:339–362
45. Peleg M (1992) On the use of the WLF model in polymers and foods. Crit Rev Food Sci Nutr 32:59–66
46. Ljubić D et al (2014) Time: temperature superposition principle: application of WLF equation in polymer analysis and composites. Zaštita materijala 55:395–400
47. ASTM E1640-4 (2009) Standard test method for assignment of the glass transition temperature by dynamic mechanical analysis. Annual book of ASTM standards, pp 1–6. https://doi.org/10.1520/E1640-18.2
48. Mai Y-W, Cotterell B (1986) On the essential work of ductile fracture in polymers. Int J Fract 32:105–125
49. Miehe C, Schänzel L-M (2014) Phase field modeling of fracture in rubbery polymers. Part I: finite elasticity coupled with brittle failure. J Mech Phys Solids 65:93–113
50. Liu K, Piggott MR (1998) Fracture failure processes in polymers. I: Mechanical tests and results. Polym Eng Sci 38:60–68
51. Lynch SP, Moutsos S (2006) A brief history of fractography. J Fail Anal Prev 6:54–69
52. Hayes MD, Edwards DB, Shah AR (2024) Fractography in failure analysis of polymers. Elsevier
53. Tan CJ, Andriyana A, Ang BC, Wong D (2020) Mechanical deformation and fracture mechanisms of polymeric fibres from the perspective of fractography–a review. Eur Polym J 137:109924
54. Fukahori Y, Liang H, Busfield JJC (2008) Criteria for crack initiation during rubber abrasion. Wear 265:387–395
55. Mostafa A, Abouel-Kasem A, Bayoumi MR, El-Sebaie MG (2009) Insight into the effect of CB loading on tension, compression, hardness and abrasion properties of SBR and NBR filled compounds. Mater Des 30:1785–1791
56. Muhr AH, Roberts AD (1992) Rubber abrasion and wear. Wear 158:213–228
57. Stadlbauer F et al (2013) Influence of experimental parameters on fatigue crack growth and heat build-up in rubber. Materials 6:5502–5516
58. Kerchman V, Shaw C (2011) Experimental study and finite element simulation of heat build-up in rubber compounds with application to fracture. Rubber Chem Technol 76:386–405
59. Tong X, Chen X, Xu J, Zheng Y, Zhi SJ (2018) The heat build-up of a polymer matrix composite under cyclic loading: experimental assessment and numerical simulation. Int J Fatigue 116:323–333
60. Ovalle Rodas C, Zaïri F, Naït-Abdelaziz M, Charrier P (2016) A thermo-visco-hyperelastic model for the heat build-up during low-cycle fatigue of filled rubbers: formulation, implementation and experimental verification. Int J Plast 79:212–236

61. Luo W et al (2018) Modeling of the heat build-up of carbon black filled rubber. Polym Test 69:116–124
62. Ruellan B et al (2018) Fatigue crack growth in natural rubber: the role of SIC investigated through post-mortem analysis of fatigue striations. Eng Fract Mech 201:353–365
63. Yeoh OH (1990) Characterization of elastic properties of carbon-black-filled rubber vulcanizates. Rubber Chem Technol 63:792–805
64. Ogden RW (1972) Large deformation isotropic elasticity: on the correlation of theory and experiment for compressible rubberlike solids. Proc Roy Soc A: Math Phys Eng Sci 328:567–583
65. Arruda EM, Boyce MC (1993) A three-dimensional constitutive model for the large stretch behavior of rubber elastic materials. J Mech Phys Solids 41:389–412
66. Bergström JS, Boyce MC (2011) Mechanical behavior of particle filled elastomers. Rubber Chem Technol 72:633–656
67. Bergström JS, Boyce MC (2000) Large strain time-dependent behavior of filled elastomers. Mech Mater 32:627–644
68. Bergström JS, Boyce MC (1998) Constitutive modeling of the large strain time-dependent behavior of elastomers. J Mech Phys Solids 46:931–954
69. Hurtado J, Lapczyk I, Govindarajan S (2013) Parallel rheological framework to model non-linear viscoelasticity, permanent set and Mullins effect in elastomers. Constitutive Models Rubber VIII 95:95–100
70. Maier PG, Goritz D (1996) Molecular interpretation of the Payne effect. KGK 49:18–21
71. Ogden RW, Roxburgh DG (1999) A pseudo–elastic model for the Mullins effect in filled rubber. Proc Roy Soc Lond Ser A: Math Phys Eng Sci 455:2861–2877

Dr. Abhay Kumar with a collaborative Ph.D. from IIT Kharagpur and Purdue University, delves into the intricate dynamics of rubber-filler interactions, unraveling structure–property relationships. His pioneering research on Mullins damage constitutive relations illuminates both recoverable and unrecoverable stress softening. Emphasizing simulation studies, he offers practical insights into filled rubber systems, bridging theory with industrial solutions. Esteemed for profound comprehension and adeptness, Dr. Kumar integrates theoretical insights to address real-world challenges in materials science.

Nikhil Kumar is a senior research fellow the Advanced Technology Development Centre, Indian Institute of Technology Kharagpur. He is specialized in fabrication of smart polymeric materials for diagnosis and therapeutics. Nikhil completed his Bachelor of Pharmacy and Master of Pharmacy in Pharmacology. His masters research work was focused on developing dressings for diabetic foot ulcers. His current research revolves around the design and application of smart polymeric biomaterials for advanced diagnosis and therapeutic interventions for cutaneous squamous cell carcinoma. He has published 12 peer-reviewed research articles. He is actively engaged in research focused on polymer characterization, with particular interest in the structural, thermal, and rheological analysis of advanced polymeric materials. Mr. Kumar has been involved in research projects that explore the application of spectroscopic techniques (FTIR, NMR), thermal analysis (DSC, TGA), to evaluate polymer performance under varied environmental conditions. Mr. Kumar is also developing a novel diagnostic tool that integrates multiple characterization techniques FTIR, Raman spectroscopy, AFM and nanoindentation to probe and differentiate pathological tissue conditions such as fibrosis and hyperkeratosis.

Manas Jain is currently working as senior research scholar, in the combustion lab of department of Aerospace Engineering, IIT Kharagpur. His area of study and research interests includes, chemically reacting flows, High speed and pressure gain combustion, Flame dynamics, Deflagration to detonation transition, Spectroscopy, Laser-optics diagnostics techniques and mechanical behaviour analysis of materials.

Chapter 6
Thermal Analysis of Polymers

Roumita Hore, Nikhil Dhore, and Komal Agarwal

Abstract In this chapter, we shall discuss the response of a polymer to temperature variation. This chapter will cover thermal characterization techniques such as Differential Scanning Calorimetry (DSC), Thermogravimetric Analysis (TGA), Dynamic Mechanical Analysis (DMA), Dilatometry, Heat Deflection Temperature (HDT), Vicat Softening Point, and Melt Flow Index (MFI). This chapter will provide information related to the thermal transition of a polymer, % crystallinity, thermal stability, degradation kinetics, and variation in dynamic properties of a polymer with temperature.

6.1 Introduction

Thermal analysis can be elaborated as a group of methods by which the physical and chemical properties of a substance, mixture, and reaction mixtures are determined as a function of temperature and time, while the sample is subjected to a controlled temperature. It includes heating or cooling (dynamic) or constant holding temperature (isothermal), or a combination. For all rubber industry products, temperature and its effects play an essential role from the production stage to the final failure of the product in the field. Thermal analysis of elastomers provides a quick way to determine the basic composition of different rubber vulcanizates. There are different thermal analytical techniques available. Some standard methods used are mentioned in Table 6.1.

Among all the above techniques, as the most used techniques in polymer research and rubber industries, Differential Scanning Calorimetry (DSC), Thermogravimetric Analysis (TGA), and Dynamic Mechanical Analysis (DMA) are discussed here.

R. Hore (✉) · N. Dhore
Rubber Technology Centre, IIT Kharagpur, Kharagpur, West Bengal, India
e-mail: horeroumita2019@gmail.com

K. Agarwal
Department of Physics and Technology, UiT-The Arctic University of Norway, Tromsø, Norway

S. Chattopadhyay and N. Kumar (eds.), *Polymer Characterization*,
https://doi.org/10.1007/978-981-95-2609-3_6

Table 6.1 Different thermal techniques

Thermal behavior based on	Characterization techniques
Energy	Differential scanning calorimetry
Mass	Thermogravimetric analysis
Mechanical property	Thermomechanical analysis
Temperature	Differential thermal analysis
Dimension	Thermodilatometry
Magnetic property	Thermomagnetometry
Electrical property	Thermoelectrometry
Acoustic	Thermosonimetry
Optical property	Thermo-optometry

6.2 Differential Scanning Calorimetry

Differential Scanning Calorimetry (DSC) is a thermo-analytical technique in which the difference in heat flow rate between a sample and an inert reference is measured as a function of time and temperature. The method was developed by Watson and O'Neill in 1962 and introduced commercially at the 1963 Pittsburgh Conference on Analytical Chemistry and Applied Spectroscopy [1].

6.2.1 Working Principle

There are different types of DSC available, such as heat flux DSC, power compensated DSC, modulated DSC, and pressure DSC [2–5].

Heat Flux DSC: In a heat flux DSC, a pan containing the sample and an empty reference pan are placed on a thermoelectric disk enclosed by a furnace. As the furnace is heated at a linear heating rate, the heat is transmitted to the sample and reference pan. The difference in temperature of the sample and reference pan due to the heat capacity (C_p) of the sample is measured by thermocouples, and the resultant heat flow is calculated using the thermal equivalence of Ohm's law. The working principle is shown in Fig. 6.1 [2].

Power-Compensated DSC: When an exothermic (heat yielded) or endothermic (heat absorbed) change occurs in the sample, power or energy is applied or removed from one or both of the calorimeters to compensate for the energy change occurring in the sample. The power-compensated DSC system is maintained in a "thermal null" state at all times. The amount of power required to keep the system in an equilibrium condition is directly proportional to the energy changes occurring in the sample. It provides an accurate measure of the calorimetric properties of the sample since the fundamental measurement with the power-compensated DSC is energy flow [3].

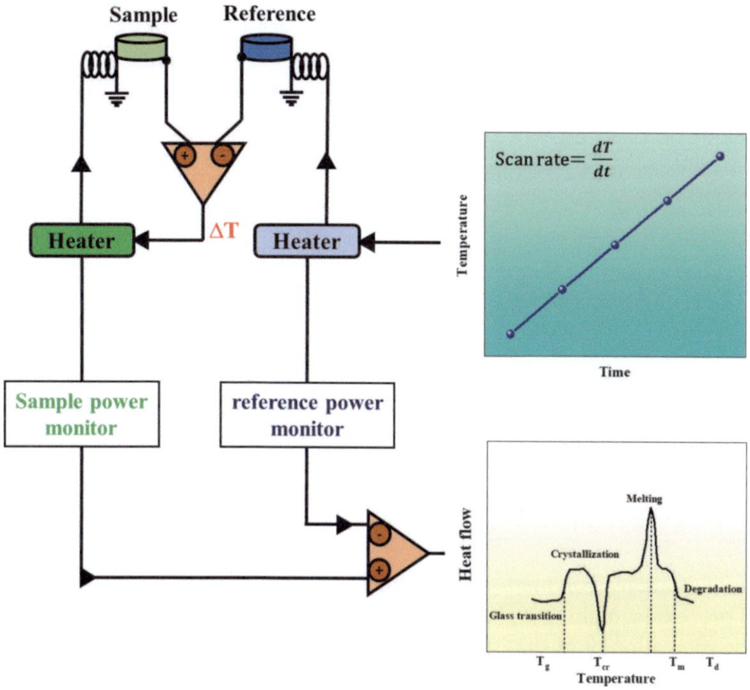

Fig. 6.1 Schematic of heat flux DSC

The difference between heat flux DSC and power-compensated DSC is tabulated in Table 6.2.

Modulated DSC (MDSC): MDSC is a technique that also measures the difference in heat flow in the sample and reference as a function of time and temperature, like a heat flux DSC. However, in MDSC, a sinusoidal temperature modulation is superposed on the linear (constant) heating profile to yield a temperature program in which the average sample temperature varies continuously in a sinusoidal manner, Fig. 6.2. A significant advantage of MTDSC is its ability to resolve overlapping

Table 6.2 Comparison between heat flux DSC and power-compensated DSC

Heat flux DSC	Power-compensated DSC
The sample and reference are kept in the same furnace	The sample and reference are kept in separate furnaces
Heat flow to the sample and reference is measured	The amount of power required to maintain the same temperature is measured
The temperature difference is converted to heat flow using mechanical calculations	Heat flow is measured directly as the amount of power required, directly proportional to the heat flow

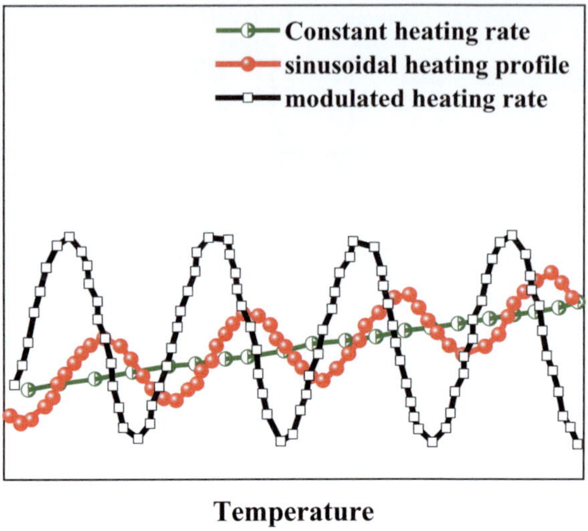

Fig. 6.2 Heating profile of modulated DSC

transitions in many polymers. In particular, the glass transition may coincide with several processes, making it challenging to analyze. These processes may include an enthalpy relaxation or hysteresis peak, a curing exotherm, or even crystallization on heating.

Another benefit of MTDSC is the increased sensitivity observed in the reversing heat flow because of the decreased noise level. The constant heating profile provides total heat flow information. The sinusoidal heating profile gives heat capacity information corresponding to the rate of temperature change. MDSC data are calculated from three signals: time, modulated heat flow, and modulated heating rate. However, raw data for MDSC is very complicated, requiring deconvolution to obtain standard data [4].

Pressure DSC: In pressure DSC, the sample can be submitted to different pressures, which allows the characterization of substances to distinguish between overlapping peaks. Applications of this technique include studies of pressure-sensitive reactions, evaluation of catalysts, and resolution of overlapping transitions [5].

6.2.2 Instrumentation

The key components of DSC include a sample chamber, a temperature control unit, and a detection system, Fig. 6.3. The sample is enclosed in a small aluminum pan (DSC pan), with a tightly closed lid. The reference is typically an inert material like alumina or an empty aluminum pan. The sample and reference pan are placed in the chamber where precise temperature control and inert atmosphere are crucial. Nitrogen or Helium is used to create an inert atmosphere. The DSC furnace increases the temperature at a constant heating rate, and a thermocouple measures the inside temperature [6]. The list of different thermocouples used in DSC is mentioned in the Table 6.3. The liquid nitrogen is a cooling agent facilitating negative temperature transitions like glass transition and crystallization. A thermopile or thermistor provides a voltage signal, which is converted to heat flow that corresponds to the thermal energy needed to maintain the temperature difference between the sample and the reference.

In DSC, heat flow is measured as a function of temperature. Endothermic transitions (melting, glass transition) cause absorption of heat; conversely, exothermic transitions (crystallization) cause release of heat, resulting in opposite heat flow signal (peak directions).

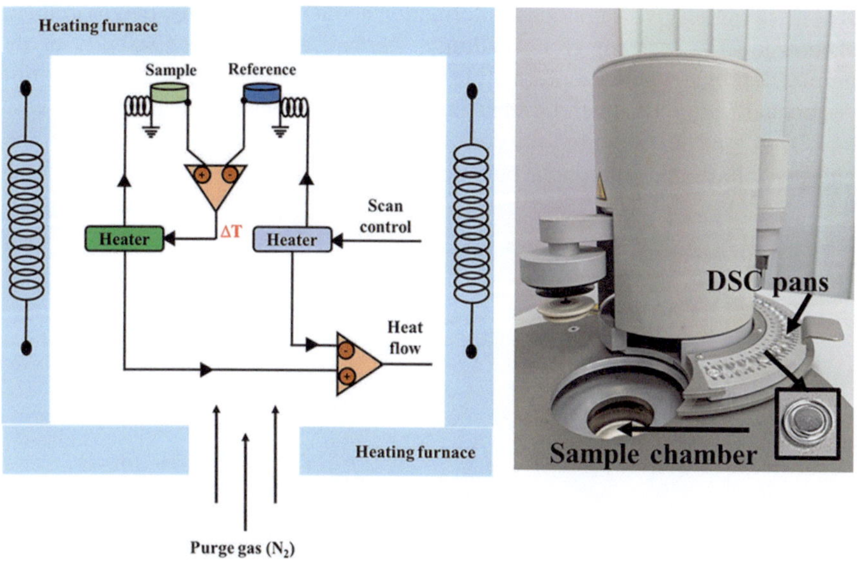

Fig. 6.3 Instrumentation of DSC

Table 6.3 Thermocouples used in DSC

Name of thermocouple	Approx max temperature (°C)
Copper-constantan	250
Iron-constantan	450
Chromel-constantan	1000
Chromel–alumel	1350
Platinum-platinum—10% rhodium	1600
Tungsten-tungsten—26% rhenium	2400
Chromel-constantan differential thermocouple	–150–600

6.2.3 Calibration

DSC is calibrated as per ASTM E967 and ASTM E968 [7, 8]. The aim of calibration and adjustment is to provide a measuring system that consistently delivers reproducible and accurate results. Indium (T_m 156.6 °C) is one of the most used materials for DSC calibration. Apart from indium, tin (T_m 231.9 °C) and zinc (T_m 419.5 °C) are also used for calibration.

6.2.4 Data Analysis

DSC thermogram gives significant information about sample purity, polymorphism, reaction kinetics, percentage crystallinity, thermal, and oxidative stability of the samples. It provides distinct information about four major transitions: glass transition (A), crystallization (B), melting transition (C), and oxidative or degradation transition (D). A typical DSC curve and corresponding thermal transitions for crystalline, semi-crystalline, and amorphous polymer are shown in Fig. 6.4.

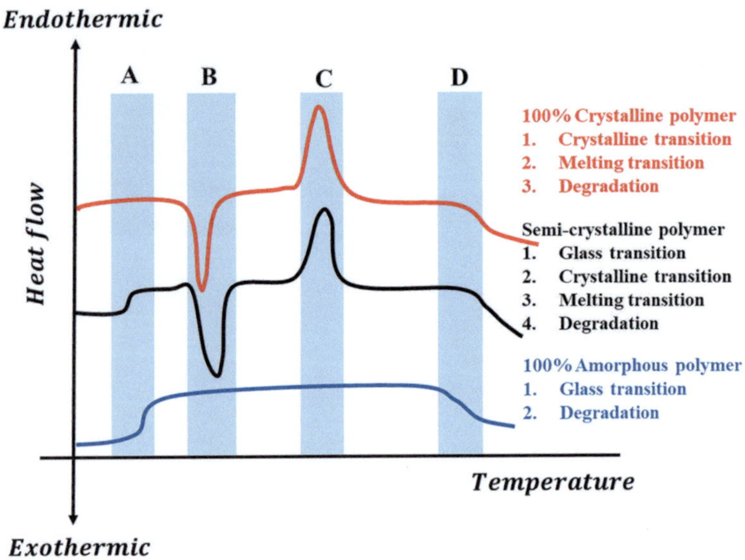

Fig. 6.4 DSC thermogram of a crystalline, semi-crystalline, and amorphous polymer

6.2.5 *First and Second Order Transitions*

Melting Transition (T_m)

The melting point is the temperature at which the crystalline phase of polymers becomes more disordered and becomes a viscous liquid. Polymers consist of very long carbon chains and are also held together by strong covalent bonds, unlike small molecules. Only crystalline materials and semi-crystalline polymers show a melting transition. Amorphous polymers do not.

It is a phase transition as it involves latent heat. The melting transition is a first-order transition [9], as the molar volume (V) and molar entropy (S), first-order derivatives of Gibbs free energy (G), change sharply at $T_{m,}$ shown in Fig. 6.5 and the following Eqs. (6.1) and (6.2), respectively.

$$\left(\frac{\partial \mathbf{G}}{\partial \mathbf{P}}\right)_{\mathbf{T}} = V \tag{6.1}$$

$$\left(\frac{\partial \mathbf{G}}{\partial \mathbf{T}}\right)_{\mathbf{P}} = S \tag{6.2}$$

Glass Transition (T_g)

The glass transition is a property of only the amorphous portion of a semi-crystalline polymer. The crystalline portion remains crystalline during the glass transition. Below the glass transition temperature, the amorphous regions of a polymer are in the glassy state. In this state, the segmental movement of the polymer chains gets frozen.

Glass transition is not a phase transition as there is no involvement of latent heat. It is a second-order transition [10] because at T_g there is a sharp change of the coefficient of thermal expansion (α) and isothermal compressibility (β), which are second-order derivatives of Gibbs free energy (G), as shown in Fig. 6.6 and the Eqs.

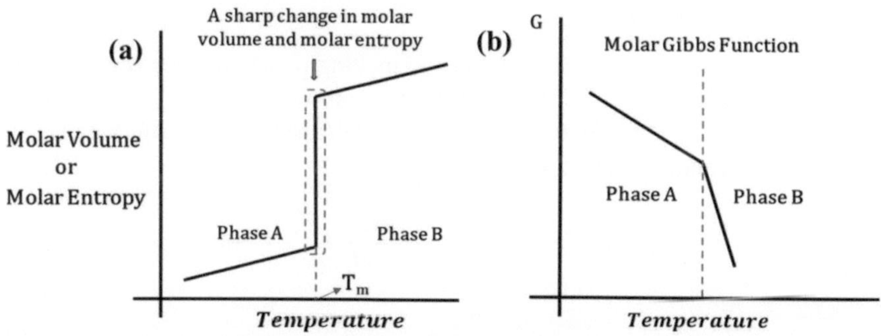

Fig. 6.5 Thermodynamics of melting **a** molar volume or entropy versus temperature, **b** Gibb's energy versus temperature

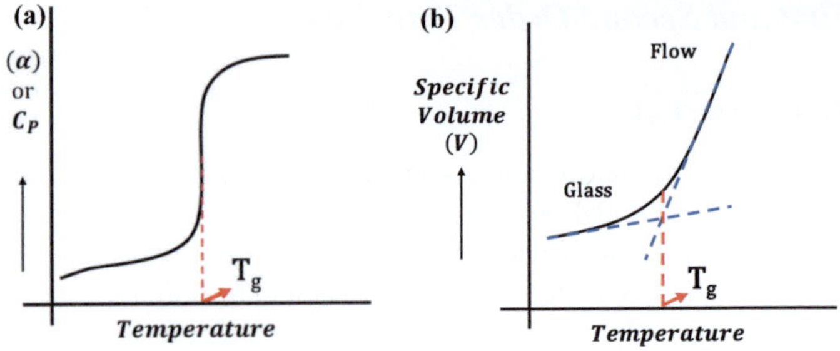

Fig. 6.6 Thermodynamics of glass transition **a** coefficient of thermal expansion or heat capacity versus temperature, **b** specific volume versus temperature

(6.3), (6.4), and (6.5).

$$\mathbf{C_P} = \left(\frac{\partial \mathbf{H}}{\partial \mathbf{T}}\right)_{\mathbf{P}} \tag{6.3}$$

$$\alpha = \frac{1}{\mathbf{V}}\left(\frac{\partial \mathbf{V}}{\partial \mathbf{T}}\right)_{\mathbf{P}} = \frac{1}{\mathbf{V}}\left(\frac{\partial^2 \mathbf{G}}{\partial \mathbf{T}\partial \mathbf{P}}\right) \tag{6.4}$$

$$\beta = -\frac{1}{\mathbf{V}}\left(\frac{\partial \mathbf{V}}{\partial \mathbf{P}}\right)_{\mathbf{T}} = -\frac{1}{\mathbf{V}}\left(\frac{\partial^2 \mathbf{G}}{\partial \mathbf{P}^2}\right)_{\mathbf{T}} \tag{6.5}$$

H_m and H_c are the enthalpies of melting and crystallization, respectively. $H_{f,100}$ is the enthalpy of melting for 100% crystalline polymer.

6.2.6 Determination of Percentage Crystallinity and Blend Miscibility

Percentage Crystallinity

The area under the curve (crystallization and melting peak) provides an idea about the extent of crystallinity. The higher area or depth under the curve corresponds to the higher degree of crystallinity, Fig. 6.7. To calculate % crystallinity precisely, the following formula, Eq. (6.5), can be used.

$$\textit{Degree of crystallinity} = \frac{H_m - H_c}{H_{f,100}} \times 100$$

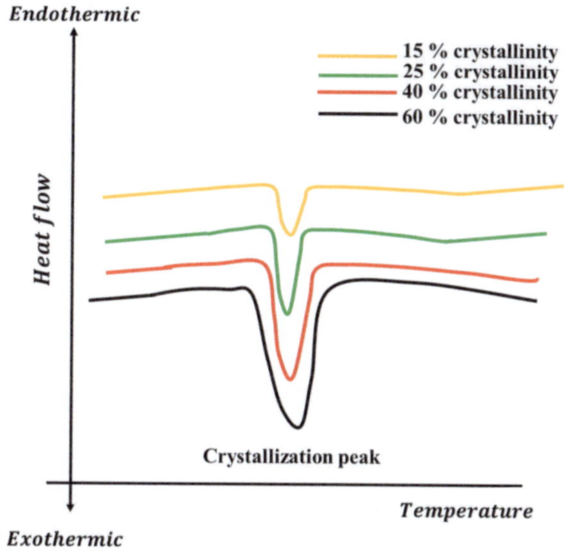

Fig. 6.7 DSC thermogram of different polymers with different crystallinity (%)

Blend Miscibility

Depending on the glass transition, DSC can differentiate between miscible and immiscible blends [11–13]. The miscible blend shows a single glass transition, but the immiscible blend shows two glass transitions. The range of T_g appears based on the blend composition, which is the ratio of the weight fraction taken. In Fig. 6.8, we can see that polymer A and polymer B have their distinct DSC thermograms. If they form a miscible blend, a single T_g (Blend) is observed. If they form an immiscible blend, two transitions as T_g (A) and T_g (B) are observed.

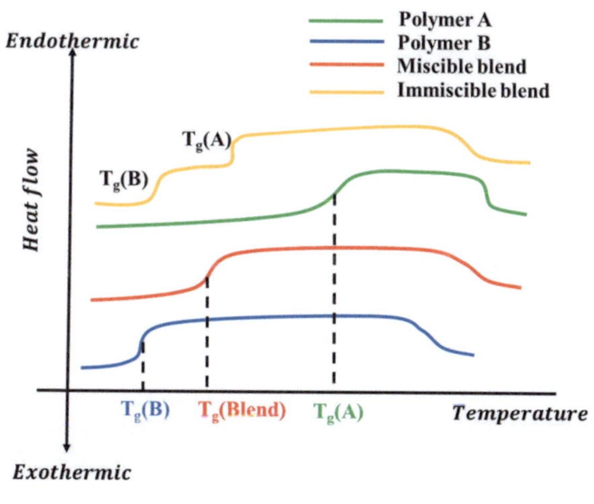

Fig. 6.8 DSC thermogram for blend miscibility

DSC offers high sensitivity and is capable of detecting small changes in heat flow, making it ideal for studying subtle thermal transitions. It provides quantitative data, such as enthalpy changes (ΔH), associated with thermal events. The technique operates over a wide temperature range, typically from −150 to 600 °C, and is versatile, applicable to various materials, including polymers, biopolymers, composites, and small molecules. DSC is also fast and reliable, delivering rapid results with high reproducibility.

6.3 Thermogravimetric Analysis

Thermogravimetric analysis or thermal gravimetric analysis (TGA) is a method of thermal analysis in which the mass of a sample is measured over time as the temperature changes [14]. The principle of thermogravimetry is based on a thermobalance, which is an instrument that continuously measures the weight changes of a substance at gradually varying temperatures, Fig. 6.9.

There are three major types of thermogravimetric methods available, as follows—

Isothermal or static thermogravimetry: A sample weight is recorded as a function of time at a constant temperature.

Quasistatic thermogravimetry: The sample temperature is raised in sequential steps separated by isothermal intervals, during which sample mass reaches stability before the start of the next temperature ramp.

Dynamic thermogravimetry: A sample is heated at a linear rate of increasing temperature.

Derivative thermogravimetric (DTG) analysis describes the rate of weight change with respect to time or temperature and is represented as $(d\omega/dt)$ or $(d\omega/dT)$, Fig. 6.9.

Fig. 6.9 TGA and DTG plot

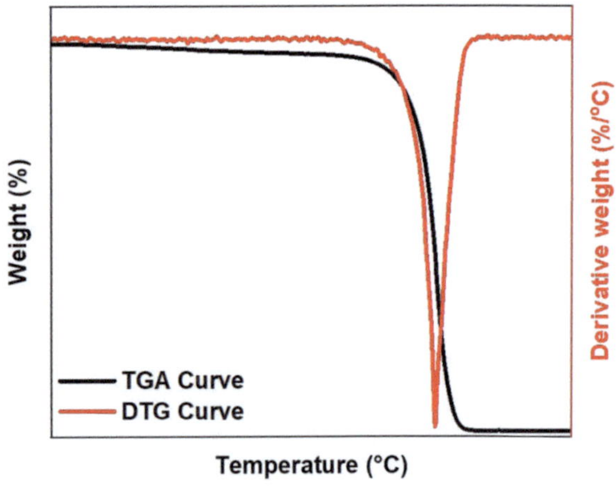

In studying multicomponent systems, the role of DTG is very important, as it easily distinguishes two or more components when the rate of weight change is reflected in the form of peaks.

6.3.1 Working Principle

TGA operates by heating a sample at a controlled rate while continuously measuring its weight. The sample is placed in a crucible, and its weight is constantly monitored using a highly sensitive balance and subjected to a temperature program (heating, or isothermal conditions). The experiment can be conducted in an inert atmosphere (e.g., nitrogen) to study decomposition or in an oxidative atmosphere (e.g., air) to study oxidation. As the temperature increases, the sample may lose weight due to evaporation, decomposition, or oxidation. The weight change (both loss and gain) of the sample is recorded as a function of temperature or time, producing a thermogravimetric (TG) curve. The thermobalance is the heart of the instrument. The balance operates on a null-balance principle. Current is then applied to the meter movement to return the balance to the null position. The amount of current applied is proportional to the weight loss or gain.

The decomposition of the sample starts with the evaporation of moisture and low molecular weight components such as process oil and plasticizer. The degradation of polymers happens in the range from 250–350 °C except for high heat-resistant polymers. Beyond 600 °C, the oxidative decomposition of carbonaceous fillers starts. Metal oxides remain unchanged as ash, Fig. 6.10.

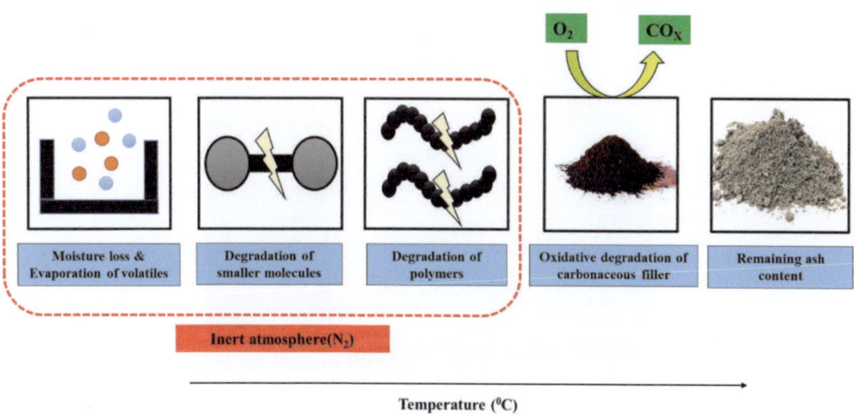

Fig. 6.10 Different types of degradation

6.3.2 Instrumentation

A TGA consists of a sample holder, a thermogravimetric balance, a temperature control unit, a gas flow system, detectors, and a data acquisition and control system, Fig. 6.11. The sample is typically placed in a high-temperature resistant metal crucible. The heart of the TGA is a highly sensitive thermobalance that continuously measures the mass of the sample. It uses a servo-controlled mechanism or force compensation method to detect small changes in mass based on an electromagnetic force. The heating furnace can raise the temperature to 1200 °C. The temperature range and heating rate are used as input to define a precise heating profile. The thermocouples placed inside the chamber sense the temperature, typically made of materials like platinum–rhodium or chromel–alumel. The atmosphere of the chamber influences thermal degradation. The gas flow system allows a constant flow of gas. Nitrogen, oxygen, or air can be used as per the requirement. The instrument is also provided with multiple gas inlets, facilitating different atmospheres at different stages of degradation. A mass loss detector continuously monitors the mass change as a function of time and temperature.

Fig. 6.11 Thermogravimetric analyzer

6.3.3 Calibration

TGA is calibrated as per ASTM E 1582—Standard Practice for Calibration of Temperature Scale for Thermogravimetry [15]. ASTM E 2040—Standard Test Method for Mass Scale Calibration of Thermogravimetric Analyzers [16] and ISO 11358 [17]. Curie temperature plays a significant role in temperature calibration. A ferromagnetic material with a known Curie point is placed in the sample holder. Materials like alumel (Curie temperature 153 °C), nickel (358 °C), $Ni_{83}Co_{17}$ (555 °C), $Ni_{63}Co_{37}$ (747 °C), $Ni_{37}Co_{63}$ (931 °C), cobalt (1116 °C) are used. At the Curie temperature, the magnetic property of these materials decreases sharply, causing an apparent weight change. Any deviation from the standard Curie point is considered for correction and reference point for calibration.

6.3.4 Data Analysis

TGA provides valuable insights into the thermal behavior and composition of both synthetic and biopolymers. It helps determine the thermal stability of a polymer by identifying the onset temperature at which decomposition begins and the temperature range over which this process occurs. For example, TGA analysis of nylon 6,6 bristles revealed that thermal degradation started at approximately 482 °C, with a total mass loss of 99%, indicating excellent thermal stability up to that point [18].

TGA also tracks the rate and extent of weight loss during decomposition, revealing multiple decomposition steps such as the loss of volatile components or polymer backbone degradation. TGA also provides data for composition analysis by quantifying fillers, additives, and volatile components within a polymer matrix, while also determining the residual mass, such as ash content, after decomposition. In the case of nylon 6,6, a small moisture content of <1% was detected at around 56 °C before major decomposition, showing the sensitivity of TGA in detecting even minor volatile losses. Another good example is the analysis of glass-filled epoxy resin, where TGA identified a 31.8% inert glass filler content after thermal decomposition and oxidative burn-off [19]. Similarly, it has been used for detecting the percentage of electrospun polyacrylonitrile fibers embedded in polyvinyl alcohol (PVA) thin films [20]. Additionally, TGA provides information on oxidative stability, indicating the temperature at which oxidation begins in an oxidative atmosphere and the extent of weight gain due to oxidation.

For quality control, TGA ensures consistency in polymer production and verifies the thermal properties of both raw materials and finished products. In formulation development, TGA is used to optimize polymer blends, composites, and additives, and to study the effects of stabilizers, plasticizers, and flame retardants [21]. An advanced application is the use of Auto Stepwise TGA to separate closely overlapping decomposition events, as seen in the characterization of ABS polymers, where the butadiene rubber and SAN copolymer decomposition could be distinguished for better compositional analysis [19].

6.3.5 Determination of Blend Miscibility, Filler Content, and Ash Content

The TGA-DTG thermogram for a miscible polymer blend shows a single but sharp mass loss, whereas the immiscible polymer blend shows multiple distinct mass losses [22, 23].

Let us consider a TGA-DTG thermogram of compounded rubber. Figure 6.12 shows three distinct peaks (DTG curve) and mass losses (TGA curve) marked as A, B, C, and D. The initial mass loss (A) happens due to moisture loss (degradation temperature, T_d 100 °C), or it can be an indication of the existence of low molecular weight entities like process oil or plasticizer (if $T_d > 100$ °C). The temperature window for rubber degradation is 250–350 °C. Thus, B is associated with the polymer loss. As single degradation happens at B, it is either a miscible rubber blend or a single rubber used as a base matrix. The third and final transition occurs at around 600 °C, which may be due to the degradation of fillers (mass loss of filler, C). The mass loss represented by D is the remaining ash. The ash is generated by zinc oxide, silica, or other metal oxides. This is how we can reverse engineer the compounded polymer [24]. To evaluate the weight% of oil, filler, and ash content in the compounded rubber, the following equations can be used—

$$\textbf{Oil content} = \frac{\mathbf{A}}{\mathbf{B}} \times \mathbf{100} \tag{6.6}$$

$$\textbf{Filler content} = \frac{C}{B} \times 100 \tag{6.7}$$

$$\textbf{Ash content} = \frac{D}{B} \times 100 \tag{6.8}$$

TGA can also distinguish between copolymers. Random copolymers show a single degradation, but block copolymers and graft copolymers respond with multiple degradation transitions, as shown in Fig. 6.13.

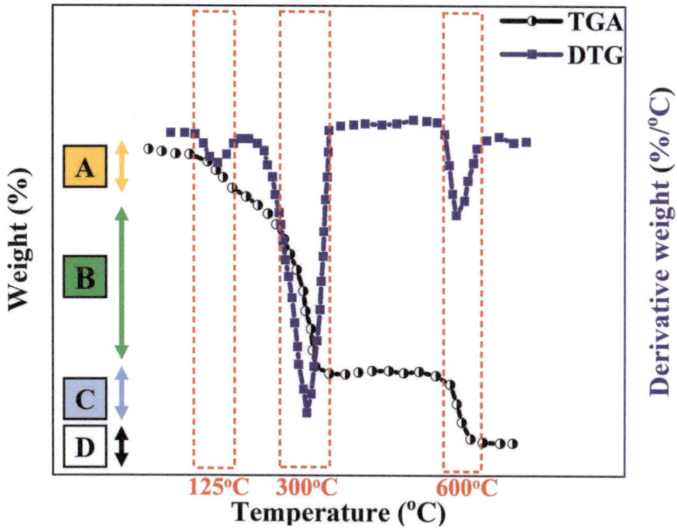

Fig. 6.12 TGA-DTG thermogram of compounded rubber

Fig. 6.13 TGA thermogram of block and random copolymers

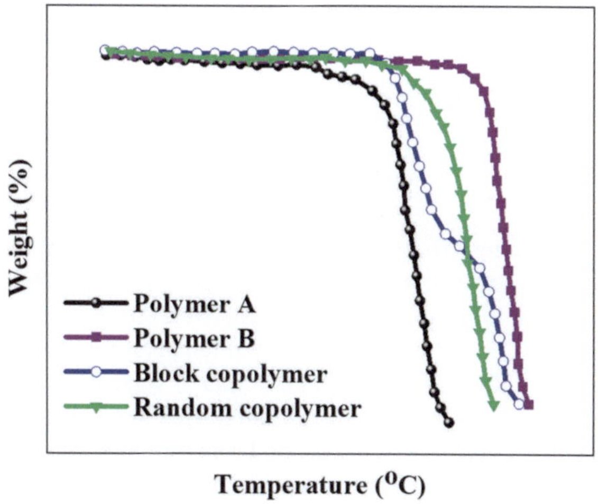

6.3.6 Degradation Kinetics

Polymer generally decomposes into solid residue and gaseous matter. There are two major categories for studying the thermal degradation kinetics of polymers: differential and integral methods. The differential method contains Kissinger [25], Carroll-Freeman [26], and Friedman methods, while Coats-Redfern [27], Doyle, and Flynn–Wall–Ozawa (F–W–O) methods [28] belong to the integral method. Coats-Redfern,

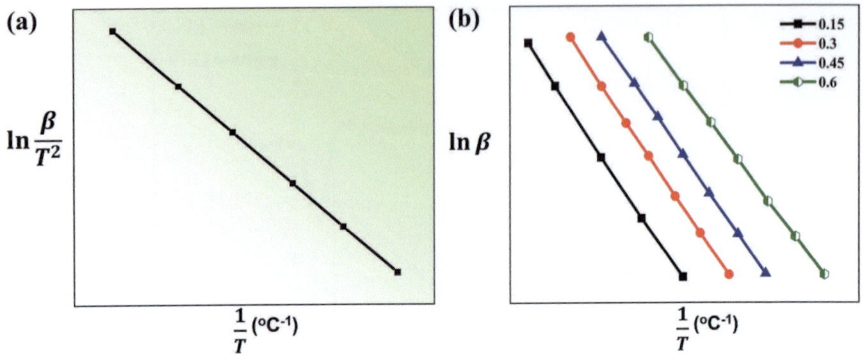

Fig. 6.14 **a** Kissinger method, varying heating rate and **b** Flynn–Wall–Ozawa method, varying degree of conversion

Flynn–Wall–Ozawa (F–W–O), and Doyle methods are usually used to investigate the thermal decomposition kinetics of polymers.

Kissinger method:

$$ln\frac{\beta}{T^2} = \ln\frac{AR}{E} - \frac{E}{R}(\frac{1}{T}) \quad (6.9)$$

Flynn–Wall–Ozawa method:

$$ln\beta = 5.3305 + \ln\frac{A}{F(\alpha)} - 1.052\frac{E}{R}(\frac{1}{T}) \quad (6.10)$$

Equations (6.9) and (6.10) are the mathematical representations of the Kissinger and Flynn–Wall–Ozawa method, where β is the heating rate, A is the pre-exponential factor, E is the activation energy, and R is the universal gas constant. $F(\alpha)$ is the degree of conversion (α), as a function of temperature (T). In the Kissinger method and the Flynn–Wall–Ozawa method, ln (β/T^2) versus 1/T and ln (β) versus $1/T$ are plotted with varying heating rate (β) and degree of conversion (α), respectively. Figure 6.14 depicts the nature of the curve. From the slope and intercept, A and E are calculated. The study of degradation kinetics provides an idea about the degradation rate, which is how fast or slow the sample undergoes degradation.

6.4 Dynamic Mechanical Analysis

Dynamic Mechanical Analysis (DMA) is a powerful analytical technique used to characterize the mechanical and viscoelastic properties of polymers and polymer composites by measuring their response to periodic (oscillating) stress or strain as a function of temperature, frequency, or time. DMA evaluates the storage modulus (E′,

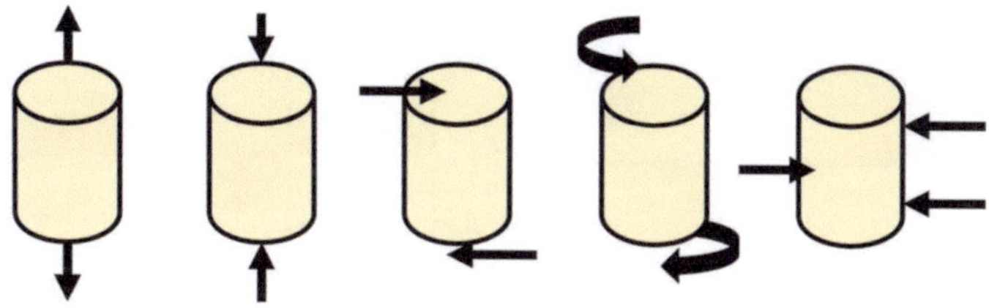

Fig. 6.15 Different types of forces and deformations

elastic response), loss modulus (E″, viscous response), and damping behavior (tanδ) of materials, providing insights into their molecular dynamics, structure–property relationships, and phase transitions [29].

The sample can be mounted in the DMA in a number of ways, depending on the geometry and end application, as shown in Fig. 6.15.

6.4.1 Working Principle

DMA operates by applying a periodic stress or strain to a sample and measuring the resulting deformation. The sinusoidal nature of the applied force allows separation of the response into storage modulus, representing stored elastic energy, and loss modulus, representing energy dissipated as heat. The phase difference (δ) between applied stress and resulting strain is used to calculate the damping factor (tanδ), indicating the material's energy dissipation characteristics. DMA is valuable for understanding viscoelastic behavior under varying temperatures and frequencies, assessing the degree of crosslinking and curing processes, and characterizing dynamic mechanical behavior.

Figure 6.16 depicts the stress–strain response of a Hookean solid, Newtonian liquid, and viscoelastic materials. Hookean solids are ideal elastic with 100% recovery; thus, the phase difference is zero. Conversely, Newtonian liquids are ideal viscous with 0% recovery. Therefore, phase lag appears as 90^0. Viscoelastic materials reside in the mid-range.

6.4.2 Instrumentation

The core components of DMA include a clamping system for mounting samples. As mentioned in Fig. 6.15, various force directions and strain responses are possible. Thus, the clamping system is designed to maintain a uniform stress distribution across the sample, Fig. 6.17. A standard clamping system includes tension, compression, and torsional fixtures. Parallel plates are used for shear or oscillation. A force or

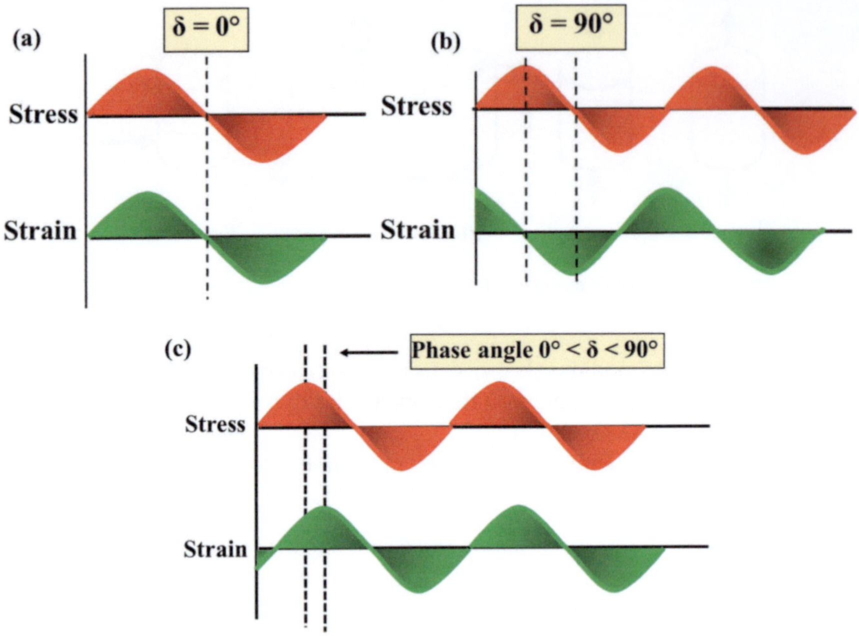

Fig. 6.16 Stress–strain response of **a** ideal elastic material, **b** ideal viscous material, **c** viscoelastic material

displacement actuator generates mechanical stress on the sample. A precise force transducer or load cell measures the force applied. The load cell typically converts the mechanical force into an electrical signal. The signal is then fed into the data acquisition system for analysis. The strain sensors used in DMA are highly sensitive. Another actuator controls the oscillation frequency to apply a periodic strain to the sample. A DMA can operate in a broad frequency range from 0.00001 to 1000 Hz. It can measure strain ranging from 0.00001 to 100%. All these ranges of measurements vary as per the instrument configuration, build-up, and user requirements. A temperature-controlled chamber with a temperature-sensing probe ensures precise, accurate, and uniform temperature distribution. Liquid nitrogen is used to experiment with the negative temperature range. The data acquisition system collects signals from the force transducer, displacement sensor, and temperature probe. The acquired data is further processed and analyzed to obtain storage modulus, loss modulus, complex modulus, and loss tangent (tanδ) as a function of amplitude, frequency, temperature, or time.

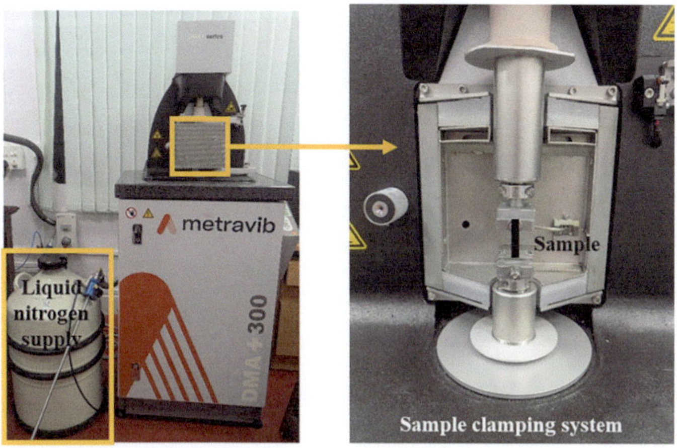

Fig. 6.17 Dynamic mechanical analyzer

6.4.3 Data Analysis

DMA is a subset of thermal techniques for characterizing viscoelastic properties as a function of strain, frequency, or temperature. The most common experiments executed in DMA include amplitude sweep, frequency sweep, and temperature sweep, which correspond to the material response as a function of varying amplitude, frequency, and temperature, respectively. DMA can investigate and quantify properties such as a glass transition (T_g), damping factor (tanδ), relaxation properties, and decomposition under specific environmental conditions or phase changes such as melt, curing, or crystallization. In the context of thermal response, this chapter describes the effect of temperature on modulus and tanδ. Other dynamic mechanical tests are discussed in Chap. 5 in detail.

Damping factor or loss tangent (tanδ) is the ratio of loss and storage modulus (Eqs. (6.11), (6.12), and 6.13), which represents the mechanical energy dissipated as heat during the loading/unloading cycle. The temperature corresponding to the maximum tanδ is considered as the glass transition temperature, shown in Fig. 6.18. A perfectly elastic material has tanδ as zero, whereas a perfectly viscous material has tanδ as infinite.

$$\boldsymbol{E}^{'} = \boldsymbol{E}^{*}\mathbf{cos}\boldsymbol{\delta} \tag{6.11}$$

$$\boldsymbol{E}^{''} = \boldsymbol{E}^{*}\mathbf{sin}\boldsymbol{\delta} \tag{6.12}$$

$$\tan\delta = \textit{Loss}\ \text{modulus (E'')/Storage modulus (E')} \tag{6.13}$$

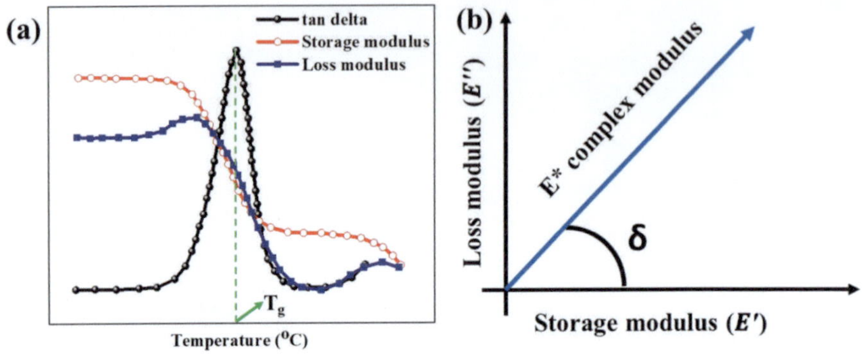

Fig. 6.18 **a** A characteristic DMA temperature sweep curve and **b** a schematic representation of complex modulus (E*)

DMA is sensitive to experimental conditions, including amplitude of applied force, temperature stability, humidity control, and sample geometry, which significantly influence results. Interpreting DMA data for complex or heterogeneous materials can be challenging due to overlapping relaxation phenomena. However, DMA offers notable advantages, including high sensitivity to subtle molecular transitions often undetected by other techniques, such as DSC, notably secondary relaxations. For synthetic polymers, DMA measurements help assess the impact of fillers and additives, characterize curing processes of thermosets, and analyze mechanical strength, durability, and stability under varying conditions. For example, DMA analysis identified the melting temperature (T_m) of semi-crystalline poly (ethylene oxide) (PEO) membranes, highlighting how lithium salts incorporation influenced thermal transitions, critical for their use as solid electrolytes in lithium-ion batteries [30].

DMA also provides insights into polymer microstructures and molecular relaxation processes. In biopolymers, DMA aids in analyzing thermal transitions, mechanical properties, hydration, and environmental effects, as well as evaluating biocompatibility and degradation behavior. Graphene oxide (GO) membranes analyzed using DMA demonstrated how intercalated molecules such as water and ethanol impact their viscoelastic properties. DMA-based stress relaxation studies showed significant differences in relaxation behavior depending on internal molecular structures, confirming that ethanol intercalation reduced cross-link density, enhancing relaxation and viscoelasticity [30].

6.4.4 Determination of Glass Transition Temperature

In Fig. 6.19, a characteristic scan of the DMA temperature sweep is presented, where tanδ of individual polymers, their blends, and copolymers are highlighted. On account of the separate existence of two blocks and phase separation in block

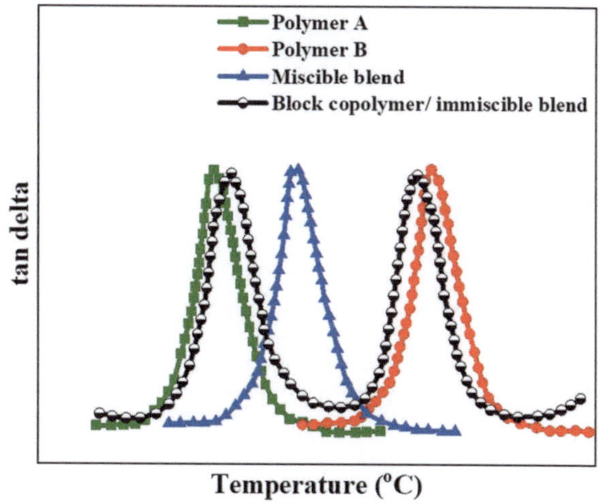

Fig. 6.19 Identification of the nature of copolymers from temperature sweep

copolymer and immiscible blends, respectively, two distinct glass transitions are observed. But miscible blends and random copolymers show a single T_g [31].

It should be noted that while analyzing properties in DMA as a function of amplitude or frequency, or temperature, the measurement is solely executed in the linear viscoelastic region (LVR).

6.4.5 Effect of Temperature on Viscoelasticity

The states of the viscoelastic transition as a function of temperature are divided into four major regions, shown in Fig. 6.19. The material shows the highest storage modulus in the glassy state. As temperature increases, the segmental movement of the polymer chains increases. In this transition zone, enough mobility exists for substantial friction between chain segments. With further increased temperature, the material enters a rubbery state where strain is accommodated by entropic changes to polymer segments between entanglements, providing a good elastic response. At the very end, the temperature reaches a value where chains overcome the constraint of entanglement and start to flow [32].

As shown in Fig. 6.20, three crossovers between storage and loss modulus are observed. Crossover points describe the switch over between the elastic and viscous component of a viscoelastic material. Storage and loss modulus represent the elastic and viscous components, respectively.

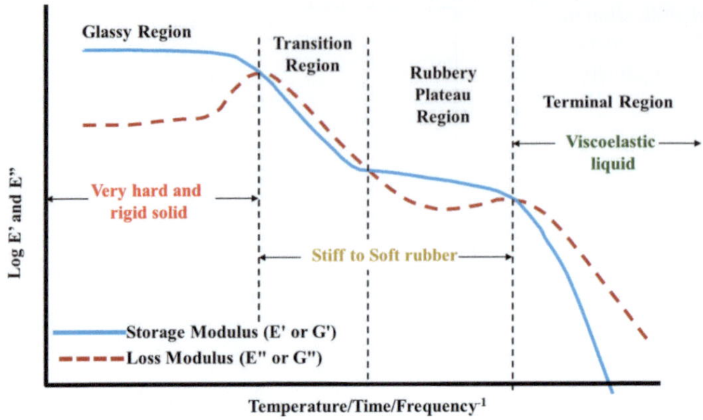

Fig. 6.20 Effect of temperature on storage and loss modulus of a viscoelastic material

6.4.6 *Effect of Crosslinking on Modulus and Glass Transition*

Introducing more crosslinks into a polymer backbone will proportionally increase the crosslink density. As the number of crosslinks increases, molecular motion in the sample gets restricted, causing a rise in the glass transition temperature. For low values of crosslink density, T_g can be found to increase linearly with the number of crosslinks. In Fig. 6.21, the order of crosslink density of polymers is in the range of A < B < C < D < E. Therefore, a sharp decrease in the storage modulus lies in the range of A < B < C < D < E. The crosslinks are responsible for better structural integrity and recovery; thus, lower crosslink density facilitates poor elasticity. The crosslink density also influences the modulus of the polymer. Higher crosslink density allows higher stiffness and more resistance toward deformation [33].

6.5 Dilatometry

Dilatometry measures the dimensional changes, i.e., a material's expansion or contraction as a function of temperature (Fig. 6.22). It's used to study various materials' thermal behavior and phase transitions, including polymers, ceramics, glasses, metals, and composites. The ASTM standards like E228 and D696 are used to analyze [34, 35]. The types are push-rod, optical, capacitance, and high-resolution laser dilatometers, each suited for different applications and materials. The sample is placed in the furnace and subjected to a controlled temperature program (heating, cooling, or isothermal conditions). As the furnace temperature changes, the sample expands or contracts. In the pushrod type, the pushrod moves up or down with the sample, and the extensometer helps to measure this movement. In optical dilatometry, the optical sensors monitor the sample's dimensions, and the software

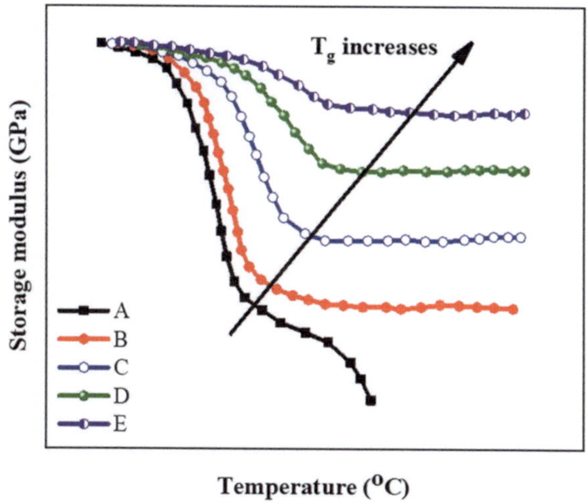

Fig. 6.21 An experiment to study the effect of crosslinking on modulus and T_g

analyzes the data. The resulting data allows for calculating the coefficient of thermal expansion (CTE), glass transition temperature (T_g), softening point, decomposition temperature, and phase transformation.

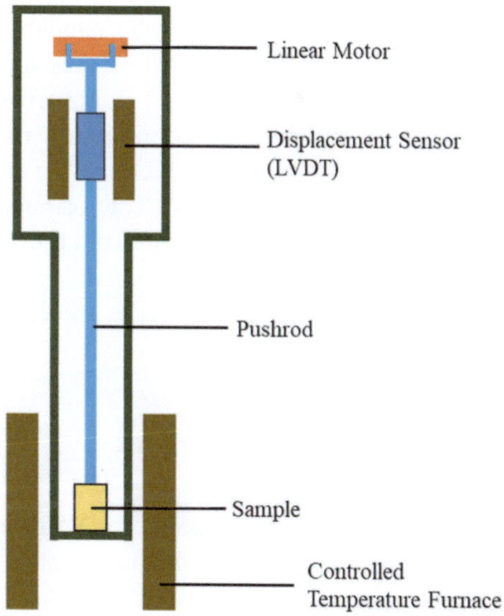

Fig. 6.22 Pushrod-type dilatometer

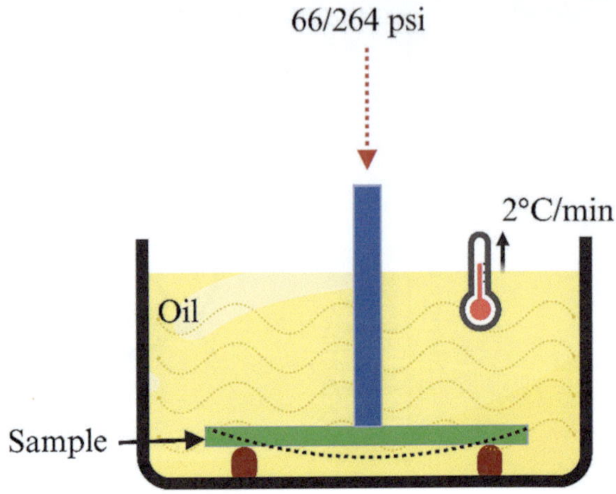

Fig. 6.23 Heat deflection temperature measurement

6.6 Heat Defection Temperature

The Heat Deflection Temperature (HDT) technique measures a material's resistance to deformation under a specific load at an elevated temperature (Fig. 6.23). It's used to evaluate the thermal resistance of materials, particularly plastics and polymers, and helps determine their suitability for applications at a range of temperatures. The test procedure for HDT is standardized by ASTM D648 [36] and ISO 75 [37]. A sample should be prepared in the standard test bar. This test bar then placed in the three-point bending device. A load was applied in the middle and supported at both ends of the test bar. A sample is immersed and heated at a controlled rate in an oil bath. The temperature continuously rises, and the deflection of the test bar is noted. The temperature at which the bar deflects by a specified amount (0.25 mm) is recorded as the HDT value. The load applied to the specimen can be 66 psi (soft plastics and other polymers for general use.) or 264 psi (stronger and more durable thermoplastics/thermosetting materials for heavy-duty use). HDI testing is mainly used in the automotive, aerospace, and consumer goods industries [38].

6.7 Vicat Softening Point

The Vicat Softening Temperature (VST) is the temperature at which a flat-ended needle penetrates the specimen to a depth of 1 mm under a specific load (Fig. 6.24). This test gauges a plastic's ability to retain its surface mechanical properties at high temperatures. This method is defined by standards like ISO 306 and ASTM D1525. A test specimen of 3- and 6.5-mm thickness and at least 10 mm in width and length is placed in the testing apparatus. A load of 10N or 50 N is applied to the specimen. Further, the specimen is immersed in an oil bath at 23 °C. The temperature (50 or

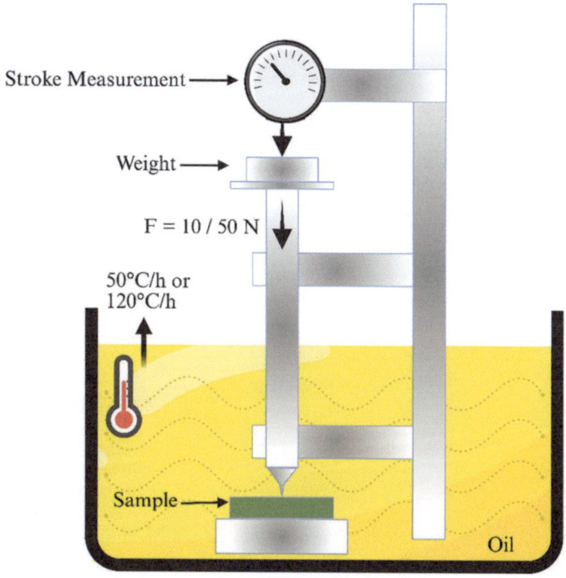

Fig. 6.24 Vicat softening point measurement

120 °C/h) of the bath is raised until the needle penetrates 1 mm into a specimen. The temperature at which the needle penetrates 1 mm is noted and known as VST [39]. While VST and HDT tests assess thermal properties, the VST test reveals the temperature at which the specimen loses its "stability-form" and softens, whereas the HDT test measures the temperature at which the specimen loses its "load-bearing" capability.

6.8 Melt Flow Index

Melt Flow Index (MFI), also known as Melt Flow Rate (MFR), measures a given polymer's ease of flow and is also known as the rheological properties in the molten state of a polymer under a known applied pressure. It is defined as the mass of polymer flowing in ten minutes through a capillary of a specific diameter and length by a pressure (Fig. 6.25). The method is described in similar standards ASTM D1238 and ISO 1133.

Approximately 4–5 gm of polymer sample is placed in the apparatus's barrel (preheated). A die with an opening of around 2 mm diameter is inserted into the apparatus's barrel. The packed material should be free of bubbles and packed properly. The apparatus also has a piston, which acts as a medium that causes extrusion of the molten polymer. The standard load on the piston is 2.16 kg, 5 kg, etc. This weight helps to exert force on the molten polymer, and immediately, it starts flowing through the die. A polymer melt sample collected and weighed accurately for the desired time period. MFI is expressed in grams of polymer per 10 min of the

Fig. 6.25 Melt flow index measurement diagram

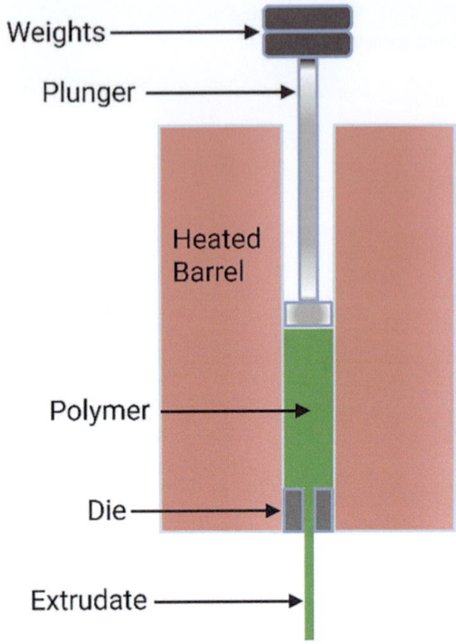

polymer flow through a capillary (g/10 min). Melt flow rate is an indirect measure of molecular weight, with a high melt flow rate corresponding to low molecular weight [40].

6.9 Thermal Conductivity

The ability of a material to transfer heat is an attribute of thermal conductivity. Thermal conduction is the method of thermal energy transfer within or between different materials. Two methods used for the measurement of thermal conductivity: the steady-state method using parallel plate geometry or coaxial cylinders, and the transient method, which is developed for materials with low thermal conductivity [41]. The unit of thermal conductivity is $Wm^{-1}\ K^{-1}$. Generally, polymers have low thermal conductivity. Incorporation of ceramic or nanofillers increases the value to some extent.

The steady-state guarded hot plate method uses a hot plate to apply a constant heat flux to the sample and a cold plate to maintain a temperature gradient. The rate of heat flow (q), sample thickness (x), area (A), and temperature gradient ($\Delta\theta$) is used to calculate the thermal conductivity using the following Eq. (6.14).

$$\kappa = \frac{q \times x}{A \times \Delta\theta} \tag{6.14}$$

In the transient method, a heat pulse is applied to the sample. Then the temperature change is monitored over time.

Acknowledgements We have used bioRender.com (student license) for making illustration.

Conflict of Interest Authors declare no competing conflict of interest.

References

1. Nur Azira T, Amin I (2016) Advances in differential scanning calorimetry for food authenticity testing. Adv Food Authenticity Test 311–335. https://doi.org/10.1016/B978-0-08-100220-9.00012-6
2. Danley RL (2002) New heat flux DSC measurement technique. Thermochim Acta 395:201–208. https://doi.org/10.1016/S0040-6031(02)00212-5
3. Zucca N, Erriu G, Onnis S, Longoni A (2004) An analytical expression of the output of a power-compensated DSC in a wide temperature range. Thermochim Acta 413:117–125. https://doi.org/10.1016/J.TCA.2003.10.006
4. Danley RL (2003) New modulated DSC measurement technique. Thermochim Acta 402:91–98. https://doi.org/10.1016/S0040-6031(02)00541-5
5. Liu H, Yu L, Dean K et al (2009) Starch gelatinization under pressure studied by high pressure DSC. Carbohydr Polym 75:395–400. https://doi.org/10.1016/J.CARBPOL.2008.07.034
6. Harvey JP, Saadatkhah N, Dumont-Vandewinkel G et al (2018) Experimental methods in chemical engineering: differential scanning calorimetry—DSC. Can J Chem Eng 96:2518–2525. https://doi.org/10.1002/CJCE.23346;WGROUP:STRING:PUBLICATION
7. (2018) Test method for temperature calibration of differential scanning calorimeters and differential thermal analyzers
8. (2025) Practice for heat flow calibration of differential scanning calorimeters
9. Karasevskii AI, Lubashenko VV (1996) Melting as a first-order phase transition in a system of defects. Phys Status Solidi B Basic Res 194:483–497. https://doi.org/10.1002/PSSB.2221940206;JOURNAL:JOURNAL:15213951;REQUESTEDJOURNAL:JOURNAL:15213951;WGROUP:STRING:PUBLICATION
10. Wu J (1999) The glassy state, ideal glass transition, and second-order phase transition. J Appl Polym Sci 71:143–150. https://doi.org/10.1002/(SICI)1097-4628(19990103)71:1%3c143::AID-APP17%3e3.0.CO;2-I
11. Zheng S, Huang J, Li Y, Guo Q (1997) A DSC study of miscibility and phase separation in crystalline polymer blends of phenolphthalein poly(ether ether sulfone) and poly(ethylene oxide). J Polym Sci B Polym Phys 35:1383–1392. https://doi.org/10.1002/(SICI)1099-0488(19970715)35:9%3c1383::AID-POLB8%3e3.0.CO;2-N
12. Aouachria K, Belhaneche-Bensemra N (2006) Miscibility of PVC/PMMA blends by vicat softening temperature, viscometry, DSC and FTIR analysis. Polym Test 25:1101–1108. https://doi.org/10.1016/J.POLYMERTESTING.2006.07.007
13. Huang JM, Yang SJ (2005) Studying the miscibility and thermal behavior of polybenzoxazine/poly(ε-caprolactone) blends using DSC, DMA, and solid state 13C NMR spectroscopy. Polymer (Guildf) 46:8068–8078. https://doi.org/10.1016/J.POLYMER.2005.06.104
14. Ng HM, Saidi NM, Omar FS et al (2018) Thermogravimetric analysis of polymers. Encycl Polym Sci Technol. https://doi.org/10.1002/0471440264.PST667
15. Interlaboratory Study to Establish Precision Statements for ASTM E1582, Standard test method for determination of low levels of antimony in carbon and low-alloy steel by graphite furnace atomic absorption spectrometry
16. (2019) Test method for mass scale calibration of thermogravimetric analyzers

17. ISO 11358-1:2022— Plastics—thermogravimetry (TG) of polymers—Part 1: general principles. https://www.iso.org/standard/79999.html. Accessed 14 May 2025
18. Hornsby PR, Wang J, Rothon R et al (1996) Thermal decomposition behaviour of polyamide fire-retardant compositions containing magnesium hydroxide filler. Polym Degrad Stab 51:235–249. https://doi.org/10.1016/0141-3910(95)00181-6
19. Characterization of polymers using TGA application note. PerkinElmer. https://www.perkinelmer.com/libraries/app_009908_01_characterization_of_polymers_using_tga. Accessed 14 May 2025
20. Sahay R, Agarwal K, Subramani A et al (2020) Helicoidally arranged polyacrylonitrile fiber-reinforced strong and impact-resistant thin polyvinyl alcohol film enabled by electrospinning-based additive manufacturing. Polymers 12:2376. https://doi.org/10.3390/POLYM12102376
21. Sengupta R, Sabharwal S, Bhowmick AK, Chaki TK (2006) Thermogravimetric studies on Polyamide-6,6 modified by electron beam irradiation and by nanofillers. Polym Degrad Stab 91:1311–1318. https://doi.org/10.1016/J.POLYMDEGRADSTAB.2005.08.012
22. Wong ACY, Lam F (2002) Study of selected thermal characteristics of polypropylene/polyethylene binary blends using DSC and TGA. Polym Test 21:691–696. https://doi.org/10.1016/S0142-9418(01)00144-1
23. Khanra S, Sreenivasan P, Das S et al (2022) Immobilization of a biobased process aid at the interface for binary silicone and fluoroelastomer based super specialty blends with silica for enhanced compatibility. J Mater Sci 57:13974–13990. https://doi.org/10.1007/S10853-022-07504-1/FIGURES/12
24. Datta S, Stoček R, Kuřitka I, Sáha P (2015) Determination of compounding formulation of cured rubber by reverse engineering. Polym Eng Sci 55:1450–1458. https://doi.org/10.1002/PEN.24092
25. Goswami S, Kiran K (2012) Application of Kissinger analysis to glass transition and study of thermal degradation kinetics of phenolic-acrylic IPNs. Bull Mater Sci 35:657–664. https://doi.org/10.1007/S12034-012-0336-0/FIGURES/10
26. Liu NA, Fan WC (1999) Critical consideration on the Freeman and Carroll method for evaluating global mass loss kinetics of polymer thermal degradation. Thermochim Acta 338:85–94. https://doi.org/10.1016/S0040-6031(99)00197-5
27. Ganeshan G, Shadangi KP, Mohanty K (2018) Degradation kinetic study of pyrolysis and co-pyrolysis of biomass with polyethylene terephthalate (PET) using Coats-Redfern method. J Therm Anal Calorim 131:1803–1816. https://doi.org/10.1007/S10973-017-6597-5/TABLES/8
28. Sun JT, Huang YD, Gong GF, Cao HL (2006) Thermal degradation kinetics of poly(methylphenylsiloxane) containing methacryloyl groups. Polym Degrad Stab 91:339–346. https://doi.org/10.1016/J.POLYMDEGRADSTAB.2005.04.037
29. Haris NIN, Hassan MZ, Ilyas RA et al (2022) Dynamic mechanical properties of natural fiber reinforced hybrid polymer composites: a review. J Market Res 19:167–182. https://doi.org/10.1016/J.JMRT.2022.04.155
30. Patra S, Ajayan PM, Narayanan TN (2020) Dynamic mechanical analysis in materials science: the Novice's tale. Oxford Open Mater Sci. https://doi.org/10.1093/OXFMAT/ITAA001
31. Khanra S, Kumar A, Ganguly D et al (2022) Effect of FKM-g-acrylamide reactive compatibilizer on mechanical, thermal and aging behaviors of fluoroelastomer (FKM)/silicone rubber (MVQ) blend. Polym Eng Sci 62:1239–1255. https://doi.org/10.1002/PEN.25922
32. Haugh EF (1959) Time, temperature, and linear viscoelasticity. J Appl Polym Sci 1:144–149. https://doi.org/10.1002/APP.1959.070010203
33. Alyes NM, Gómez Ribelles JL, Gómez Tejedor JA, Mano JF (2004) Viscoelastic behavior of poly(methyl methacrylate) networks with different cross-linking degrees. Macromolecules 37:3735–3744. https://doi.org/10.1021/MA035626Z/ASSET/IMAGES/MEDIUM/MA-2003-05626Z_0022.GIF
34. (2022) Test method for linear thermal expansion of solid materials with a push-rod dilatometer
35. (2024) Test method for coefficient of linear thermal expansion of plastics between 30C and 30C with a vitreous silica dilatometer

36. (2018) Test method for deflection temperature of plastics under flexural load in the edgewise position
37. ISO 75-1:2020—plastics—determination of temperature of deflection under load—part 1: general test method. https://www.iso.org/standard/77576.html. Accessed 14 May 2025
38. All about heat deflection temperature. Xometry. https://www.xometry.com/resources/3d-printing/heat-deflection-temperature/. Accessed 14 May 2025
39. Stephenson CE, Willbourn AH (1959) The vicat softening point test method: correlations, uses, and variants
40. Shenoy AV, Saini DR (1986) Melt flow index: more than just a quality control rheological parameter. Part I. Adv Polym Technol 6:1–58. https://doi.org/10.1002/ADV.1986.060060101; WGROUP:STRING:PUBLICATION
41. Subir R, Goswami R, Singh V et al (2022) Construction and validation of a low cost thermal conductivity measurement device. Lect Notes Mech Eng. https://doi.org/10.1007/978-981-16-9613-8_38/TABLES/2

Roumita Hore is currently a research scholar at the Rubber Technology Centre, Indian Institute of Technology, Kharagpur, India. She has been awarded the prestigious Prime Minister Research Fellowship in 2022. She has completed her M. Tech (2021) and B. Tech (2017) in Polymer Science and Technology and B. Sc (2014) in Chemistry (Hons) from Calcutta University. Roumita has served as an executive R & D in CICO Technologies limited for 2 years (2017–2019). Her area of research includes the fabrication of sustainable new-generation rubber composites for the application of passive vibration isolation in electric vehicles. Roumita is experienced in polymer characterization, particularly in thermal and chemical analysis of polymeric systems.

Nikhil Dhore is pursuing a Ph.D. at the Indian Institute of Technology Kharagpur, focusing on the development of sustainable polymers for decorative and protective applications. Nikhil has completed his bachelor's in chemical technology and a master's specializing in paint technology and has hands-on experience of 3 + years in the coating field research. His current research on the synthesis of water-based polyurethane polymer, non-isocyanate polyurethane, and biobased epoxy amine systems. More than 6 + peer-reviewed international journal articles in his account.

Dr. Komal Agarwal is a Principal Investigator at UiT The Arctic University of Norway, with expertise in polymer thin films and biomaterials for advanced functional applications. Her research spans piezoelectric and impact-resistant polymers, bio-inspired adhesives, and polymeric scaffolds for tissue engineering. She brings a strong innovation background with contributions to synthetic polymer design, biomimetic systems, and applied materials science.

Chapter 7
Chromatographic Techniques for Polymer Analysis

Nikhil Kumar, Payel Maity, Nikhil Dhore, Homen Dahal, and Priya Devi

Abstract Chromatography, a fundamental polymer characterization technique, is covered extensively in this chapter. This subsection provides an in-depth exploration of chromatographic methods tailored specifically for polymer analysis. It discusses the principles behind chromatography, including the separation mechanism based on differences in molecular interactions with the stationary phase. Various chromatographic techniques are elucidated, along with their applications in characterizing polymers based on molecular weight distribution, chemical composition, and structural heterogeneity. Additionally, the subsection addresses recent advancements and emerging trends in chromatographic techniques for polymer characterization, highlighting their significance in advancing our understanding of polymer properties and behaviour.

Keywords Gel permeation chromatography · HPLC · Affinity chromatography · Ion-exchange chromatography

N. Kumar (✉)
Advanced Technology Development Centre, Indian Institute of Technology Kharagpur, Kharagpur, India
e-mail: nikhilsingh25694@gmail.com

P. Maity
School of Nanoscience and Technology, Indian Institute of Technology Kharagpur, Kharagpur, India

N. Dhore
Rubber Technology Centre, Indian Institute of Technology Kharagpur, Kharagpur, India

H. Dahal
Department of Chemistry, Indian Institute of Technology Kharagpur, Kharagpur, India

P. Devi
Department of Chemical Sciences, Tezpur University, Tezpur, India

S. Chattopadhyay and N. Kumar (eds.), *Polymer Characterization*,
https://doi.org/10.1007/978-981-95-2609-3_7

7.1 Introduction

In polymer research, chromatography is a useful analytical method extensively used for macromolecule separation, identification, and characterization. Since polymers having diverse molecular architectures, molecular weight distributions, and compositions, chromatographic techniques offer a robust platform for resolving these attributes with high precision. It also helps to define polymer performance and processing behaviour.

The term chromatography originates from the Greek word '*Chroma*' meaning colour '*Graphy*' meaning writing or recording. Chromatography is a physical method of separation of chemical compounds. The components to be separated are distributed between two phases, one of which is stationary phase while the other is the mobile phase that moves in a definite direction. Originally, chromatography was developed as a technique for separating plant pigments, where different coloured components were separated along a stationary phase. The Nobel Prize in Chemistry for the development of chromatography was awarded to Archer John Porter Martin and Richard Laurence Millington Synge in 1952 for their invention of partition chromatography; essentially, the foundational work behind modern chromatography techniques (Fig. 7.1). Over time, the method evolved into a powerful analytical tool used across various scientific fields, including polymer characterization, pharmaceuticals, and biochemistry [1].

The historical evolution of chromatography in polymer science dates back to the mid-twentieth century with the development of Gel Permeation Chromatography (GPC) by Moore in 1964, which provided an effective method for molecular weight determination [2]. Since then, significant advancements, including multi-detector approaches integrating light scattering, viscometry, and mass spectrometry, have enhanced the accuracy of chromatographic polymer characterization. The advent

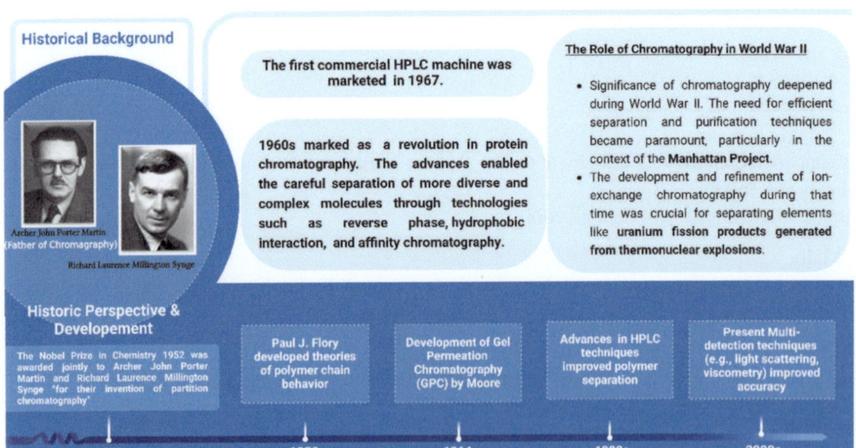

Fig. 7.1 A historical overview of chromatography, highlighting key milestones from its invention and World War II significance to the development of HPLC and modern polymer analysis techniques

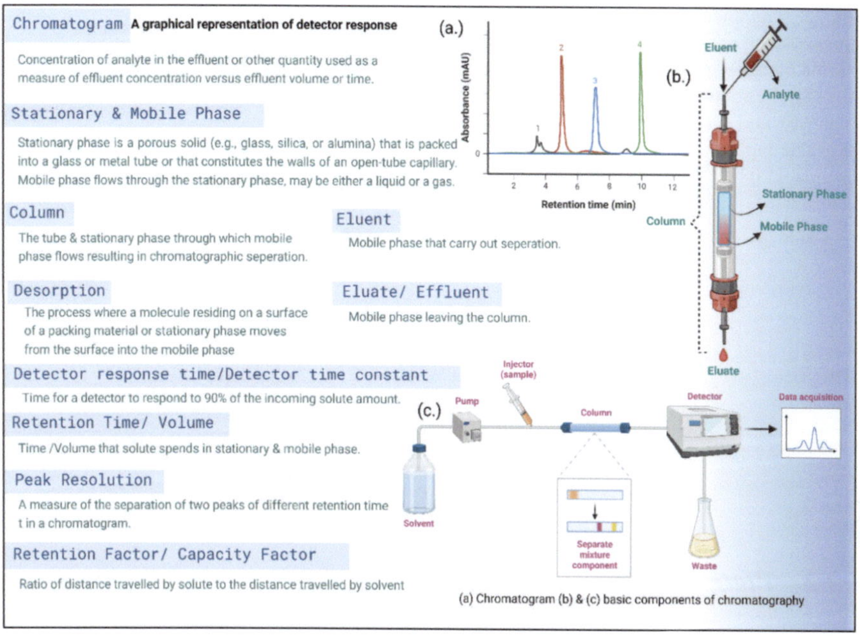

Fig. 7.2 Key concepts and components of chromatography, including definitions (e.g. stationary phase, eluate), detector response, and system setup with a sample chromatogram

of two-dimensional chromatography (2D-LC) and field-flow fractionation (FFF) has further expanded the analytical capabilities, enabling the characterization of complex polymer systems such as branched, hyperbranched, and block copolymers [3]. This chapter deals with an in-depth exploration of chromatographic methodologies utilized in polymer characterization, highlighting fundamental principles, instrumental aspects, to underscore its significance in polymer research. The key concepts and components of chromatography are as shown in Fig. 7.2.

7.2 Terminologies Related to Chromatography

7.2.1 Retention

In chromatographic techniques, the relative affinity of a chemical compound for the stationary and mobile phases determines the rate of migration. The solutes that are more strongly attracted to the stationary phase are retained back in contrast to the solutes that are more strongly attracted to the mobile phase (Fig. 7.3).

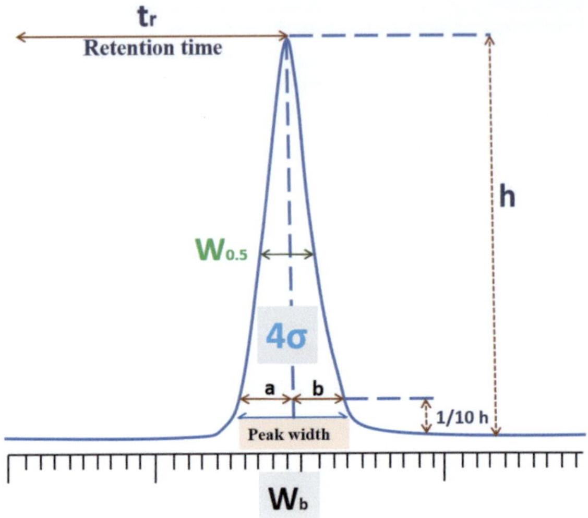

Fig. 7.3 Graphical representation of a chromatographic peak showing retention time, peak height (h), and peak width W for evaluating chromatographic performance

7.2.2 Plate Height

Plate height refers to the 'height equivalent to a theoretical plate' (HETP), which is a measure of column efficiency. Peak width is directly proportional to the square root of the distance that the peak has migrated. Mathematically, the square of the standard deviation is equivalent to constant time the distance travelled. The proportionality constant between the standard deviation and the distance travelled is known as the height equivalent to a theoretical plate (HETP) [4].

$$\mathrm{HETP} = \sigma^2 / \mathrm{L} \tag{7.i}$$

where σ represents standard deviation and L is the distance travelled.

In elution chromatography, in which the peak develops on a time scale, so the derived form of the above equation will be:

$$\mathrm{HETP} = \mathrm{L}\sigma_t^2 / t_r^2 \tag{7.ii}$$

where L is column length, t_r is time of retention of the peak by the column, and σ_t is standard deviation of the peak measured in units of time.

Fundamental procedures that contribute HETP includes:

- '*Molecular diffusion/longitudinal diffusion*' where solute molecules randomly diffuse outward from the centre of the solute band due to their thermal energy, causing peak broadening. This effect is inversely proportional to the flow velocity,

u. This rapid flow reduces the time for diffusion and mathematically, plate height of this factor is expressed as **B/u**, in which B is a constant related to the diffusion coefficient of the solute.

- '*Eddy diffusion*' is phenomenon where solute molecules in a chromatography column travel at different speeds due to the uneven flow paths created by the packing particles, essentially 'wandering' through the column's granular bed, causing peak broadening. This contributes to the plate height as a constant factor, **'A'**, independent of velocity [the Van Deemter equation (Eq. 7.iii)].
- '*Non-equilibrium or mass transfer*' causes peak broadening due to slowness of diffusion rate (in/out) at which a solute molecule can move between the mobile phase (liquid flowing through the column) and the stationary phase (particles within the column), leading to a delay in reaching equilibrium and affecting the overall separation efficiency. This effect is proportional to velocity and is expressed as Csu and Cmu, in which Cs and Cm are constants relating to the stationary and mobile phases, respectively.

The above three factors can be assembled to calculate HETP using given two equations:

$$HETP = A + \frac{B}{u} + C_S u \tag{7.iii}$$

First, we need to evaluate B, A, Cm, and Cs, in terms of underlying diffusivity and flow processes then assemble them into a total plate height equation [5, 6]. This equation is valid for low velocities. However, for higher velocities or flow rate Eq. (7.v) will be applicable:

$$HETP = \left(\frac{1}{A} + \frac{1}{C_m u}\right)^{-1} + \frac{B}{u} + c_s u \tag{7.iv}$$

7.2.3 *Efficiency and Resolution*

The peak maxima and width are essential in determining the chromatogram's efficiency and resolution (separate solute zones). The first, peak maxima, denotes the maximum concentration of a peak. The maxima of two adjacent peaks must be separated to achieve better resolution. Such separation depends on the purity of the solute and the selectivity of the stationary and mobile phases. The second peak width, the distance between peak maxima (the highest point of each peak), might be large enough to appear separated; if the peaks are very wide, the solutes will still not be considered fully resolved in chromatography [7].

7.2.4 Column Efficiency

The efficiency of a column is indicated as the number of theoretical plates per column. The larger the number of theoretical plates per column, the sharper the peak [8]. The number of theoretical plates is calculated using the following equation:

$$\begin{aligned}&\text{Number of theoretical plates per column} \times 100/\text{length of HPLC column (cm)}\\&= \text{Number of theoretical plates per metre}\end{aligned} \tag{7.v}$$

$$N = 16\left(\frac{t_r}{w}\right)^2 \tag{7.vi}$$

$$N = 5.54\left(\frac{t_r}{w_{0.5}}\right)^2 \tag{7.vii}$$

$$N = \left(\frac{t_r}{\sigma}\right)^2 \tag{7.viii}$$

In the above equations, tr is the retention time, and w is the peak width obtained by drawing tangents to the sides of the Gaussian curve at the inflection points and extrapolating the tangents to intercept the baseline. The distance between the intercepts is the peak width. From Eq. (7.vi), we can calculate the number of theoretical plates (N) by drawing the tangents.

An alternate way to calculate N using Eq. (7.vii) is to use the peak width at half the peak height, $w_{0.5.}$ The half-height method is more straightforward to determine directly from the system because there is no need to draw tangents. It also enables the determination of N if the peak is not entirely separated from a neighbouring peak, as long as the valley between the peaks is lower than the half-height.

The plate number can be associated to the statistical broadening of a peak, that tangents drawn to the sides of a Gaussian distribution intersect the baseline at ±2 standard deviations (σ). A well-behaved chromatographic peak should be Gaussian in shape, so we can conclude that the peak width at baseline is 4σ. If you substitute 4σ for w in Eq. (7.vi), N will be the square of the retention time divided by σ (Eq. 7.viii).

- The number of theoretical plates measures the 'accuracy' of the column.
- Plate numbers may range from 100 to 10^6.
- The plate number depends on the length of the column. The extreme value of 10^6 plates was obtained with an open tubular gas chromatographic column 1.6 kms (1 mile) long.

7.2.5 Resolution

Resolution is the ability to separate two chromatographic peaks [9]. Resolution, R, is given by:

$$R = \frac{t_{r2} - t_{r1}}{0.5(w_1 + w_2)} \tag{7.ix}$$

where t_{r1} and t_{r2} and w_1 and w_2 are the times and widths of the two immediately adjacent peaks. If the peaks are sufficiently close, which is the pertinent problem, w is nearly the same for both peaks, and the resolution may be expressed as:

$$R = \frac{t_{r2} - t_{r1}}{4\sigma} \tag{7.x}$$

If the distance between the peaks is 4σ, then R is 1 and 2.5% of the area of the first peak overlaps 2.5% of the area of the second peak.

7.2.6 Peak Asymmetry Factor (A_s)

It is a measure of peak asymmetry in chromatography. It calculated as the ratio of the peak width after the peak maximum to the peak width before the peak maximum. It is represented as A_s and calculated using following Eq. (7.xi).

$$A_s = b/a \tag{7.ix}$$

where b is the distance from the peak midpoint (*perpendicular from the peak highest point*) to the trailing edge of the peak measured at 10% of peak height and a is the distance from the leading edge of the peak to the peak midpoint (*perpendicular from the peak highest point*) measured at 10% of peak height [10].

- *If* $A_s = 1.0$ indicate a symmetrical peak.
- *If* $A_s > 1.0$ indicate a tailing (where the peak tail is wider than the front).
- *If* $A_s < 1.0$ indicate a fronting (where the peak front is wider than the tail).

7.2.7 Tailing Factor (T_f)

The Tailing Factor is the USP coefficient that measures peak symmetry in a chromatogram. It is represented as T_f and calculated using the following Eq. (7.xii), where a and b are same as Eq. (7.xi) [11].

$$T_f = (a + b)/2a \tag{7.xii}$$

- *If* $\boldsymbol{T_f = 1.0}$ indicate peak with a Gaussian distribution.
- *If* $\boldsymbol{T_f > 1.0}$ indicate a tailing peak.
- *If* $\boldsymbol{T_f < 1.0}$ indicate a fronting peak.

7.3 Chromatography Techniques

In chromatography, column is a key in analytical and preparative methods for polymer characterization. It enables the separation of polymer constituents based on their molecular weight, polarity, or specific interactions.

7.3.1 High-Performance Liquid Chromatography

High-Performance Liquid Chromatography (HPLC) is used to separate, identify, and quantify components of a mixture dissolved in a liquid. It relies on a mobile phase and a stationary phase to separate the different components of the sample (Figs. 7.4 and 7.5). The separation is achieved by the differential migration of the sample components through the column under high pressure. Depending on the type of interaction between the polymer and the stationary phase, different modes of chromatography are used, such as size exclusion chromatography (SEC), adsorption chromatography, ion-exchange chromatography, and affinity chromatography. Among these, SEC/gel permeation chromatography (GPC) widely in determining polymer molecular weight distribution and polydispersity. It also helps in characterizing copolymer compositions, purifying synthetic and natural polymers, analysing degradation products, and studying functional group modifications [12].

7.3.2 Gel Permeation Chromatography (GPC)

Gel permeation chromatography (GPC) is a category of size-exclusion chromatography (SEC) that is commonly used to determine the molecular weight distribution (MWD) of polymers, depending on their size or diameter (Fig. 7.6). This method was first invented by Lathe and Ruthven in 1955. The terminology '*gel permeation chromatography*' was given by J.C. Moore of the Dow Chemical Company, who further explored the method in 1964. Later, this patented column technique was certified by Waters Corporation, which marketed it in 1964. GPC systems and consumables

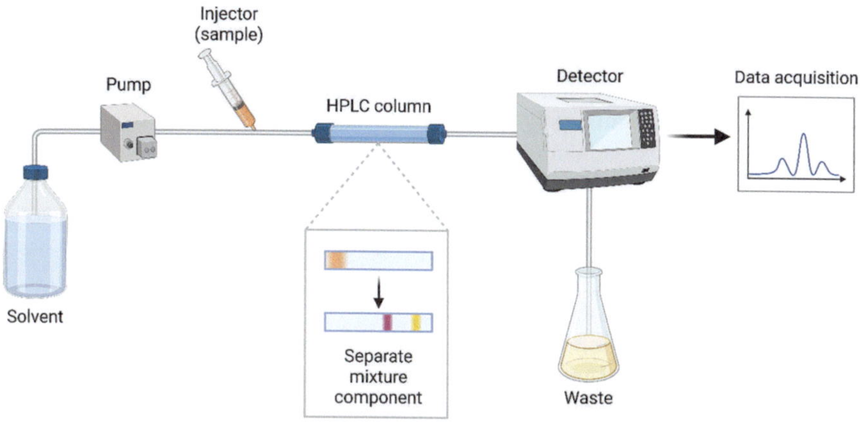

Fig. 7.4 Schematic of a chromatography system showing solvent delivery, sample injection, column separation, detection, and data acquisition

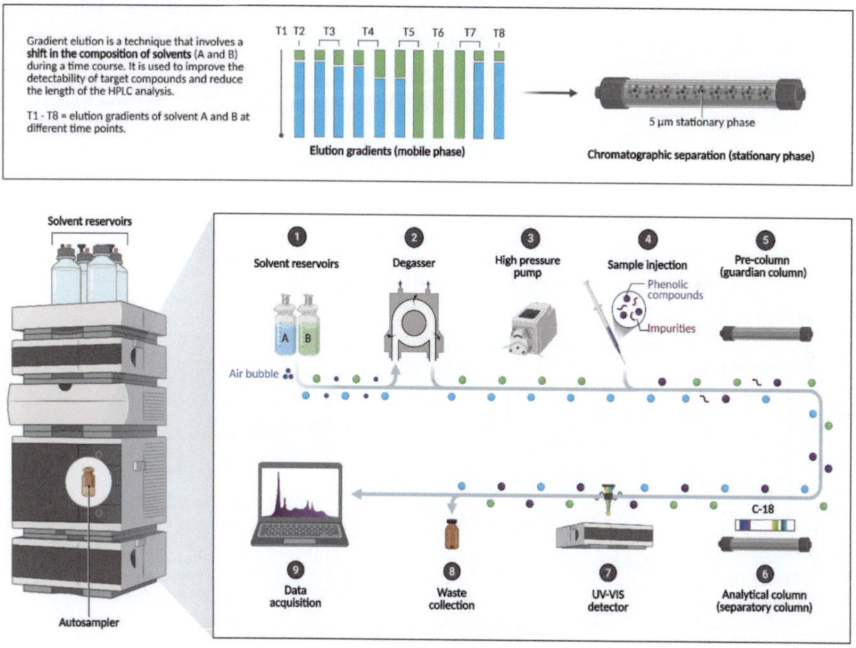

Fig. 7.5 Illustration of gradient elution in HPLC, showing changes in mobile phase composition over time and the stepwise process from solvent reservoirs to data acquisition in a chromatographic system

are currently accessible from several manufacturers. Separating polymers is essential for both their analysis and purification of the aimed product. When describing polymers, examining their size distribution, dispersity (Đ), and molecular weight is essential [13]. Generally, polymers can be described through various molecular weight measurements, including number average molecular weight (Mn), weight average molecular weight (Mw), size average molecular weight (Mz), and viscosity average molecular weight (Mv). GPC simplifies the calculation of Đ and Mv, whereas Mn, Mw, and Mz can be derived from further data.

Principle

Generally, maximum chromatographic techniques separate the analytes based on the chemical or physical interactions between the mobile and stationary phases, whereas in GPC, separation is based on the size or hydrodynamic diameter of the analyte. Separation happens through porous gel beads packed inside a column, with the principle of separation based on differential exclusion or inclusion of macromolecules. Bigger particles come first out of the column, whereas smaller molecules come through the porous gel beads present in the column and take a longer time to come out. Larger molecules avoid those columns and elute faster. The whole process happens without any interaction between the analytes and the stationary phase surface. Smaller analytes can permeate the pores and spend more time inside the

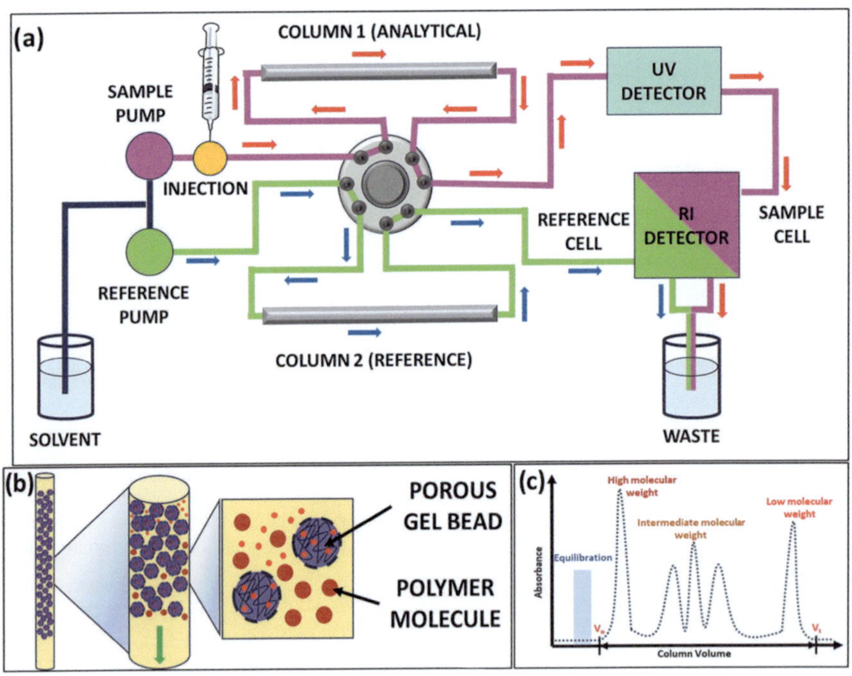

Fig. 7.6 Gel permeation chromatography instrumentation

gel particles, increasing their retention time. Each column has a series of molecular weights that can be separated based on pore sizes [14].

The total volume of a column packed with porous gel beads has been swelled by eluent is given by:

$$\mathbf{V_t} = \mathbf{V}_0 + \mathbf{V_m} + \mathbf{V_i} \quad (7.i)$$

where V_t is the total column volume, V_m is the volume occupied by the gel matrix, V_o is the void volume of the mobile phase, i.e. the volume of liquid outside the gel matrix, and V_i is the volume of the liquid inside the matrix.

Each column can separate a specific range of molecular weights. So, the porous gel beads also have been chosen within the range of molecular weight of analytes. For polymer separations, pore sizes should be the order of the analysed polymers.

Dependence on Molecular Conformation

GPC separates the polymers depending on their hydrodynamic volume, which is controlled by the molecular conformation, not only the molecular weight. Polymers with the same molecular weights can elute at different times if they have different shapes in solution. For example, linear polymers tend to have larger hydrodynamic volumes than branched or compact structures and therefore elute earlier. Conversely, star-shaped, or highly branched polymers appear smaller in size and elute later, leading to an underestimation of molecular weight if linear standards are used for calibration. Solvent quality also impacts conformation. In good solvents, polymers expand into coil-like structures, while in poor solvents, they collapse, reducing hydrodynamic volume and shifting elution profiles. Temperature changes can further alter conformation, especially in thermoresponsive polymers, affecting the GPC results. Therefore, molecular conformation directly affects the accuracy of molecular weight determination in GPC. To correct for these effects, universal calibration (using intrinsic viscosity and molecular weight) or multi-detector GPC (e.g. combining refractive index with light scattering or viscometry) is often employed. Understanding polymer conformation is essential for accurate interpretation of GPC data, especially for non-linear or responsive polymers.

Instrumentation

GPC is also like other chromatographic techniques. Here, organic solvents dissolve the samples and are injected into a column for separation [15]. A pump continuously supplies fresh eluent to the column, and at last, multiple detectors are often used to obtain additional information about the polymer samples, as shown in Fig. 7.6 and Table 7.1.

Data Analysis

GPC is a widely used technique for examining the molecular weight and MWD of polymer samples. It can be used as a universal calibration parameter, where the hydrodynamic volume (V_η) is proportional to the product of the intrinsic viscosity of

Table 7.1 Key components of gel permeation chromatography (GPC)

Component	Description
Gel (Stationary Phase)	Gels are used as stationary phases in GPC, offering precise control of pore size for effective separation. They are preferred for their lack of ionizing groups and low affinity for the analytes. They must be mechanically stable and chemically inert. Common commercial gels include dextran gel, acrylamide gel, and agarose gel
Column	The GPC column is packed with microporous gels. The total penetration volume refers to the maximum volume the analyte can permeate without retention on the stationary phase. Columns typically have a much larger volume than the injected sample
Eluent (Mobile Phase)	The eluent should dissolve the polymer, be inert to interactions with the polymer, and wet the gel surface.[5,6] Common eluents include: • THF for general polymer use at room temperature • o-Dichlorobenzene, trichlorobenzene for polyalkynes (130–150 °C) • HFIP for polyamides and polyesters • Water/buffers for proteins and polysaccharides
Pump	Pumps are critical for delivering a constant flow of eluent. Two types are commonly used: • Piston pumps and • Peristaltic pumps. A consistent flow rate is vital for accurate molecular weight calibration.[9]
Detector	Two types of detectors are commonly used: • Concentration-dependent: UV–VIS absorption, DRI, RI detectors • Molecular weight-dependent: LALLS, MALS Multiple detectors are often used in series to improve accuracy. UV and RI detectors are most commonly used for polymer composition analysis in GPC

the polymer [η] in the SEC eluent and molecular weight [M]. If the Mark-Houwink-Sakurada constants [K,α] are known, a plot of log [η]M versus elution volume or elution time provides a universal calibration curve for any polymer in that solvent [16]. A calibration curve can be attained by plotting the logarithm of molecular weight versus retention time or volume. This allows for the determination of molecular weights and MWD for unknown polymers.

Determination of Absolute Mw Using Triple Detection GPC

Triple detection GPC is a technique used to determine the **absolute molecular weight (Mw)** of polymers, independent of calibration standards [17]. It combines three detectors:

- **Refractive Index (RI) Detector**

The RI detector response is proportional to the product of the concentration (c) of the polymer and the refractive index increment (d_n/d_c):

$$\boldsymbol{RI} \propto \left(\frac{dn}{dc}\right) \times \boldsymbol{c} \tag{7.ii}$$

Here, d_n/d_c represents the change in refractive index with respect to concentration and depends on both the nature of the polymer and the solvent used.

- **Light Scattering Detector (Typically MALS—Multi-Angle Light Scattering)**

In light scattering (LS) detection used in GPC, the detector signal is proportional to the product of the weight-average molecular weight (M_w), the square of the refractive index increment (dn/dc), and the polymer concentration (c). This relationship is expressed as:

$$\boldsymbol{LS} \propto \boldsymbol{Mw} \times \left(\frac{dn}{dc}\right)^2 \times \boldsymbol{c} \tag{7.iii}$$

In this context, M_W reflects the average molecular weight of the polymer chains weighted by their mass, and dn/dc is the refractive index increment, indicating how strongly the polymer alters the refractive index of the solution. Because the LS signal depends on (dn/dc), even small inaccuracies in measuring dn/dc can lead to significant errors in molecular weight determination. Therefore, obtaining an accurate dn/dc specific to the polymer–solvent system is crucial for reliable light scattering analysis. This detector is sensitive to **molecular size**, especially for high-molecular-weight species.

- **Viscometer Detector**

In GPC with a viscometer detector, the signal is proportional to the product of the intrinsic viscosity [η] and the polymer concentration c, expressed as,

$$\boldsymbol{Viscosity} \propto [\eta] \times \boldsymbol{c} \tag{7.iv}$$

Intrinsic viscosity [η] is a measure of a polymer's hydrodynamic volume in solution—it reflects how much space a polymer occupies and how it affects the viscosity of the solvent. It is influenced by the polymer's molecular weight, shape, and interaction with the solvent. Since the viscometer detects changes in flow resistance as polymer chains pass through, the signal increases with higher intrinsic viscosity and concentration. This relationship is key for understanding polymer conformation and branching when used alongside light scattering and refractive index detectors in triple detection GPC.

Triple detection GPC provides a **comprehensive analysis** of polymers by combining concentration, molecular weight, and structural information in a single experiment, making it especially useful for complex or non-linear polymer architectures.

Applications

- **Molecular weight distribution in polyolefins:** High-temperature SEC is routinely used to profile the MWD of polyolefin resins (polyethylene, polypropylene), since properties like processability and mechanical strength are highly sensitive to MWD. For example, recent studies on polyethylene with controlled unimodal vs. bimodal MWD have demonstrated that bimodal distributions (two distinct molecular weight fractions) yield simultaneously improved processability and higher tensile strength compared to unimodal distributions of similar average molecular weight. SEC reveals these multi-modal distributions (often appearing as multiple peaks in the chromatogram), allowing manufacturers to tailor catalysts or reactor conditions to produce polymers (e.g. bimodal HDPE) with optimized performance characteristics [18].

- **Copolymer composition heterogeneity:** Many copolymers (e.g. random or block copolymers) exhibit variation in composition from chain to chain—a chemical composition distribution (CCD)—in addition to MWD. Conventional SEC will elute all chains of similar size together, even if their comonomer content differs, so composition differences are not resolved. For instance, a styrene–butadiene copolymer with varying styrene content would show a single broad SEC peak, as all chains span a similar size range. Detecting such heterogeneity requires either specialized detectors or multidimensional methods (described below). SEC alone cannot separate molecules by composition, even with advanced detection, highlighting the need for multi-dimensional techniques to fully characterize copolymers [19–22].
- **Polymer additive and oligomer analysis:** SEC is useful for separating high molecular weight polymer from low molecular weight additives, residual monomers, or oligomeric by products in a formulation. Because large polymer chains elute first and small molecules elute later, one can isolate fractions of small compounds for further analysis. For example, in a plastic formulation containing antioxidant or plasticizer additives, SEC can cleanly separate the polymer (eluting in the high-molecular-weight region) from the additive compounds (eluting at much larger elution volume). This makes it possible to detect and quantify additives in polymers by coupling SEC with appropriate detectors. In fact, SEC is commonly used as a sample cleanup step: the polymer matrix is eluted and discarded, while the small-molecule fraction is collected for identification by techniques like GC–MS. One notable study demonstrated an offline SEC–GC method to simultaneously determine a polymer's MWD and the concentration of its additives in each fraction, illustrating the power of hyphenated use of SEC for formulation analysis [23–25].
- **End-group or functionality analysis:** The distribution of functional end-groups in polymers (or the presence of telechelic oligomers) can sometimes be probed by SEC in combination with selective detection or derivatization. While SEC can't separate on functionality alone, one approach is to use dual detectors—for example, a refractive index detector in parallel with a UV detector tuned to a chain end chromophore. Differences between the chromatogram signals can reveal how a UV-active end-group is distributed across the MWD (higher signals at low MW indicate higher end-group concentration per mass, etc.). This allowed quantification of acid end-group content as a function of molecular size—essentially mapping functionality across the MWD. In more advanced setups, one can perform liquid chromatography at critical conditions (LCCC) as

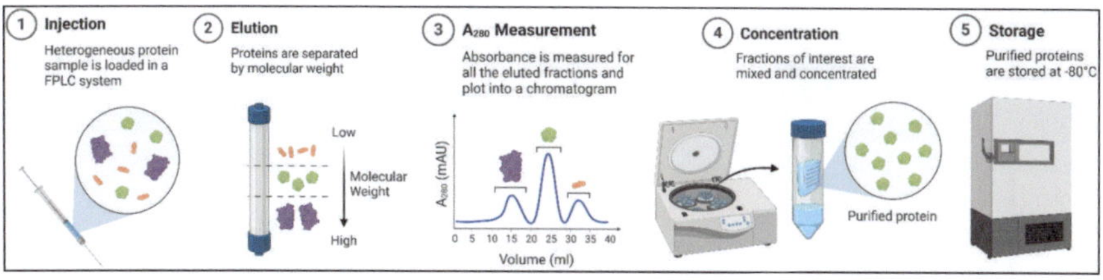

Fig. 7.7 Workflow of FPLC protein purification: (1) sample injection, (2) size-based elution, (3) absorbance monitoring for peak detection, (4) concentration of target fractions, and (5) storage of purified proteins at −80 °C

one dimension (where separation is based on end-group type or a specific block segment), followed by SEC as a second dimension, to directly separate polymers by end-group while simultaneously measuring their molecular weight.

- **Degradation and stability profiling:** Monitoring changes in molecular weight via SEC is a key method for studying polymer degradation (oxidative, thermal, or biodegradation). As polymer chains break over time, SEC traces show a shift of the molecular weight distribution towards lower values, and often the emergence of low-MW fragments. Such SEC analysis provides a quantitative measure of degradation progress. In one study, SEC confirmed a significant drop in PLA molecular weight during the early stages of hydrolysis, signalling the onset of degradation well before any weight loss was observed. Similarly, in thermal or photo-degradation studies, SEC can reveal if a polymer's high molecular weight tail is truncating or if a bimodal distribution develops (indicating a fraction of the material has depolymerized). This information is crucial for understanding polymer lifetime and for comparing how different additives or environmental conditions affect stability (Fig. 7.7).

7.3.3 *Supercritical Fluid Chromatography (SFC)*

Supercritical fluid chromatography (SFC) is a technique that uses a liquid or gas in a state above critical temperature or critical pressure, allowing gases and liquids to coexist. Lovelock first put forward the idea of SFC in 1958. Klesper first employed it in 1962. It is used to analyse and purify molecules of low to moderate molecular weight, thermally liable molecules, as well as separate chiral compounds. Supercritical fluid like carbon dioxide is used as the mobile phase. As the supercritical phase represents a state in which liquid and gas properties converge, the SFC technique is also known as 'Convergence chromatography' [26].

Principle

SFC uses a supercritical fluid as the mobile phase. A supercritical fluid is a material that shows both gaseous and liquid characteristics at a certain temperature and pressure higher than its critical point. Carbon dioxide is commonly used due to its

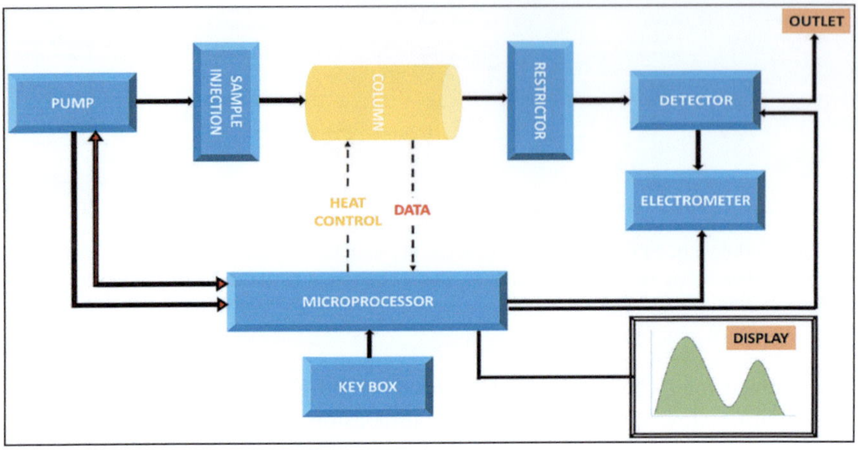

Fig. 7.8 Supercritical fluid chromatography instrumentation

non-toxicity, low cost, and suitable critical parameters. The separation in SFC is based on the principles of adsorption and partition chromatography. Compounds in the sample interact differently with the stationary phase (usually a column packed with silica or other materials) and the supercritical fluid mobile phase, leading to their separation. When a mixture of compounds is introduced at different temperatures, they interact differently with the stationary phase and the mobile phase, allowing them to pass through the column at different rates and separate from each other.

Instrumentation

Key components of SFC are high-pressure pumps, an oven, an injection system, columns packed with a stationary phase, a backpressure regulator, and detectors for compound analysis, as shown in Fig. 7.8.

Stationary Phase

SFC columns are like HPLC columns in terms of coating ingredients. The two most common types of columns used in SFC are open-tubular columns and packed columns. Open-tubular columns are well-known because they are similar to fused-silica columns of HPLC. This column is distinguished by the presence of an interior coating consisting of a stationary phase composed of cross-linked siloxane material. Coating thickness ranges between 0.05 to 1.0 μm. The column has a length ranging from 10 to 20 m.

Mobile Phase

In SFC, a mobile phase can be made of a large variety of different substances. The mobile phase can be chosen from the solvent groups consisting of inorganic solvents, hydrocarbons, alcohols, ethers, and halides; alternatively, it can be one of the following: acetone, acetonitrile, pyridine, and so on. Carbon dioxide is the most commonly used supercritical fluid due to its excellent properties like low critical temperature and pressure requirements, inexpensive, easy to get, unaffected by UV radiation, nontoxic, and it is a good solvent for non-polar compounds. Besides carbon dioxide, the following compounds can also be used: ethane, n-butane, N_2O, dichlorodifluoromethane, diethyl ether, ammonia, and tetrahydrofuran.

Pumps

Syringe pumps are used for capillary SFC to maintain consistent pressure while reciprocating pumps are used for packed columns to facilitate mixing of the mobile phase or the introduction of modifier fluids.

Injectors

In capillary SFC, a tiny amount of sample should be immediately injected into the column. Hence, pneumatically controlled valves are used. A typical injection valve is commonly used for packed SFC.

Columns

In SFC, two types of analytical columns are employed: packed columns and capillary columns. Packed columns consist of tiny deactivated particles on which the stationary phase attaches. These are made of stainless steel. Capillary columns are open tubular columns of fused silica with modest interior diameters.

Detectors

One of the most notable advantages of SFC over HPLC is the wide range of detectors. The flame ionization detector (FID), which is typically part of the equipment for GC, can be used to configure SFC. Since FID is an extremely sensitive detector, a detector of this kind can provide a positive contribution to the quality of SFC analyses. A mass spectrometer, a UV–visible spectrometer, or an infrared spectrometer can be coupled to an SFC system more easily than an HPLC system can be related to these types of spectrometers. Thermionic detectors and fluorescence emission spectrometers are two examples of additional detectors that can be connected to SFC. These detectors are typically used in conjunction with HPLC.

Application

It has advantage over other traditional HPLC chromatographic techniques, includes high efficiency, quick separation, and less solvent usage. SFC is widely used in separating oligomers, unreacted monomers, and degradation products, which is required for ensuring quality, purity and also for analysing functional or end groups (such as in PEGylated or carboxylated oligomers).

Additionally, SFC has good resolving power for stereoisomers and chiral polymers. In the biopolymers, SFC is used to analyse polymer-drug conjugates, allows separation and quantification of free drugs, polymer carriers, and conjugated species. Various detectors, including UV–Vis, Evaporative Light Scattering Detector (ELSD), and mass spectrometry (MS), integrated SFC is used to boost the sensitivity of detection (especially for non-UV absorbing or complex polymer systems).

7.3.4 Ion Exchange Chromatography

Ion-exchange chromatography (IEC) can be classified into cationic and anionic chromatography. This technique is helpful for polymers with ionic functional groups like sulphonic acid, carboxylic acid, or amine groups. Anion-exchange chromatography (AEC) focuses on separating negatively charged polymers by utilizing a positively charged stationary phase. Cation-exchange chromatography (CEC) works the opposite manner, separating positively charged polymers with a negatively charged stationary phase. The retention of polymers in the column is influenced by the strength and quantity of ionic interactions, and they are eluted through a salt or pH gradient (Fig. 7.9).

Ion-exchange chromatography is used to determine degree of ionization or charge density and functional group analysis. This technique is also effective in separating polymers monomer or by-products based on ionic heterogenicity. Furthermore, it is also used in characterization of polyanions-based polymers (polyelectrolytes) like poly (acrylic acid) and sulphonated polymers, as well as polycations such as poly (ethylene imine) and quaternized polymers. This technique is also employed for analysing charged polymer–protein conjugates or nucleic acid–polymer complexes (Bioconjugate Polymers), especially in the context of drug delivery research.

Example: Cationic polymers, such as quaternized poly (vinylpyridine), are examined on cation-exchange columns to evaluate their degree of quaternization. Anionic

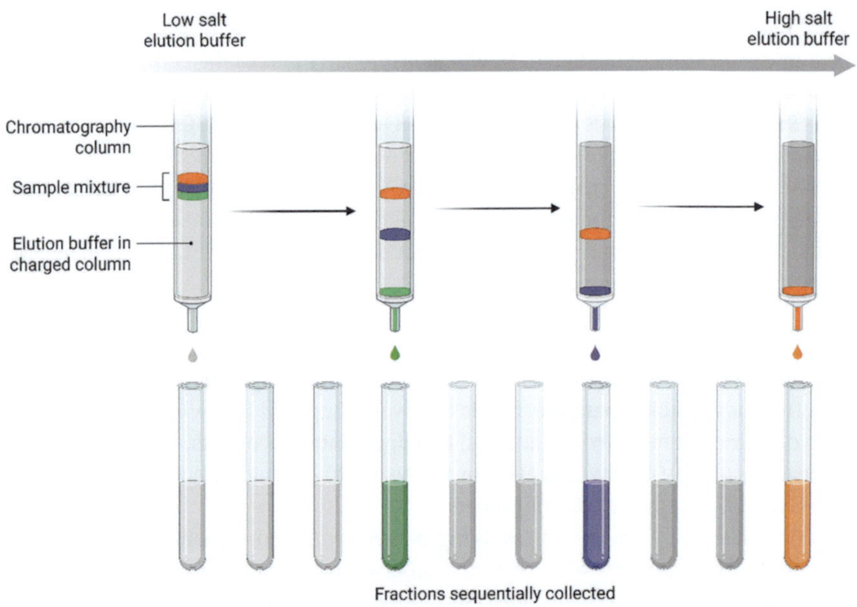

Fig. 7.9 Stepwise visualization of column chromatography: a sample mixture is loaded onto the column, and components are separated based on their affinities to the stationary phase and collected in fractions over time

polysaccharides like heparin or alginate are analysed through anion-exchange chromatography to assess their molecular charge density and purity [27, 28].

7.3.5 *Affinity Chromatography*

Affinity chromatography is a separation technique that plays important role in biochemistry, particularly to purifying proteins, enzymes, or nucleic acids through specific binding interactions (Fig. 7.10). Apart from separation of biopolymers, it is also used in the field of modified polymer systems, especially polymers with particular functional groups or bioactive ligands. Affinity chromatography works by separating them based on the specific interactions that occur between the functional groups on the polymer and the immobilized ligands on the stationary phase. These interactions can be categorized into antibody-antigen, enzyme–substrate, receptor-ligand, and metal–ligand (histidine-tagged polymers interacting with immobilized metal ions) interactions [29, 30].

Applications of affinity chromatography involves identification of functional group in polymers like biotin, PEG, or peptides, that can be characterized by examining their binding affinity to various ligands. This technique is also useful for distinguishing between functionalized and non-functionalized polymers. Affinity columns can effectively retain functional polymers while allowing the non-functional to washed away. Furthermore, it also helps in analysing biopolymers (such as polysaccharides and polypeptides) that can interact with biological ligands. It can be instrumental in evaluating the biological activity or binding specificity of polymer conjugates. This method can also separate conjugated from unconjugated proteins

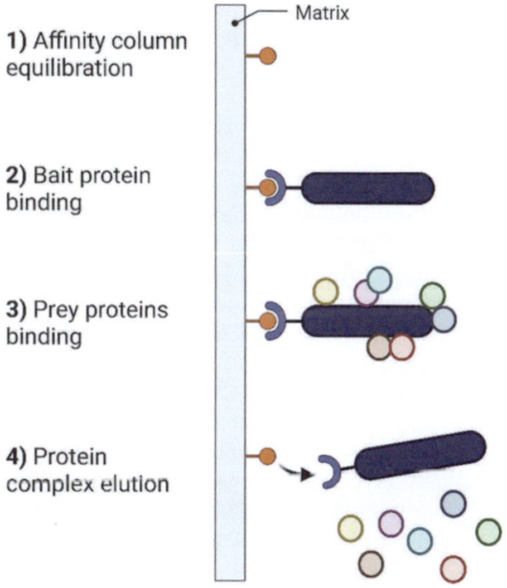

Fig. 7.10 Stepwise visualization of affinity chromatography

or polymers, which is crucial for characterizing polymer-drug or polymer-protein conjugates. For Example, PEGylated polymer that has a terminal biotin. You can characterize it using a streptavidin-immobilized resin. Only the polymers with biotin will bind, which clearly indicates the success of the biotinylation process.

7.3.6 Two-Dimensional Gel Permeation Chromatography (2D-GPC)

The two-dimensional (2D) correlation technique has been effectively used to evaluate changes in different molecular spectra like IR, Raman, and NIR. This technique is also useful for understanding the reaction dynamics and mechanisms in polymerization processes [8–10]. 2D-GPC combines two different separation techniques. The first separates molecules based on their size, and the second might focus on chemical properties or other factors. 2D-GPC has been introduced to study the fine intricacies of the polymerization of simple silane coupling agents [31, 32]. *(Discussed in* Chap. 11*advanced characterization of polymers).*

Acknowledgements We used the Biorender student license (https://BioRender.com) for the illustrations in this article.

Competing Interest The authors declare no competing interests.

References

1. The birth of partition chromatography. https://www.researchgate.net/publication/285954771_The_birth_of_partition_chromatography. Accessed 15 May 2025
2. Jain AV (2005) Analysis of organophosphate and carbamate pesticides and anticholinesterase therapeutic agents. Toxicology of organophosphate and carbamate compounds, 681–701. https://doi.org/10.1016/B978-012088523-7/50048-X
3. DM Meunier JH Wade M Janco 2021 Recent advances in separation-based techniques for synthetic polymer characterization ACS Publications 93 273 294 https://doi.org/10.1021/ACS.ANALCHEM.0C04352
4. Majors R, America JH-L-GN (2013) Undefined The chromatography and sample preparation terminology guide. go.gale.com
5. GS-A Chemistry 1998 Undefined (1998) Height equivalent to a theoretical plate theory for rectangular GC columns ACS Publications 70 4805 4816 https://doi.org/10.1021/AC980328Z
6. Colmsjö A, A ME-J of C (1987) Undefined Assessment of the height equivalent to a theoretical plate in liquid chromatography. Elsevier
7. O Bousquet AP Foucault GF Le 1991 Efficiency and resolution in countercurrrent chromatography Taylor & Francis 14 3343 3363 https://doi.org/10.1080/01483919108049395
8. F Gritti G Guiochon 2013 Perspectives on the evolution of the column efficiency in liquid chromatography ACS Publications 85 3017 3035 https://doi.org/10.1021/AC3033307
9. Carda-Broch S, … JT-L-A chimica (1999) Undefined Evaluation of several global resolution functions for liquid chromatography. Elsevier

10. Pápai Z, A TP-J of C (2002) Undefined Analysis of peak asymmetry in chromatography. Elsevier
11. JG-A Chemistry 1963 Undefined (1963) kinetic origin of tailing in chromatography ACS Publications 35 1999 2002 https://doi.org/10.1021/AC60206A008
12. Bioanalysis AG-P and P of, 2000 undefined Basic HPLC theory and practice. taylorfrancis.com. https://doi.org/10.4324/9780203449479-9/BASIC-HPLC-THEORY-PRACTICE-ANDY-GRAY
13. Leonard RG, Quigley K (2004) Adhesives and sealants. Encyclopedia of Analytical Science: Second Edition, 28–37. https://doi.org/10.1016/B0-12-369397-7/00005-4
14. Lee B, Park T, Technology CC-PS and (1994) Undefined (1970) Principles and Application of GPC/SEC. koreascience.kr 5:393–402. https://doi.org/10.1080/00372367008068438
15. GPC INSTRUMENTATION PROCESS d.o.o. https://gpcip.eu/en/. Accessed 15 May 2025
16. CJB Dobbin A Rudin MF Tchir 1980 Evaluation of GPC methods for estimation of Mark–Houwink–Sakurada constants Wiley Online Library 25 2985 2992 https://doi.org/10.1002/APP.1980.070251229
17. Z Lu Z Wang F Lu H Li 2013 Determination of molecular weight and distribution of polycarboxylate superplasticizer by triple detection GPC Adv Mat Res 687 461 465 https://doi.org/10.4028/WWW.SCIENTIFIC.NET/AMR.687.461
18. E Prasad Paradhakshina N Dhore R Arukula CRK Rao 2024 Water soluble redox-active polyurethanes as efficient corrosive inhibitors for mild steel J Appl Polym Sci 141 e56029https://doi.org/10.1002/APP.56029
19. Z Fodor A Fodor JP Kennedy 1992 A GPC method to determine the composition of two component copolymers Polym Bull 29 689 696 https://doi.org/10.1007/BF01041156/METRICS
20. I Suárez MJ Caballero B Coto 2010 Composition effects on ethylene/propylene copolymers studied by GPC-MALS and GPC-IR Eur Polym J 46 42 49 https://doi.org/10.1016/J.EURPOLYMJ.2009.09.005
21. I Suá Rez J Caballero B Coto 2011 A fast and reliable procedure to determine the copolymer composition by GPC-IR: application to ethylene/propylene copolymers and comparison with 13C NMR Polym Eng Sci 51 317 322 https://doi.org/10.1002/PEN.21811
22. JM Sebastian RA Register 2001 Block copolymer molecular weight determination via gel permeation chromatography: choosing a combining rule J Appl Polym Sci 82 2056 2069 https://doi.org/10.1002/APP.2051;WGROUP:STRING:PUBLICATION
23. AA Cuthbertson C Lincoln J Miscall 2024 Characterization of polymer properties and identification of additives in commercially available research plastics Green Chem 26 7067 7090 https://doi.org/10.1039/D4GC00659C
24. A Ortín E López P Hierro del 2018 Simplified robust triple detection methods for high temperature GPC analysis of polyolefins Macromol Symp 377 1700044 https://doi.org/10.1002/MASY.201700044;PAGEGROUP:STRING:PUBLICATION
25. WJ Wang S Kharchenko K Migler S Zhu 2004 Triple-detector GPC characterization and processing behavior of long-chain-branched polyethylene prepared by solution polymerization with constrained geometry catalyst Polymer (Guildf) 45 6495 6505 https://doi.org/10.1016/J.POLYMER.2004.07.035
26. West C (2018) Current trends in supercritical fluid chromatography. Analyt Bioanalyt Chem 410:25 410:6441–6457. https://doi.org/10.1007/S00216-018-1267-4
27. A Nordborg EF Hilder 2009 Recent advances in polymer monoliths for ion-exchange chromatography Anal Bioanal Chem 394 71 84 https://doi.org/10.1007/S00216-009-2636-9/TABLES/2
28. NS Terefe O Glagovskaia K Silva De R Stockmann 2014 Application of stimuli responsive polymers for sustainable ion exchange chromatography Food Bioprod Process 92 208 225 https://doi.org/10.1016/J.FBP.2014.02.003
29. Urh M, Simpson D, Zhao K (2009) Chapter 26 affinity chromatography: general methods. Methods Enzymol 463:417–438. https://doi.org/10.1016/S0076-6879(09)63026-3

30. NA Samojlova MA Krayukhina IA Yamskov 2004 Use of the affinity chromatography principle in creating new thromboresistant materials J Chromatogr B 800 263 269 https://doi.org/10.1016/J.JCHROMB.2003.10.049
31. K Izawa T Ogasawara H Masuda 2002 2D gel permeation chromatography (2D GPC) correlation studies of the growth process for perfluoro-octyltriethoxysilane polymer aggregates Phys Chem Chem Phys 4 1053 1061 https://doi.org/10.1039/B107556J
32. K Izawa T Ogasawara H Masuda 2002 Two-dimensional correlation gel permeation chromatography (2D GPC) study of 1H,1H,2H,2H-perfiuorooctyltriethoxysilane sol-gel polymerization process J Phys Chem B 106 2867 2874 https://doi.org/10.1021/JP012878Y/ASSET/IMAGES/MEDIUM/JP012878YU00004A.GIF

Nikhil Kumar is a senior research fellow the Advanced Technology Development Centre, Indian Institute of Technology Kharagpur. He is specialized in fabrication of smart polymeric materials for diagnosis and therapeutics. Nikhil completed his Bachelor of Pharmacy and Master of Pharmacy in Pharmacology. His masters research work was focused on developing dressings for diabetic foot ulcers. His current research revolves around the design and application of smart polymeric biomaterials for advanced diagnosis and therapeutic interventions for cutaneous squamous cell carcinoma. He has published 12 peer-reviewed research articles. He is actively engaged in research focused on polymer characterization, with particular interest in the structural, thermal, and rheological analysis of advanced polymeric materials. Mr. Kumar has been involved in research projects that explore the application of spectroscopic techniques (FTIR, NMR), thermal analysis (DSC, TGA), to evaluate polymer performance under varied environmental conditions. Mr. Kumar is also developing a novel diagnostic tool that integrates multiple characterization techniques FTIR, Raman spectroscopy, AFM and nanoindentation to probe and differentiate pathological tissue conditions such as fibrosis and hyperkeratosis.

Payel Maity is a Ph.D. candidate at the Indian Institute of Technology Kharagpur, specializing in metalpolymer composites for environmental pollution control. With a solid academic background in Chemistry, she completed both her Bachelor's and Master's degrees in Chemistry, where her research focused on the study of polymer structures and their diverse metal composites. Her current research centres on designing and applying polymer-containing metal-organic frameworks, particularly in advancing water safety and addressing severe environmental pollution-related challenges.

Nikhil Dhore is pursuing his Ph.D. at the Indian Institute of Technology Kharagpur, focusing on the development of sustainable polymers for decorative and protective applications. Nikhil has completed his bachelor's in chemical technology and a master's specializing in paint technology and has hands-on experience of 3+ years in the coating field research. His current research is on the synthesis of water-based polyurethane polymer, non-isocyanate polyurethane, and biobased epoxy amine systems. He has to his credit more than 6+ peer-reviewed international journal articles in his account.

Homen Dahal holds a strong foundation in chemistry, with research centred on cationic Gemini surfactants and their structure–property relationships. His work encompasses the synthesis and detailed characterization of organic molecules using a range of spectroscopic techniques. Through his studies, he aims to connect molecular design with practical biomedical applications.

Priya Devi holds an academic background in chemistry, with research focused on enhancing the mechanical strength of degraded cellulosic paper through polymer-based reinforcement coatings. Her research employs spectroscopic techniques to characterize lignocellulosic biomass and evaluate the physicochemical effects of polymer coatings on paper substrates.

Chapter 8
Spectroscopical Analysis of Polymers: Techniques and Applications

Homen Dahal, Priya Devi, Nikhil Dhore, Manas Jain, and Nikhil Kumar

Abstract This chapter proposes a thorough overview of various spectroscopic methods used in the analysis of polymers. Understanding the molecular structures and composition of polymers during synthesis remains a key challenge in polymer science. To address this, the chapter provides a detailed perspective on methods such as Raman Spectroscopy, Mass Spectrometry (MS), Nuclear Magnetic Resonance (NMR), Ultraviolet–Visible (UV–Vis) Spectroscopy, and Fourier Transform Infrared (FTIR) Spectroscopy. This chapter aims to provide conceptual understating of spectroscopic methods for polymer characterization.

Keywords Spectroscopy · NMR · Raman. FTIR · Mass spectroscopy

8.1 Introduction

Electromagnetic radiation (EMR) is generated when a subatomic particle, such as an electron, is propelled by an electric field. The movement generates oscillating electric and magnetic fields that propagate perpendicularly to one another as a collection of

---

H. Dahal
Department of Chemistry, Indian Institute of Technology Kharagpur, Kharagpur, India

P. Devi
Department of Chemical Sciences, Tezpur University, Tezpur, Assam, India

N. Dhore
Rubber Technology Centre, Indian Institute of Technology Kharagpur, Kharagpur, India

M. Jain
Department of Aerospace Engineering, Indian Institute of Technology Kharagpur, Kharagpur, India

N. Kumar (✉)
Advanced Technology Development Centre, Indian Institute of Technology Kharagpur, Kharagpur, India
e-mail: nikhilsingh25694@gmail.com

S. Chattopadhyay and N. Kumar (eds.), *Polymer Characterization*,
https://doi.org/10.1007/978-981-95-2609-3_8

tiny energy particles known as photons. Electromagnetic radiation is defined by its energy, wave number, wavelength, and frequency (Fig. 8.1) [1].

Points to Remember

- Wavelength (λ): It is the distance between two consecutive peaks of a wave (crests). Unit = m
- Frequency (ν): It is the number of waves that are formed in a given length of time. Unit = number of wave cycles per second or hertz (Hz).
- Wave number ($\bar{\nu}$): Reciprocal of wavelength. ($\bar{\nu} = 1/\lambda$) Unit = cm^{-1}
- Energy of EMR (E): EM particles having small pockets of energy called photons.

$$E = h\nu \, or \, E = \frac{hc}{\lambda} \quad \text{(i)}$$

  where h is planks constant ((6.62 × 10^{-34} Js) ν is frequency and c is speed of light.
- EMR interacts with matter by either absorption or emission. Absorption or emission of EMR causes change in given energy and in molecular spectroscopy we measure these energy change.
- Total energy of molecules = Translational (Overall movement of molecule) + rotational (Microwave spectroscopy) + vibrational (IR spectroscopy) + electronic (Uv–vis spectroscopy)

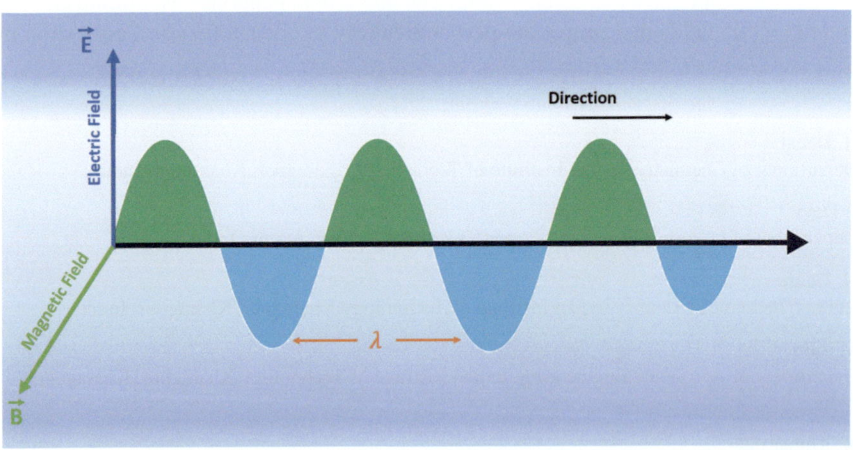

Fig. 8.1 Schematic representation of an electromagnetic wave propagating through space, illustrating the mutually perpendicular oscillations of the electric field (E, in blue) and magnetic field (B, in green), which are both perpendicular to the direction of wave

- Molecular spectra arise from the absorption or emission of electromagnetic radiation as molecules transition between quantized energy states.
- Absorbance and transmittance are two methods to characterize the interaction of light with a substance. Transmittance indicates the quantity of light that traverses a sample, whereas absorbance indicates the quantity of light that is absorbed by the sample. Their relationship is inverse; as one rises, the other lowers.

Spectroscopy is the examination of the interaction between matter (atoms and molecules) and EMR. This interaction produces two categories of spectra: atomic and molecular spectra. The movement of electrons between energy levels in an atom, resulting from energy changes, is termed atomic spectra, whereas the transition of electrons among rotational and vibrational energy levels, in addition to electronic transitions, corresponds to molecular spectra (Fig. 8.2). Atomic spectrum is relatively simple, whereas the molecular spectrum is considerably more complex [2]. Spectroscopy offers vital understanding of the behavior of diatomic and polyatomic molecules by means of analysis of their energy, wave number, wavelength, frequency, and related molecular processes. It is a useful tool for computing several structural features including molecular bond dissociation energies, bond strengths, bond lengths, bond angles, and size and shape. Consequently, thorough knowledge on the composition and structure of compound depends on molecular spectroscopy [3].

Moreover, spectroscopy has been pivotal in advancing polymer science, evolving alongside synthetic materials from the nineteenth century till present. Early visible and UV–Vis techniques analyzed natural polymers like rubber, while mid-twentieth-century IR spectroscopy decoded synthetic polymers like nylon and PVC by identifying functional groups and crystallinity. In 1960s–70s, NMR and Raman spectroscopy were used for understanding microstructures and symmetrical bonds.

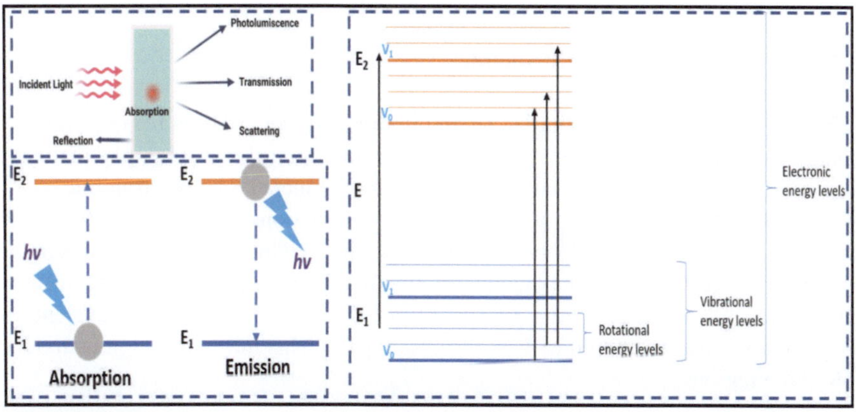

Fig. 8.2 Concept of spectroscopy and energy-level diagram

Late twentieth-century innovations like FTIR, XPS, and hybrid techniques enabled detailed real-time analysis of polymer samples. In the modern era, hyperspectral and in-situ techniques like Raman and AFM-IR optimize polymer processing, biodegradable materials, and nanocomposite, ensuring quality, sustainability, and performance. Hence, from vulcanized rubber to Kevlar®, spectroscopy remains the backbone of polymer characterization. Furthermore, spectroscopic techniques are significant in the dominion of polymer characterization, offering profound insights into the molecular architecture, composition, and behavior of polymers, which are vital for designing and experimental optimization [4].

8.2 Classification of Spectroscopic Techniques

Spectroscopic techniques used in polymer characterization can be classified based on their ability to probe specific molecular features of polymers: vibrational spectroscopy (FTIR and Raman) for analyzing molecular bond vibrations and functional groups, electronic absorption spectroscopy (UV–Vis) for studying conjugated systems and electronic transitions, nuclear magnetic resonance spectroscopy (NMR) for determining molecular structure and dynamics, and mass spectrometry (MS) for measuring molecular weights and fragmentation patterns. These complementary techniques provide a comprehensive understanding of polymer composition, microstructure, and properties, enabling researchers to correlate molecular features with material performance for both fundamental studies and industrial applications [5].

8.2.1 Fourier Transform Infrared Spectroscopy (FTIR) for Polymer Characterization

8.2.1.1 Theory and Basic Principle

Infrared spectroscopy a qualitative technique for polymer analysis for over seventy years. An infrared spectrum provides as an unique fingerprint of a sample, featuring absorption peaks that correspond to the vibrational frequencies between atomic bonds (Fig. 8.4). Different polymers possess distinct combinations of atoms, resulting in no two molecules generating identical infrared spectra. The intensity of the peaks in the spectrum directly indicates the quantity of functional groups present [6]. The infrared (IR) region of the electromagnetic spectrum (Fig. 8.3) encompasses wavelengths from approximately 780 nm to 1 mm (or 1000 μm). This infrared region is divided into three regions:

(a) **Near IR (NIR)**: Wavelength: 700 nm to 2500 nm (or 0.7 to 2.5 μm) and Wavenumber: 14,000 to 4,000 cm^{-1}.

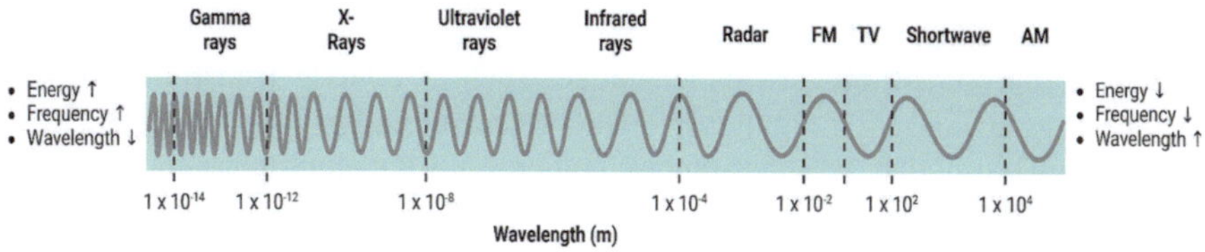

Fig. 8.3 Electromagnetic spectrum

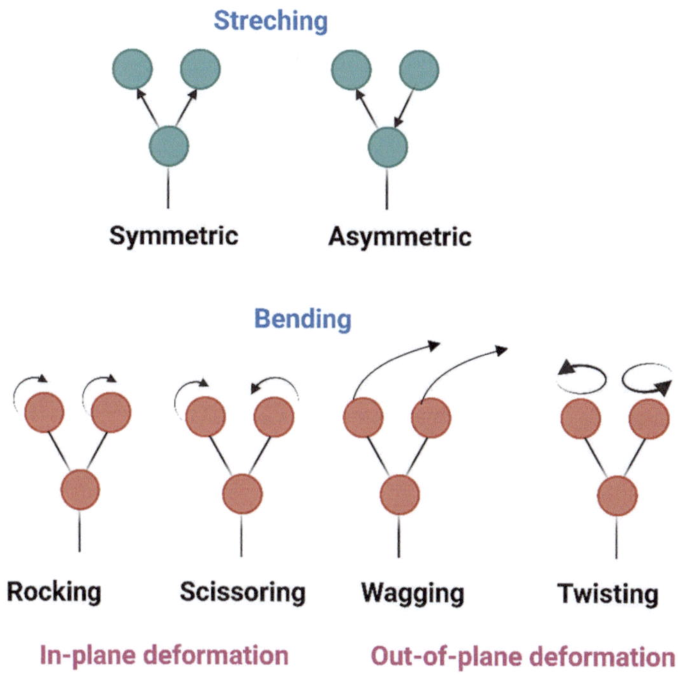

Fig. 8.4 Different types of vibrations

(b) **Mid IR: Wavelength**: 700 nm to 2500 nm (or 0.7 to 2.5 μm) and Wavenumber: 14,000 to 4,000 cm^{-1}.
(c) **Far IR: Wavelength**: 700 nm to 2500 nm (or 0.7 to 2.5 μm) and Wavenumber: 14,000 to 4,000 cm^{-1}.

The fingerprint zone (1450–500 cm^{-1}) of an infrared spectrum is termed the fingerprint region because, similar to a human fingerprint, this segment of the spectrum is nearly unique for each specific molecule [7].

Previously used infrared instruments were of the dispersive and destructive type. These instruments had separate components for the infrared prism or grating and the detector. The prism or grating isolated different colors (frequencies) of energy released from the source, while the detector assessed the energy at each frequency that intersected the sample. This formed a spectrum, which is a graph of intensity against frequency. In contrast, modern IR techniques such as Fourier Transform Infrared Spectroscopy (FTIR) offer significant advantages over earlier instruments. FTIR is simple, non-destructive, provides precise and highly significant measurements, has increased scan speed.

In FTIR analysis, infrared light from the source transverse through a Michelson interferometer, which simultaneously measures all infrared frequencies across the optical path. The Michelson interferometer comprises a beam splitter, a movable mirror, and a stationary mirror. The beam splitter splits the input infrared beam into two optical beams. One optical beam reflects off a fixed flat mirror, while the other reflects off a flat mirror mounted on a mechanism that allows it to move a very short distance—typically a few millimeters—away from the beam splitter. After reflection, the two beams are recombined at the beam splitter. Because one beam travels a fixed path and the other travels a path that changes as its mirror moves, the combined signal that exits the interferometer results from the interference of these two beams. The resultant signal is termed an interferogram, possessing the distinctive characteristic that each data point, as a function of the moving mirror's position, combines information regarding every infrared frequency produced by the source (Fig. 8.5) [8].

FTIR spectroscopy is based on the absorption of infrared (IR) radiation by a sample. The molecular bonds of a polymer sample absorb specific frequencies when IR radiation travels through it, causing them to vibrate (Hook's Law). The absorbed

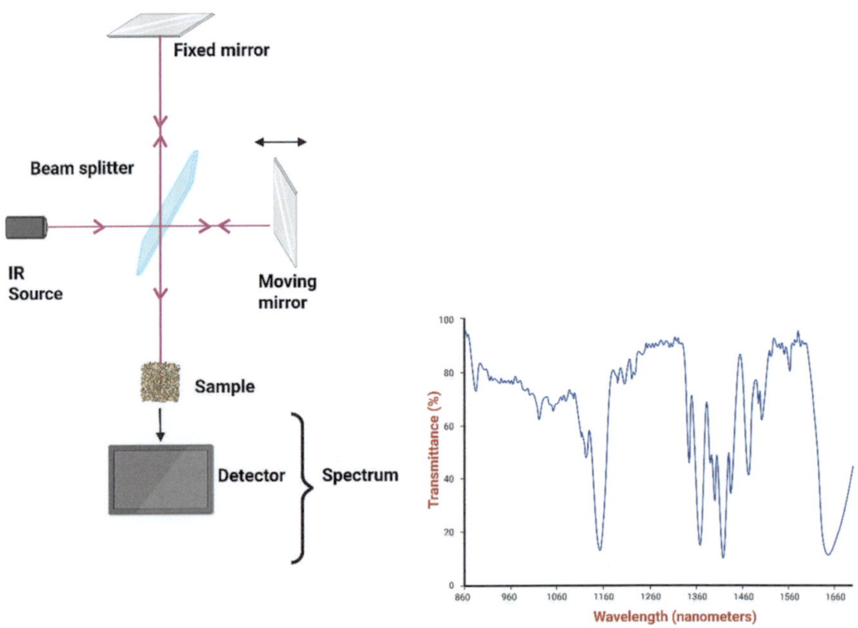

Fig. 8.5 Basic instrumentation of FTIR

frequencies correspond to the vibrational modes of the chemical bonds within the polymer. Molecular vibrations are primarily of two types: stretching (symmetric and asymmetric) and bending (scissoring, rocking, wagging, and twisting). These vibrations occur due to the absorption of IR radiation, leading to change in dipole moment. Stretching involves bond length variations, while bending involves bond angle changes. The resulting spectrum is a plot of the intensity of the transmitted or absorbed IR radiation as a function of the wavelength or wavenumber (cm^{-1}). The Fourier Transform aspect of FTIR refers to the mathematical process (DFT) used to convert the raw data (interferogram) into a spectrum.

FTIR spectroscopy is based on three main principles:

(a) Diffuse Reflectance Method (adhering samples—powders)
(b) High Sensitivity Reflection Measurement (metal sheet)
(c) Attenuated Total Reflectance (ATR) (solid/liquid samples).

8.2.1.2 Sample Preparation for FTIR Analysis

FTIR spectroscopy offers multiple sampling techniques to analyze polymers in different forms. In transmission mode, the sample is prepared as a thin film or KBr pellet, and the IR beam passes through it, measuring the transmitted radiation. Attenuated Total Reflectance (ATR) is a surface-sensitive method where the sample is pressed against a high-refractive-index crystal (e.g., diamond or zinc selenide), and the IR beam undergoes internal reflection, with the evanescent wave interacting with the sample. For reflective or opaque samples, reflection mode is used, where the IR beam reflects off the sample surface for analysis. Meanwhile, diffuse reflectance is ideal for powdered samples, where the IR beam scatters off the sample, and the diffusely reflected light is collected. Each technique provides unique advantages depending on the sample's physical state and analytical requirements.

8.2.1.3 Interpretation of FTIR Spectra

The FTIR spectrum of a polymer is a fingerprint that provides information about its chemical structure. Different functional groups absorb IR radiation at characteristic wavenumbers (Table 8.1). Below is a table listing common functional groups found in polymers, along with their characteristic FTIR absorption bands (wavenumbers in cm^{-1}) [9].

8.2.1.4 Applications of FTIR in Polymer Characterization

FTIR is widely used to identify unknown polymers by comparing their spectra with reference spectra in databases. FTIR provides detailed information about the chemical structure of polymers, including the presence of specific functional groups and the type of polymerization (e.g., addition, condensation) (Table 8.2).

Table 8.1 Functional groups in polymers and their characteristic FTIR absorption bands

Functional group	Type of vibration	Wavenumber range (cm^{-1})	Comments
C–H Stretching	Alkane (sp^3)	2850–2960	Strong bands for saturated hydrocarbons (e.g., polyethylene)
	Alkene (sp^2)	3000–3100	Weaker than alkane C–H stretching (e.g., polypropylene)
	Aromatic (sp^2)	3050–3150	Sharp bands for aromatic rings (e.g., polystyrene)
O–H Stretching	Alcohol/Phenol	3200–3600	Broad band due to hydrogen bonding (e.g., polyvinyl alcohol)
	Carboxylic Acid	2500–3300	Very broad band due to strong hydrogen bonding (e.g., polyacrylic acid)
N–H Stretching	Amine/Amide	3300–3500	Sharp bands for primary amines, broader for secondary amines (e.g., nylon)
C=O Stretching	Ketone	1700–1750	Strong band (e.g., poly(methyl methacrylate))
	Aldehyde	1720–1740	Slightly higher than ketones
	Carboxylic Acid	1700–1725	Often accompanied by O–H stretching
	Ester	1735–1750	Strong band (e.g., polyethylene terephthalate)
	Amide (Amide I)	1640–1690	Strong band in polyamides (e.g., nylon)
C=C Stretching	Alkene	1620–1680	Medium intensity (e.g., polybutadiene)
	Aromatic	1450–1600	Multiple bands in this range (e.g., polystyrene)
C–O Stretching	Ether	1000–1300	Strong band (e.g., polyethers like polyethylene oxide)
C-N Stretching	Ester	1050–1300	Often overlaps with C–O–C stretching
	Amine/Amide	1020–1230	Medium intensity (e.g., polyamides)
C–Cl Stretching	Alkyl Halide	600–800	Strong band (e.g., polyvinyl chloride)

(continued)

Table 8.1 (continued)

Functional group	Type of vibration	Wavenumber range (cm^{-1})	Comments
C-F Stretching	Fluorocarbon	1000–1400	Strong bands (e.g., polytetrafluoroethylene)
C≡N Stretching	Nitrile	2200–2300	Sharp band (e.g., polyacrylonitrile)
S–H Stretching	Thiol	2550–2600	Weak band (e.g., thiol-functionalized polymers)
S=O Stretching	Sulfoxide/Sulfonate	1030–1060 (S=O)	Strong bands (e.g., sulfonated polymers)
		1120–1200 (SO_2)	
N=O Stretching	Nitro	1500–1600 (asymmetric)	Strong bands (e.g., nitrocellulose)
		1300–1400 (symmetric)	
Si–O Stretching	Siloxane	1000–1100	Strong band (e.g., polydimethylsiloxane)
C=C (Aromatic)	Aromatic Ring	1450–1600	Multiple bands (e.g., polystyrene)
C–H Bending	Alkane (sp^3)	1350–1480	Scissoring, rocking, and twisting vibrations
	Alkene (sp^2)	900–1000	Out-of-plane bending (e.g., polypropylene)
	Aromatic (sp^2)	700–900	Out-of-plane bending (e.g., polystyrene)

8.2.2 Raman Spectroscopy

Raman spectroscopy is based on the Raman effect, a phenomenon of inelastic scattering of light discovered by Sir C.V. Raman in 1928. Raman spectroscopy (absorption) widely used analytic technique for characterization of organic and inorganic compounds due to its ability to provide detailed molecular fingerprints without extensive sample preparation. Raman spectroscopy provides information on both rotational and vibrational levels.

8.2.2.1 Theory and Basic Principle

When a beam of light passes the sample, the majority of scattered light proceeds through the sample without interaction. This indicates that the detector will absorb energy at the same frequency as the excitation source, resulting in no alteration of the system's total kinetic energy. This phenomenon is referred to as Rayleigh or elastic

Table 8.2 FTIR spectral features for common polymers, highlighting their characteristic absorption bands and functional groups

Polymer	Key FTIR absorption bands (cm^{-1})	Functional groups/ characteristics
Polyethylene (PE)	2850–2960 (C–H stretch)	Strong C–H stretching and bending (1465, 720 cm^{-1})
	1465, 720 (C–H bend)	Simple hydrocarbon structure
Polypropylene (PP)	2850–2960 (C–H stretch)	C–H stretching and bending (1375, 1450 cm^{-1})
	1375, 1450 (C–H bend)	CH_3 asymmetric deformation (1375 cm^{-1})
Polystyrene (PS)	3000–3100 (aromatic C–H stretch)	Aromatic C–H stretching and C=C stretching (1450–1600 cm^{-1})
	1450–1600 (C=C stretch)	Monosubstituted benzene ring (700–750, 1800–2000 cm^{-1})
Poly (methyl methacrylate) (PMMA)	1730 (C=O stretch)	Strong C=O stretching (ester) and C–O stretching (1150–1250 cm^{-1})
	1150–1250 (C–O stretch)	C–H stretching (2850–3000 cm^{-1})
Polyethylene Terephthalate (PET)	1720 (C=O stretch)	C=O stretching (ester) and C–O stretching (1250 cm^{-1})
	1250 (C–O stretch)	Aromatic C=C stretching (1500–1600 cm^{-1})
Nylon (Polyamide)	3300 (N–H stretch)	N–H stretching and C=O stretching (amide I, 1640–1690 cm^{-1})
	1640–1690 (C=O stretch, amide I)	N–H bending (amide II, 1550 cm^{-1})
Polyvinyl Chloride (PVC)	2900–3000 (C–H stretch)	C–H stretching and C–Cl stretching (600–800 cm^{-1})
	600–800 (C–Cl stretch)	Strong C–Cl band indicates halogenated polymer
Polytetrafluoroethylene (PTFE)	1000–1400 (C–F stretch)	Strong C–F stretching bands. No C–H bands due to fluorination

scattering. The minimal quantity of dispersed light indicates a variation in the total kinetic energy of the system (~1 in 10^7), known as Raman or stokes or red shift (Raman scattering/inelastic scattering). Radiation having less frequency or energy than that of the incident beam is termed "Stokes" radiation, whereas radiation with a greater frequency or energy is known as "anti-Stokes" (blue shift) radiation (Fig. 8.6). At ambient temperature, the anti-Stokes-shifted Raman energy is weaker than the Stokes-shifted energy; hence, it is typically neglected and eliminated by filters. This

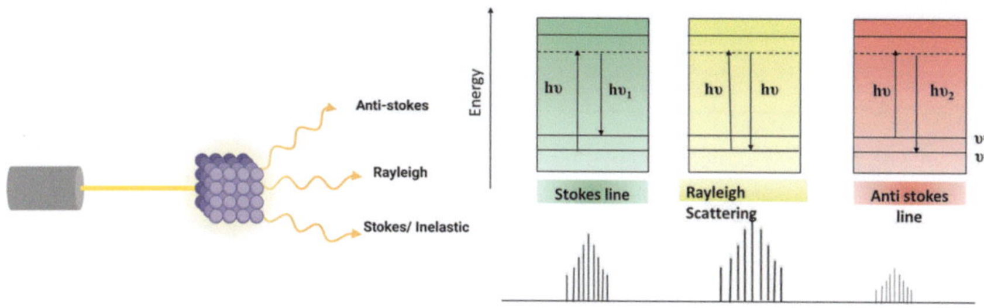

Fig. 8.6 Principle of Raman spectroscopy

results from the interactions between the incident electromagnetic waves and the vibrational energy levels of the sample's molecules. By measuring the inelastic scattering of light, Raman spectroscopy reveals vibrational modes of chemical bonds, enabling the identification of polymer structures, additives, and degradation product.

Its high spatial resolution (~1 μm) and non-destructive nature and make it particularly valuable for studying crystallinity, phase transitions, and copolymer composition. Additionally, Raman spectroscopy excels in industrial applications, such as quality control of multilayer films and real-time monitoring of polymerization processes. Unlike infrared (IR) spectroscopy, Raman is less affected by water and can detect symmetric bonds (e.g., C=C, S–S), making it a complementary tool for comprehensive polymer analysis. This technique's versatility and sensitivity to both bulk and surface properties underscore its importance in materials science and industrial polymer research. This chapter explores the principles, applications, and future potential of Raman spectroscopy, highlighting its transformative role in advancing polymer research and industrial applications [10, 11].

8.2.2.2 Instrumentation of Raman Spectrometer

The key components of a Raman spectrometer work together to generate, detect, and analyze the Raman signal. Raman spectroscopy instrumentation consists of several key components: a laser source (e.g., argon-ion, helium–neon, or diode lasers) that provides monochromatic excitation while minimizing fluorescence; a sample illumination system with lenses and filters to focus the laser and block unwanted wavelengths; a spectrometer (dispersive or FT-based) with a diffraction grating to separate Raman-shifted light; a sensitive detector (CCD, PMT, or InGaAs) to capture the scattered signal; and optical filters (notch or edge filters) to suppress Rayleigh scattering. A data processing system with specialized software analyzes spectra, enabling peak identification and quantification. For high-resolution analysis, a microscope may be integrated to focus the laser on micron-scale sample areas, while a cooling system (liquid nitrogen or Peltier) can enhance detector sensitivity by reducing thermal noise. Together, these components enable precise molecular characterization of materials such as polymers, composites, and biological samples (Fig. 8.7) [12].

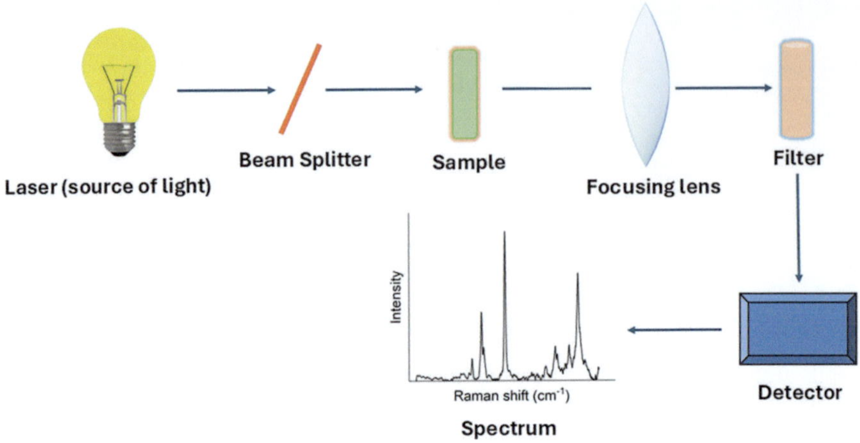

Fig. 8.7 Instrumentation of Raman spectroscopy

8.2.2.3 Interpretation of Raman Spectra

To interpret Raman spectroscopy data, analyze the Raman spectrum, which plots intensity against Raman shift (measured in wavenumbers, cm^{-1}). Each peak in the spectrum corresponds to unique molecular vibrations, revealing information on the chemical structure, composition, and interactions within the sample (Table 8.3) [13].

8.2.2.4 Applications of Raman Spectroscopy in Polymer Characterization

Raman spectroscopy detects specific vibrational modes of chemical bonds, allowing the identification of functional groups (e.g., C=O, C–H, C=C) in polymers. By comparing Raman spectra with reference databases, unknown polymers can be identified based on their unique spectral fingerprints (Table 8.4). Raman spectroscopy can track changes in chemical bonds during polymerization, providing real-time insights into reaction kinetics and mechanisms [14, 15].

Raman spectroscopy also helps in crystallinity assessment (sharp vs. broad peaks) and phase transition studies (melting, crystallization) in polymeric samples. It probes molecular interactions (H-bonding, π-π stacking) and chain conformations (gauche/trans), while characterizing polymer blends (miscibility) and composites (filler dispersion, interfacial effects). It also monitors degradation (oxidation, environmental aging), surface/interface properties (coatings, adhesion), and micro/nanoscale features (mapping, nanostructures). Advanced techniques like SERS, resonance Raman, and time-resolved Raman further enhance sensitivity for trace analysis, conjugated polymers, and dynamic processes [14].

Table 8.3 Detailed breakdown of the key aspects of Raman data interpretation in tabular form

Aspect of interpretation	Description	Examples/notes
Raman shift (cm^{-1})	The position of peaks in the spectrum, indicating the energy difference between incident and scattered light	–C–H stretch: ~2800–3000 cm^{-1} – C=C stretch: ~1600–1680 cm^{-1} – C–C stretch: ~1000–1200 cm^{-1}
Peak intensity	The height of the peaks, proportional to the concentration of the vibrating species	– Strong peaks indicate abundant or highly polarizable bonds – Weak peaks may indicate low concentration or weak Raman activity
Peak width	The breadth of the peaks, influenced by molecular environment and crystallinity	– Narrow peaks: Crystalline or ordered structures – Broad peaks: Amorphous or disordered regions
Peak shape	The symmetry and profile of the peaks, providing insights into molecular interactions	– Symmetrical peaks: Isolated vibrations – Asymmetrical peaks: Overlapping vibrations or complex interactions
Functional group identification	Assigning peaks to specific functional groups based on known Raman shifts	– C=O stretch: ~1700–1750 cm^{-1} – O–H stretch: ~3200–3600 cm^{-1} – N–H stretch: ~3300–3500 cm^{-1}
Molecular symmetry	Analyzing peak patterns to determine molecular symmetry and structure	– Highly symmetric molecules (e.g., benzene) show fewer but intense peaks – Asymmetric molecules show more complex spectra
Crystallinity and phase analysis	Using peak positions and widths to assess crystallinity and phase transitions	– Sharp peaks: Crystalline phase – Broad peaks: Amorphous phase – Peak shifts: Phase transitions (e.g., melting, crystallization)
Molecular interactions	Detecting changes in peak positions or intensities due to intermolecular interactions	– Hydrogen bonding: Shifts in O–H or N–H peaks – π-π stacking: Shifts in aromatic ring vibrations
Quantitative analysis	Using peak intensities or areas to quantify the concentration of specific components	– Calibration curves are often required for accurate quantification – Internal standards may be used for normalization

(continued)

Table 8.3 (continued)

Aspect of interpretation	Description	Examples/notes
Multivariate analysis	Applying statistical methods to analyze complex spectra with overlapping peaks	– Principal Component Analysis (PCA): Identifies patterns and reduces data dimensionality – Partial Least Squares (PLS): Correlates spectral data with sample properties
Baseline correction	Removing background signals to improve peak visibility and accuracy	– Fluorescence background is common in polymers and can obscure Raman peaks – Software tools are used for baseline subtraction
Peak deconvolution	Separating overlapping peaks into individual components for detailed analysis	– Gaussian or Lorentzian fitting is used to model peak shapes – Helps resolve complex spectra with overlapping vibrations
Spectral mapping	Creating 2D or 3D maps of Raman spectra to study spatial distribution of components	– Used in micro-Raman spectroscopy to analyze heterogeneous samples (e.g., polymer blends, composites)
Temperature and pressure effects	Analyzing changes in spectra due to variations in temperature or pressure	– Peak shifts: Indicate changes in molecular vibrations – Peak broadening: Suggests increased molecular motion
Degradation and aging studies	Monitoring changes in spectra over time to study degradation mechanisms	– Appearance of new peaks: Formation of degradation products – Disappearance of peaks: Loss of specific functional groups

8.2.3 *Ultraviolet–Visible (UV–Vis)*

Ultraviolet–visible (UV–Vis) spectroscopy is commonly used for studying the electronic transitions in polymers, providing valuable information about their optical properties, chemical structure, and functional groups. It measures the absorption of ultraviolet (UV) and visible light by a polymer sample, which occurs when electrons in the polymer are excited from lower to higher energy states [16].

8.2.3.1 Basic Principle and Theory

UV–visible spectroscopy is based on the concept of photon absorption by molecules, resulting in electronic transitions among various energy levels. Molecules absorb specific wavelengths of UV (190–400 nm) or visible light (400–800 nm), corresponding to the energy differences between their ground state and excited electronic states, thereby promoting electrons from bonding (π, σ) or non-bonding (n) orbitals

Table 8.4 Table summarizing the Raman spectra of several well-known polymers, including their characteristic Raman shifts (cm^{-1}), peak assignments, and interpretations

Polymer	Raman shift (cm^{-1})	Peak assignment	Interpretation
Polyethylene (PE)	1060, 1130, 1295, 1440, 2880	C–C stretching, CH_2 bending, CH_2 stretching	– 1060, 1130 cm^{-1}: C–C backbone vibrations – 1295 cm^{-1}: CH_2 twisting – 1440 cm^{-1}: CH_2 bending – 2880 cm^{-1}: CH_2 symmetric and asymmetric stretching
Polypropylene (PP)	810, 840, 970, 1160, 1450, 2880	CH_3 rocking, C–C stretching, CH_3 bending, CH stretching	– 810, 840 cm^{-1}: CH_3 rocking – 970 cm^{-1}: C–C backbone vibrations – 1160 cm^{-1}: CH_3 bending – 1450 cm^{-1}: CH_3 symmetric bending – 2880 cm^{-1}: CH stretching
Polystyrene (PS)	620, 1000, 1030, 1600, 3050	Phenyl ring vibrations, C–H bending, C=C stretching	– 620, 1000, 1030 cm^{-1}: Phenyl ring vibrations – 1600 cm^{-1}: C=C stretching in the aromatic ring – 3050 cm^{-1}: C–H stretching in the aromatic ring
Polyethylene Terephthalate (PET)	860, 1090, 1280, 1615, 1725, 3070	C–C stretching, C=O stretching, phenyl ring vibrations	– 860 cm^{-1}: C–C stretching – 1090 cm^{-1}: C–O stretching – 1280 cm^{-1}: C–C stretching – 1615 cm^{-1}: Phenyl ring vibrations – 1725 cm^{-1}: C=O stretching (ester group) – 3070 cm^{-1}: C–H stretching in the aromatic ring

(continued)

Table 8.4 (continued)

Polymer	Raman shift (cm^{-1})	Peak assignment	Interpretation
Polyvinyl Chloride (PVC)	635, 695, 960, 1100, 1435, 2900	C–Cl stretching, CH_2 bending, CH stretching	– 635, 695 cm^{-1}: C–Cl stretching – 960 cm^{-1}: CH_2 rocking – 1100 cm^{-1}: C–C stretching – 1435 cm^{-1}: CH_2 bending – 2900 cm^{-1}: CH stretching
Polymethyl Methacrylate (PMMA)	600, 810, 1450, 1725, 2950	C=O stretching, CH_3 bending, C–C stretching	– 600 cm^{-1}: C–C–O bending – 810 cm^{-1}: CH_2 rocking – 1450 cm^{-1}: CH_3 bending – 1725 cm^{-1}: C=O stretching (ester group) – 2950 cm^{-1}: CH_3 stretching
Polycarbonate (PC)	700, 880, 1110, 1600, 1775, 3050	Phenyl ring vibrations, C=O stretching, C–H stretching	– 700, 880 cm^{-1}: Phenyl ring vibrations – 1110 cm^{-1}: C–O–C stretching – 1600 cm^{-1}: Phenyl ring vibrations – 1775 cm^{-1}: C=O stretching (carbonate group) – 3050 cm^{-1}: C–H stretching in the aromatic ring
Nylon 6 (Polyamide)	930, 1080, 1300, 1440, 1640, 2900	C–C stretching, C=O stretching, CH_2 bending	– 930 cm^{-1}: C–C stretching – 1080 cm^{-1}: C-N stretching – 1300 cm^{-1}: CH_2 twisting – 1440 cm^{-1}: CH_2 bending – 1640 cm^{-1}: C=O stretching (amide I) – 2900 cm^{-1}: CH_2 stretching

(continued)

Table 8.4 (continued)

Polymer	Raman shift (cm^{-1})	Peak assignment	Interpretation
Polytetrafluoroethylene (PTFE)	290, 385, 730, 1380, 2900	CF_2 stretching, CF_2 bending	– 290, 385 cm^{-1}: CF_2 bending – 730 cm^{-1}: CF_2 symmetric stretching – 1380 cm^{-1}: CF_2 asymmetric stretching – 2900 cm^{-1}: CH stretching (if impurities are present)
Polybutadiene (PB)	1000, 1300, 1440, 1650, 2900	C=C stretching, CH_2 bending, CH stretching	– 1000 cm^{-1}: C–C stretching – 1300 cm^{-1}: CH_2 twisting – 1440 cm^{-1}: CH_2 bending – 1650 cm^{-1}: C=C stretching (double bond) – 2900 cm^{-1}: CH stretching

to higher-energy antibonding orbitals (π^*, σ^*). The transitions comprising $\pi \rightarrow \pi^*$ (typical of conjugated systems), $n \rightarrow \pi^*$ (pertaining to lone pair electrons), and $\sigma \rightarrow \sigma^*$ (associated with saturated compounds) [17]. This transition is quantified by the Beer-Lambert Law "At a given wavelength, absorption is proportional to the concentration of absorbing molecules and the path length of the light through the sample":

$$A = \varepsilon c l \tag{i}$$

wherein absorbance (A) is contingent upon molar absorptivity (ε), concentration (c), and path length (l). The molar absorptivity, or extinction coefficient, is unique to the compound at a specific wavelength. Molar absorptivity can be significantly high for strongly absorbing compounds, exceeding 10,000, while it may be relatively low for weak absorbers, ranging from 10 to 100. Absence of absorption results in $\varepsilon = 0$.

A chromophore is the molecular component responsible for the absorption of ultraviolet or visible light. The addition of substituent groups to a fundamental chromophore structure alters both the position and intensity of its absorption band. Substituents that enhance the intensity of absorption and shift the wavelength are referred to as autochromes. These substituents have different effects on absorption intensity spectra (Fig. 8.8) [18]:

(a) Bathochromic shift (red shift) a shift to longer lower energy.

Fig. 8.8 Changes in atomic spectra

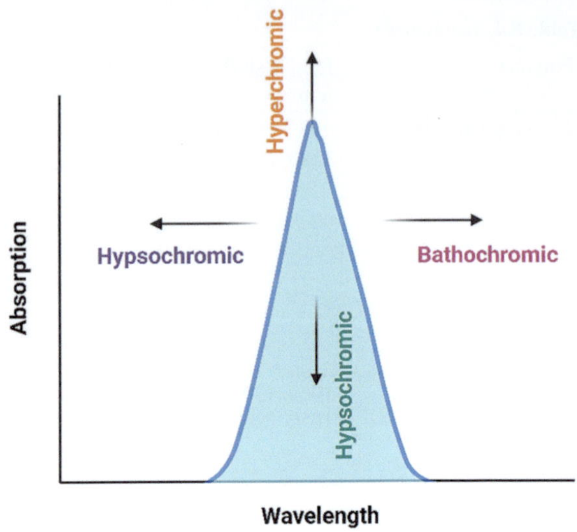

(b) Hypsochromic shift (blue shift)—shift to shorter higher energy.
(c) Hyperchromic effect—an increase in intensity.
(d) Hypochromic effect—a decrease in intensity.

8.2.3.2 Instrumentation

The light source generates a wide range of visible and ultraviolet light; it usually uses a tungsten-halogen lamp (350–2500 nm) for visible light, a deuterium lamp (160–375 nm) for UV light, or a xenon lamp (200–1000 nm) for a combined range. The monochromator then uses an exit slit to choose the desired wavelength, a diffraction grating or prism to scatter light, and an entrance slit to isolate particular wavelengths. This ensures only the intended wavelength reaches the sample, placed in cuvette. Once the light has passed through the sample, it is measured by a detector, which could be a charge-coupled device (CCD) for higher resolution, a photomultiplier tube (PMT) for high sensitivity, or a photodiode array (PDA) for quick multi-wavelength detection. The signal processor uses the Beer-Lambert to determine absorbance after the detector transforms light intensity into an electrical signal. The absorption spectrum (absorbance vs. wavelength) is then shown for analysis on a display unit. In double-beam UV–Vis instruments, a reference beam compensates for light source variation by splitting the beam into sample and reference paths which improves the accuracy of the system. UV–visible spectrometers are available in two configurations: single-beam systems, which require manual baseline correction, and double-beam systems, which provide more stability by measuring the sample and reference beams at the same time. This technique allows for a variety of applications, including material characterization, kinetic studies, and quantitative analysis, by examining how molecules absorb light at particular wavelengths [19].

8.2.3.3 UV–Visible Spectroscopy for Polymer Characterization

In polymers chromophore groups are present that absorb ultraviolet or visible light. UV–visible spectroscopy serves as an essential analytical technique for investigating these chromophores, offering valuable information about electronic transitions, molecular structure, and functional group interactions within polymeric systems. Among the most significant chromophores are conjugated π-systems found in polymers like polyacetylene and polythiophene, where extended conjugation leads to absorption at longer wavelengths. Aromatic structures, prevalent in polymers such as polystyrene and polyethylene terephthalate, exhibit characteristic absorption around 250–280 nm due to $\pi \rightarrow \pi^*$ transitions of their benzene rings. Carbonyl functionalities present in polyesters and acrylics demonstrate distinct $n \rightarrow \pi^*$ and $\pi \rightarrow \pi^*$ transitions in the UV region, while isolated and conjugated double bonds in rubbers and diene polymers show progressively red-shifted absorption. Heteroatom-containing groups in nylons and polyurethanes exhibit $n \rightarrow \pi^*$ transitions, and charge-transfer complexes in doped conductive polymers often produce visible-range absorption. UV–visible spectroscopy provides deep insights in investigating polymer degradation, tracking reactions, characterizing conjugated systems (Table 8.5) [20, 21].

8.2.3.4 Woodward-Fieser Rules: An Empirical Tool for Predicting UV–Visible Absorption Maxima

The Woodward-Fieser rules used to predict the λ_{max} (maximum wavelength of absorption) of conjugated organic compounds, particularly dienes, enones, α, β-unsaturated carbonyl compounds, and aromatic systems. The rules were initially proposed by Robert Burns Woodward in the 1940s for calculating λ_{max} values in conjugated dienes. They were later extended and refined by Louis Fieser, who incorporated additional structural features that influence the absorption. Thus, the combined system is known as the Woodward-Fieser Rules. This rule is based on the ideas that certain chromophores (light-absorbing groups) and auxochromes (substituents that alter light absorption) contribute incrementally to the λ_{max} of a molecule [22]. The total λ_{max} is determined by beginning with a base value for the core chromophore and subsequently adding increments for each substituent or structural modification. The Woodward-Fieser Rules are based on the observation that:

(a) Increased conjugation shifts λ_{max} to longer wavelengths (bathochromic shift).
(b) Substituents and ring constraints influence the energy gap between HOMO and LUMO.
(c) Empirical increments allow for the estimation of λ_{max} based on structural features.

Major limitations of this rule include:

(a) Solvent Effects: The rules assume ethanol as solvent; polar solvents can cause shifts.

Table 8.5 Table summarizing the peak positions of important chromophores in UV–visible spectroscopy

Chromophore	Example	Transition type	Peak position (nm)	Notes
Isolated C=C	Ethylene ($CH_2 = CH_2$)	$\pi \rightarrow \pi^*$	170–200	Absorbs in the far UV region
Conjugated C=C	Polyacetylene ($(-CH = CH-)_n$)	$\pi \rightarrow \pi^*$	200–400	Longer conjugation shifts absorption to longer wavelengths
Aromatic Rings	Benzene (C_6H_6)	$\pi \rightarrow \pi^*$	250–280	Strong absorption due to aromatic π-system
Carbonyl Group (C=O)	Acetone (CH_3COCH_3)	$n \rightarrow \pi^*$	270–300	Weak absorption; $\pi \rightarrow \pi^*$ transition around 180–200 nm
Nitro Group (NO_2)	Nitrobenzene ($C_6H_5NO_2$)	$n \rightarrow \pi^*$, $\pi \rightarrow \pi^*$	260–280	Strong absorption due to electron-withdrawing nature
Azo Group (N=N)	Azobenzene ($C_6H_5N = NC_6H_5$)	$\pi \rightarrow \pi^*$	400–500	Intense absorption in the visible region; used in dyes
Conjugated Polymers	Polypyrrole, Polyaniline	$\pi \rightarrow \pi^*$	300–800	Absorption depends on the extent of conjugation and doping
Peptide Bonds	Proteins (–CONH–)	$n \rightarrow \pi^*$, $\pi \rightarrow \pi^*$	210–220	Peptide bond absorption in the far UV region
Aromatic Amino Acids	Tryptophan, Tyrosine	$\pi \rightarrow \pi^*$	280	Aromatic side chains in proteins absorb at 280 nm
Metal Complexes	Hemoglobin (Fe^{2+} complex)	d-d, charge-transfer	400–600	Absorption in the visible region gives distinct colors (e.g., red for Fe^{2+})
Polyenes	Carotenoids (e.g., lycopene)	$\pi \rightarrow \pi^*$	400–500	Long conjugated systems absorb in the visible region
Charge-Transfer Complexes	Iodine-doped polyacetylene	Charge-transfer	400–700	Absorption depends on the donor–acceptor interaction

(b) Steric Hindrance: Non-planar structures due to steric crowding may deviate from predictions.

8.2.4 *Mass Spectrometry (MS)*

Mass spectrometry, a potent analytical tool, can determine the molecular weight, chemical structure, and composition of polymers. It provides detailed information about the distribution of molecular weights, the types of end groups, the repeating units, and how the molecules break apart by turning polymer molecules into ions and sorting them based on their mass-to-charge ratio (m/z). Mass spectrometry provides sensitivity, accuracy, and the capacity to examine a broad range of molecular weights; thus, it is especially useful for characterizing complicated polymer systems, including synthetic polymers, biopolymers, and polymer blends. In polymer engineering, mass spectrometry techniques like Matrix-Assisted Laser Desorption/Ionization (MALDI) and Electrospray Ionization (ESI) are used to scrutinize polymerization processes, identify contaminants and additives. Further improving mass spectrometry's capacity is coupling it with separation methods such as liquid chromatography (LC) or size exclusion chromatography (SEC), which allows understanding of complicated polymer composites [23].

8.2.4.1 Basic Principle and Theory

Mass spectrometry deals with the examination of characteristics fragments and arising from the breakdown of molecules, through three stages, i.e., ionization, mass separation, and detection. The spectrum is plotted as mass-to-charge ratio (m/z) of ionized molecules. Neutral molecules are first turned into ions using methods like electron ionization (EI), chemical ionization (CI), Matrix-Assisted Laser Desorption/ Ionization (MALDI), or electrospray ionization (ESI), which changes them from neutral to charged particles in the gas phase (Fig. 8.9). Mass analyzers such as time-of-flight (TOF), quadrupole, ion trap, magnetic sector, or Orbitrap systems each providing unique resolution and accuracy and separates these ions depending on their m/z values. This spectrum plots ion intensity against m/z, provides information including molecular weight (via molecular ion peaks), structural details (through fragmentation patterns), and isotopic distributions (Fig. 8.10) [24].

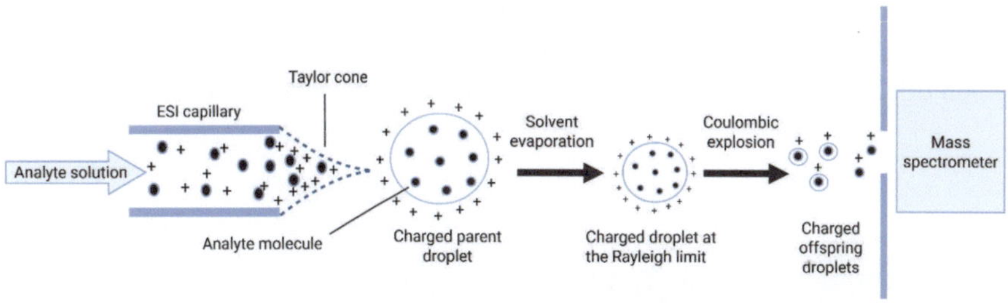

Fig. 8.9 Electrospray ionization technique

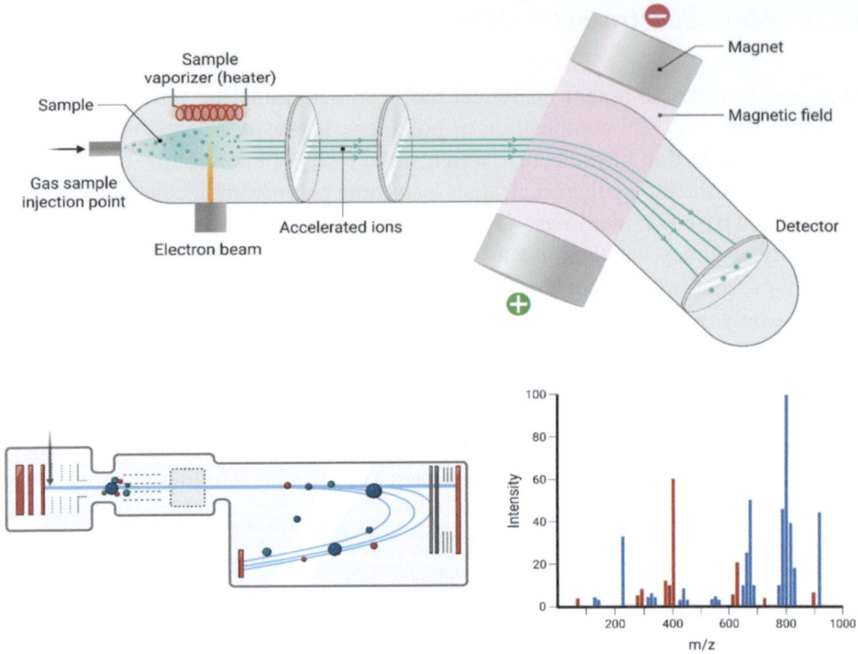

Fig. 8.10 Instrumentation of mass spectroscopy, time of flight analyzer, and mass spectra

8.2.4.2 Applications of Mass Spectroscopy in Polymer Characterization

Mass spectrometry (MS) techniques such as Matrix-Assisted Laser Desorption/Ionization-Time-of-Flight-Mass Spectrometry (MALDI-TOF–MS) and electrospray ionization-mass spectrometry (ESI–MS) are used in polymer analysis. These methods are especially helpful for determination of the molecular weight of polymers like polyethylene glycol (PEG) and polystyrene, both number-average (Mn) and weight-average (Mw) molecular weights also be determined. Additionally, MALDI-TOF MS and ESI–MS are utilized in end-group analysis to identify functional groups at polymer chain ends, as seen in polymers like polycaprolactone and polyurethane. Tandem MS (MS/MS) and ESI–MS are powerful tools for analyzing the sequence in copolymers or block copolymers, such as poly(lactic-co-glycolic acid) (PLGA), which offers comprehensive knowledge of polymer architecture [25, 26].

For the detection of additives, such as plasticizers, stabilizers, and contaminants in polymers like plasticized PVC and polypropylene, Gas Chromatography-Mass Spectrometry (GC–MS) and Liquid Chromatography-Mass Spectrometry (LC–MS) are commonly used [27–29]. Pyrolysis-GC–MS and ESI–MS are effective in identifying by-products generated during the thermal or oxidative degradation of polymers like nylon and polyethylene terephthalate (PET). Moreover, MALDI-TOF MS and ESI–MS facilitate the determination of monomer ratio and distribution in copolymers such as styrene-butadiene rubber (SBR) and ethylene–vinyl acetate (EVA) [30].

Common Mass Spectroscopy Techniques Used in Polymer Characterization Are as Follows

(a) MALDI-TOF MS (large, non-volatile polymers).
(b) ESI–MS: Electrospray Ionization MS (polar and ionic polymers).
(c) MALDI-TOF MS and ESI–MS (end group analysis).
(d) Pyrolysis-GC–MS (thermal decomposition studies of complex polymers).
(e) TOF–MS, Q-TOF MS, and Orbitrap (High-resolution advanced characterization).

8.2.5 Nuclear Magnetic Resonance (NMR) Spectroscopy

NMR spectroscopy analyzes molecular structures by studying nuclear responses to an oscillating radio-frequency electromagnetic field in the presence of a strong external magnetic field. Nuclei are components of atoms that create molecules, allowing NMR to clarify intricate aspects of the structure and dynamic behavior of chemical compounds.

8.2.5.1 Basic Principle and Theory

Nuclei are positively charged and spin on an axis, creating a tiny magnetic field. However, all nuclei are not suitable for NMR spectroscopy. ^{1}H and ^{13}C are the most important NMR-active nuclei. The natural abundance of ^{1}H is 99.9%, while ^{13}C has an abundance of 1.1%. In contrast, ^{12}C, which makes up 98.9%, is not NMR-active. Without an external magnetic field, the nuclear magnetic moments of nuclei exhibit random orientation. In the presence of an external magnetic field (B_0), nuclear magnetic moments may align parallel or antiparallel to the field. The energy difference between these two orientations is typically minimal and is depends upon the strength of B_0. When radiofrequency (RF) radiation of the high energy is applied, it causes the nuclear spin to flip. This condition is called resonance, and the nuclei are said to be in resonance with the external magnetic field. Whether a nucleus is NMR-active depends on its nuclear spin quantum number (I), which is related to its atomic mass and atomic number. The number of spin states is determined by $2I + 1$, indicates the number of possible energy levels.

Various nuclei absorb electromagnetic radiation at distinct wavelengths, corresponding to the energy necessary to achieve resonance. Nuclei of a specific type resonate at varying energies based on their chemical and electronic environments. The NMR signal's position, referred to as the chemical shift (δ), along with its pattern of splitting or multiplicity, shows information regarding the chemical environment of the nuclei (Fig. 8.11, Table 8.6). The chemical shift represents the specific magnetic field strength (expressed in parts per million, ppm) at which a nucleus comes into resonance, relative to a reference standard, typically tetramethyl silane (TMS). Electron clouds around nuclei shield them from the full strength of the external magnetic field, causing the nuclei to absorb EM radiation at slightly higher energies. This effect is known as shielding. Shielding is influenced by neighboring functional groups, which alter the electron density around the nucleus, thereby affecting its chemical shift. In proton, NMR spectroscopy spin–spin coupling (splitting) occurs when protons on adjacent carbon interacts and split into smaller peaks and forms a multiplets (n + 1 rule) following Pascal triangle. The n + 1 rule states that one proton on an adjacent carbon will result in a doublet, while two protons on adjacent carbons will produce a triplet with a 1:1 intensity ratio, and this pattern continues accordingly. The resonance of a proton interacting with "n" equivalent protons on the adjacent carbon will exhibit "splitting" into n + 1 peaks, characterized by a coupling constant J. The coupling constant refers to the interval between peaks in a split pattern, measured in Hertz (Hz) [31].

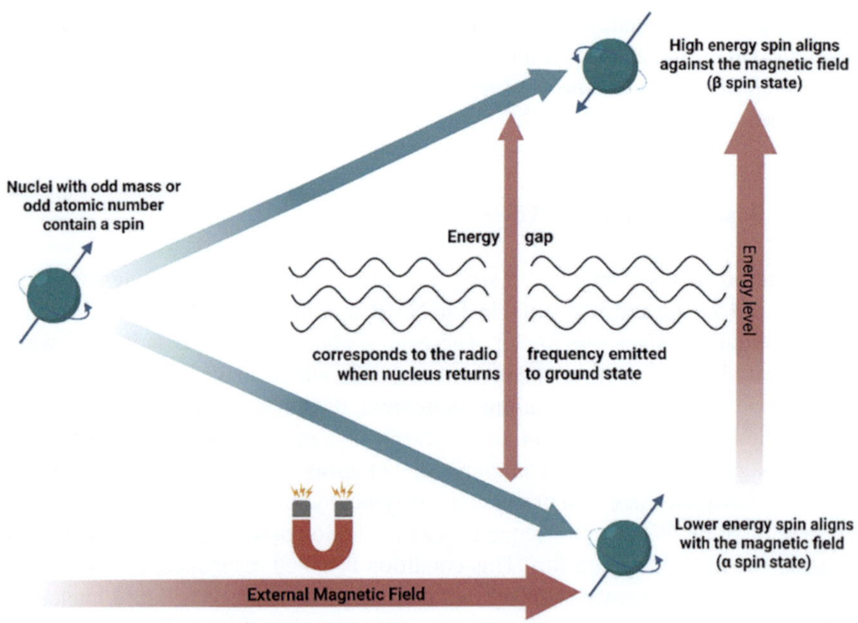

Fig. 8.11 Basic principle of NMR spectroscopy

Table 8.6 Relationship between the number of adjacent equivalent protons (n) and multiplicity in proton NMR spectra. According to the (n + 1) rule, a proton coupled to n equivalent neighboring protons shows a signal split into (n + 1) peaks with relative intensities following Pascal's triangle

Number of adjacent equivalent protons (n)	Multiplicity	Number of peaks	Relative intensities
0	Singlet (s)	1	–
1	Doublet (d)	2	1:1
2	Triplet (t)	3	1:2:1
3	Quartet (q)	4	1:3:3:1
4	Quintet (quint)	5	1:4:6:4:1

The choice of solvent in NMR spectroscopy is critical for polymer characterization, as it must fully dissolve the sample while minimizing signal interference. Deuterated solvents like chloroform-d ($CDCl_3$) and dimethyl sulfoxide-d_6 (DMSO-d_6) are commonly used for organic-soluble and polar polymers, respectively, with their residual proton peaks (7.26 ppm for $CDCl_3$, 2.50 ppm for DMSO-d_6) serving as convenient internal references. For aqueous polymers, deuterated water (D_2O) is preferred, though the residual HOD peak at 4.79 ppm must be accounted for. Chemical shifts are typically calibrated against tetramethylsilane (TMS, 0 ppm) for organic solvents or sodium 3-(trimethylsilyl)propionate-d_4 (TSP) for aqueous solutions. In solid-state NMR, where solvents are unnecessary, magic-angle spinning (MAS) is employed to resolve rigid polymer structures (Fig. 8.12) [32].

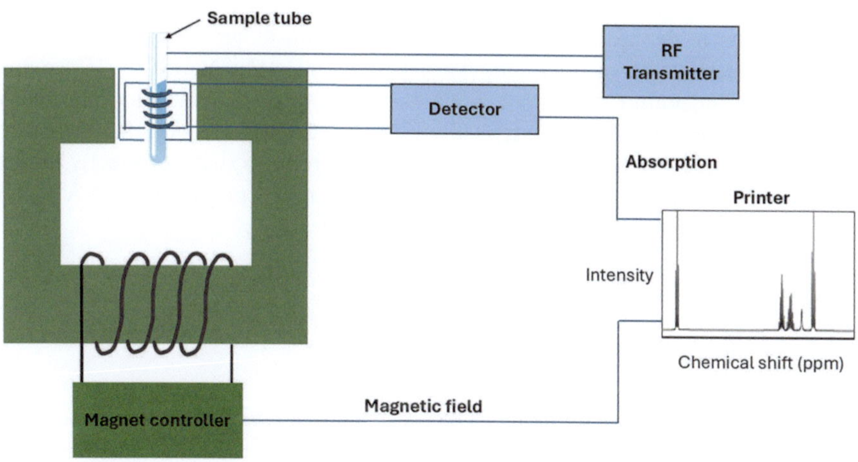

Fig. 8.12 Instrumentation of NMR Spectroscopy

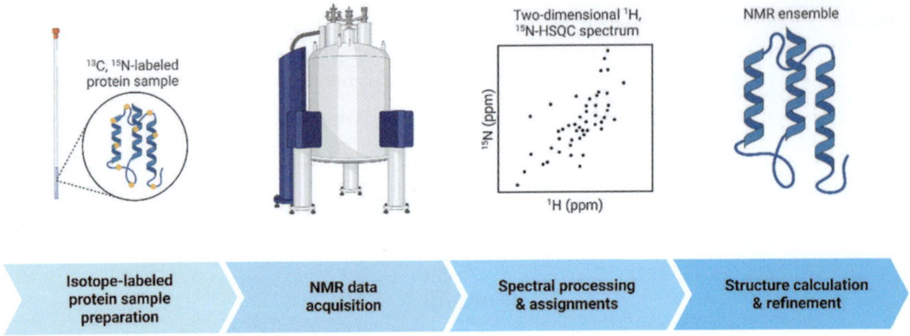

Fig. 8.13 Procedure of structural insights of protein samples using NMR spectroscopy

8.2.5.2 NMR in Polymer Characterization

NMR techniques in polymer are used for the structural and dynamic characterization to get detailed information about composition, microstructure, tacticity, branching, end-group functionality, and molecular interactions in polymer. High-resolution solution-state NMR is commonly employed for soluble polymers, while solid-state NMR techniques, such as cross-polarization magic-angle spinning (CP/MAS), are essential for insoluble or cross-linked polymer systems. With advancements in multi-dimensional NMR (e.g., COSY, HSQC, DOSY) and high-field instruments, NMR has become indispensable for polymer scientists, facilitating the design of advanced materials with tailored properties for applications in coatings, biomaterials, plastics, and composites (Fig. 8.13) [33].

In addition to the aforementioned techniques, multi-dimensional NMR is used in the characterization of polymers. This includes 2D-^{1}H-^{15}N HSQC (Heteronuclear Single Quantum Correlation), a 2D experiment that correlates a proton with its directly bound heteronuclei, typically ^{15}N or ^{13}C. Furthermore, the 3D^{1}H^{15}N NOESY-HSQC (Protein NMR) correlates an amide proton (x) with its directly bound nitrogen (y), with the third dimension (z) correlating to any proton within 5 Å of the amide proton (Table 8.7) [34].

8.2.6 X-Ray Photoelectron Spectroscopy

X-Ray Photoelectron Spectroscopy (XPS) is an analytical technique that focuses on the surface characterization of polymeric materials. The process involves bombarding a material with X-rays, usually from sources like Al Kα or Mg Kα. This interaction causes photoelectrons to be ejected from the surface. Through the analysis of the kinetic energy of emitted electrons, XPS can determine the binding energies of electrons in different elements, which gives information about the elemental composition

Table 8.7 NMR peaks of some common chemical groups in polymer

Functional group	Polymer example	^{1}H NMR (δ ppm)	^{13}C NMR (δ ppm)	Remarks
$-CH_3$ (Methyl)	Polypropylene	0.8–1.2	10–25	Backbone or side chain
$-CH_2-$ (Methylene)	Polyethylene, polystyrene	1.2–1.6	20–40	Chain methylene groups
–CH– (Methine)	Polystyrene, PMMA	1.4–2.5	30–55	May vary with environment
Aromatic ring	Polystyrene	6.3–7.5	125–145	Multiplets in ^{1}H NMR
$O-CH_3$ (Methoxy)	PMMA	3.3–3.8	50–60	Common in esters
$O-CH_2-$ (Ether linkage)	Poly(ethylene glycol)	3.3–4.0	60–75	PEG, PEO, etc.
C=O (Carbonyl, ester)	PMMA, polyesters	–	165–180	Only in ^{13}C NMR
C≡N (Nitrile)	Polyacrylonitrile	–	110–120	Sharp peak in ^{13}C NMR
–COOH (Carboxylic acid)	Poly(acrylic acid)	10–12 (broad)	170–185	Broad singlet in ^{1}H
–OH (Hydroxyl)	PVA, polyols	1–5 (broad, exchangeable)	~60 (for –C–OH carbon)	^{1}H peak varies with H-bonding
Amide (–CONH–)	Nylon, polyacrylamide	7–9 (NH), 2–4 (CH)	160–180 (C=O), 35–55 (CH)	^{1}H NH often broad
CH=CH (Alkene)	Polybutadiene, natural rubber	4.5–6.5	110–140	Cis/trans splitting
CH≡CH (Alkyne)	Polydiacetylenes	1.8–3.0	70–90	Less common

and chemical states present within 1–10 nm of a polymer surface. From XPS, quantification of the atomic percentages of elements like carbon, oxygen, nitrogen, and fluorine on the polymer surface can be determined. This is useful for identifying additives, fillers in the materials.

It can differentiate between different chemical bonds in a polymer, such as C–C, C–O, C=O, and O–C=O. This allows researchers to investigate the different types of functional groups present and track the chemical changes that occur due to processing or exposure to the environment. Additionally, XPS is instrumental in identifying surface contamination. During the processing or handling of polymers, various impurities like residual solvents, lubricants, or adventitious carbon can accumulate on the surface. XPS can detect these contaminants, which is crucial for applications that require high purity. Another significant application of XPS is in the analysis of thin films and surface coatings. By employing techniques like angle-resolved XPS (ARXPS), researchers can explore compositional gradients, measure the thickness of ultra-thin films, and assess the effectiveness of surface treatments such as plasma modification or chemical grafting.

8.3 Conclusions

Spectroscopic techniques are indispensable tools in the comprehensive characterization of polymers, offering insights into their molecular structure, composition, and functional properties. This chapter has outlined the fundamental principles and applications of key spectroscopic methods—NMR, MS, Raman, FTIR, UV–Vis, and XPS spectroscopy highlighting their role in monitoring polymer synthesis and structural analysis. A clear conceptual understanding of these techniques enables researchers to make informed choices when analyzing complex polymer systems. As polymer science continues to evolve, the integration of multiple spectroscopic approaches will remain essential for advancing material design, quality control, and innovation in polymer-based technologies.

Acknowledgements We have used bioRender.com (student license) for making illustration.

Conflict of Interest Authors declare no competing conflict of interest.

References

1. Yousif ME (2014) The electromagnetic radiation mechanism. Int J Fund Phys Sc 4:72–79. https://doi.org/10.14331/IJFPS.2014.330068
2. Svanberg S, Svanberg S (2001) Atomic and molecular spectroscopy
3. Hertel I V., Schulz C-P (2015) Molecular spectroscopy, 289–381. https://doi.org/10.1007/978-3-642-54313-5_5
4. Danielsen SPO, Beech HK, Wang S et al (2021) Molecular Characterization of polymer networks. Chem Rev 121:5042–5092. https://doi.org/10.1021/ACS.CHEMREV.0C01304/ASSET/IMAGES/LARGE/CR0C01304_0028.JPEG
5. Sivaji C, … SC-J on A (2022) Undefined a review on spectroscopy and its classification. researchgate.net 1:2583–7125. https://doi.org/10.46632/jacp/1/1/5
6. Berthomieu C, Hienerwadel R (2009) Fourier transform infrared (FTIR) spectroscopy. Photosynth Res 101:157–170. https://doi.org/10.1007/S11120-009-9439-X/FIGURES/4
7. Rosenberg Petersen C, Møller U, Kubat I et al (2014) Mid-infrared supercontinuum covering the 1.4–13.3 μm molecular fingerprint region using ultra-high NA chalcogenide step-index fibre. nature.com. https://doi.org/10.1038/NPHOTON.2014.213
8. Hipp M, Woisetschläger J, Reiterer P et al, Digital evaluation of interferograms. Elsevier
9. Khan SA, Khan SB, Khan LU et al (2018) Fourier transform infrared spectroscopy: fundamentals and application in functional groups and nanomaterials characterization. Springer 317–344. https://doi.org/10.1007/978-3-319-92955-2_9
10. Prochazka M, Spectroscopy MP-S-ER 2016 undefined (2016) Basics of Raman scattering (RS) spectroscopy, Springer 7–19. https://doi.org/10.1007/978-3-319-23992-7_2
11. Cordero E (2018) In-vivo Raman spectroscopy: from basics to applications. J Biomed Opt 23:1. https://doi.org/10.1117/1.JBO.23.7.071210

12. Dubessy J, Caumon MC, Rull F, Sharma S (2012) Instrumentation in Raman spectroscopy: elementary theory and practice. Earth Sci Cultural Heritage, 14–16
13. Ewen S, Geoffrey D (2019) Modern Raman spectroscopy : a practical approach
14. Gerrard D, Reviews WM-AS (1986) Undefined (1986) Polymer characterization by Raman spectroscopy. Taylor & Francis 22:251–334. https://doi.org/10.1080/05704928608070179
15. Edwards HGM, Johnson AF, Lewis IR (1993) Applications of Raman spectroscopy to the study of polymers and polymerization processes. Wiley Online Library 24:475–483. https://doi.org/10.1002/JRS.1250240803
16. Picollo M, Aceto M, Vitorino T (2019) UV-Vis spectroscopy. Phys Sci Rev, 4. https://doi.org/10.1515/PSR-2018-0008/MACHINEREADABLECITATION/RIS
17. Akash M, Rehman K, analysis MA-… of pharmaceutical, 2020 undefined (2020) Ultraviolet-visible (UV-VIS) spectroscopy. Springer, 29–56. https://doi.org/10.1007/978-981-15-1547-7_3
18. Liu X, Xu Z, Cole JM (2013) Vis absorption and emission properties in organic fluorophores: toward larger bathochromic shifts, enhanced molar extinction coefficients, and greater stokes shifts. ACS Publ 117:16584–16595. https://doi.org/10.1021/JP404170W
19. Das A (2021) Portable UV–visible spectroscopy—instrumentation, technology, and applications. Portable spectroscopy and spectrometry, 179–207. https://doi.org/10.1002/9781119636489.CH8
20. Blends MA-C of P, 2014 undefined (2015) Characterization of polymer blends using UV-visible spectroscopy. Wiley Online Library 9783527331536:789–820. https://doi.org/10.1002/9783527645602.CH24
21. Probe GM-MC and O (1997) Undefined UV-vis spectroscopy as an analytical tool for the characterization of polymers. spiedigitallibrary.org. https://doi.org/10.1117/12.279857.SHORT
22. Simons JW, Hase WL, Phillips RJ et al (1978) Conformational analysis. 128. The Woodward-Fieser rules and. alpha., beta.-unsaturated ketones. ACS Publications 100:1068–1073. https://doi.org/10.1021/JA00472A008
23. Duckworth H, Barber R, Venkatasubramanian V (1986) Mass spectroscopy
24. Rajawat J, Jhingan G (2019) Mass spectroscopy. Data Processing handbook for complex biological data sources, 1–20. https://doi.org/10.1016/B978-0-12-816548-5.00001-0
25. Peacock PM, McEwen CN (2004) Mass spectrometry of synthetic polymers. ACS Publications 76:3417–3428. https://doi.org/10.1021/AC040064I
26. Weidner SM, Trimpin S (2010) Mass spectrometry of synthetic polymers. ACS Publications 82:4811–4829. https://doi.org/10.1021/AC101080N
27. Aminlashgari N, Polymers–New MH-MS of, 2011 undefined (2012) Emerging mass spectrometric tools for analysis of polymers and polymer additives. Springer 248:1–37. https://doi.org/10.1007/12_2011_152
28. Chemistry JS-CA, 2008 undefined Additive Analysis. Elsevier
29. Akoueson F, Chbib C, Monchy S, et al Identification and quantification of plastic additives using pyrolysis-GC/MS: A review. Elsevier
30. Smith PB, Pasztor AJ, Mckelvy ML et al (1997) Analysis of synthetic polymers and rubbers. ACS Publ 69. https://doi.org/10.1021/A19700020
31. Wong KC (2014) Review of NMR spectroscopy: basic principles, concepts and applications in chemistry. J Chem Educ 91:1103–1104. https://doi.org/10.1021/ED500324W
32. Yang R (2018) Analytical methods for polymer characterization. Analytical methods for polymer characterization. https://doi.org/10.1201/9781351213158
33. Characterization HC-MM of P, 1991 undefined NMR characterization of polymers. books.google.com
34. Izunobi JU, Higginbotham CL (2011) Polymer molecular weight analysis by 1H NMR spectroscopy. ACS Publications 88:1098–1104. https://doi.org/10.1021/ED100461V

Homen Dahal holds a strong foundation in chemistry, with research centred on cationic Gemini surfactants and their structure–property relationships. His work encompasses the synthesis and detailed characterization of organic molecules using a range of spectroscopic techniques. Through his studies, he aims to connect molecular design with practical biomedical applications.

Priya Devi holds an academic background in chemistry, with research focused on enhancing the mechanical strength of degraded cellulosic paper through polymer-based reinforcement coatings. Her research employs spectroscopic techniques to characterize lignocellulosic biomass and evaluate the physicochemical effects of polymer coatings on paper substrates.

Nikhil Dhore is pursuing a Ph.D. at the Indian Institute of Technology Kharagpur, focusing on the development of sustainable polymers for decorative and protective applications. Nikhil has completed his bachelor's in chemical technology (2018) and a master's specializing in paint technology (2020). and has hands-on experience of 3+ years in the coating field research. He worked as a project associate at CSIR-Indian Institute of Chemical Technology, Hyderabad, where he gained 3+ years of hands-on experience in developing sustainable polymers for coating and foam applications and their characterization on FTIR, TGA, DSC, UTM, DLS, electrochemical workstations, salt spray test, weathering tests, and goniophotometry. He was a part of the technology transfer of the "tin-free and solvent-free UV-curable polyurethane acrylate" to Sai Life Sciences, Hyderabad. His current research on the synthesis of water-based polyurethane polymer, non-isocyanate polyurethane, and biobased epoxy amine systems, and hydrogel. More than 6+ peer-reviewed international journal articles in his account till 2024.

Manas Jain is currently working as a senior research scholar, in the combustion lab of department of Aerospace Engineering, IIT Kharagpur. His area of study and research interests includes, chemically reacting flows, High speed and pressure gain combustion, Flame dynamics, Deflagration to detonation transition, Spectroscopy, Laser-optics diagnostics techniques and mechanical behaviour analysis of materials.

Nikhil Kumar is a senior research fellow the Advanced Technology Development Centre, Indian Institute of Technology Kharagpur. He is specialized in fabrication of smart polymeric materials for diagnosis and therapeutics. Nikhil completed his Bachelor of Pharmacy and Master of Pharmacy in Pharmacology. His masters research work was focused on developing dressings for diabetic foot ulcers. His current research revolves around the design and application of smart polymeric biomaterials for advanced diagnosis and therapeutic interventions for cutaneous squamous cell carcinoma. He has published 12 peer-reviewed research articles. He is actively engaged in research focused on polymer characterization, with particular interest in the structural, thermal, and rheological analysis of advanced polymeric materials. Mr. Kumar has been involved in research projects that explore the application of spectroscopic techniques (FTIR, NMR), thermal analysis (DSC, TGA), to evaluate polymer performance under varied environmental conditions. Mr. Kumar is also developing a novel diagnostic tool that integrates multiple characterization techniques FTIR, Raman spectroscopy, AFM and nanoindentation to probe and differentiate pathological tissue conditions such as fibrosis and hyperkeratosis.

Chapter 9
Microscopic Techniques and Structural Insights

Komal Agarwal, Nikhil Kumar, and Biswajoy Ghosh

Abstract This chapter will provide a comprehensive overview, investigating into the complex world of polymer structures at a microscopic level. Microscopy techniques are pivotal in elucidating morphology, composition, and behaviour, offering insights crucial for their characterization and application of polymeric structures. From traditional optical to advanced electron microscopy methods, this chapter explores the principles, instrumentation, and applications of various microscopy techniques in the study of polymers, enabling an understanding of structure–property relationships.

Keywords Microscopy · Polymers · Electron and atomic force microscopy · Structure–property relationships

9.1 Introduction

Eyes!! most fascinating part in human body allowing us to see fascinating things–colours, shapes, movements of the beautiful world. Ever wondered how your brain creates images from the light that hits your eyes? or how you perceive color. And then there's the question of what the human eye can actually detect. Sesame seeds, about 1 mm in size? No problem. A strand of hair, even as fine as 20–40 m micrometers? You'd probably spot it instantly—especially if it's lying in corner of the room or sink. A housefly, roughly 8 mm long? Both visible and audible. But what about something smaller, like E. coli bacteria (~2 to 6 μm), or the influenza virus (~100 nm)? Even red blood cells and sperm cells fall well below the threshold of what the naked eye can distinguish. So how do we explore this hidden microscopic world? Simple we turn to magnification. Microscopes and advanced imaging technologies open up a universe far beyond what our eyes alone can see.

K. Agarwal · B. Ghosh (✉)
Department of Physics and Technology, UiT-The Arctic University of Norway, Tromsø, Norway
e-mail: biswajoy.ghosh@uit.no

N. Kumar
Advanced Technology Development Centre, Indian Institute of Technology Kharagpur, Kharagpur, India

S. Chattopadhyay and N. Kumar (eds.), *Polymer Characterization*,
https://doi.org/10.1007/978-981-95-2609-3_9

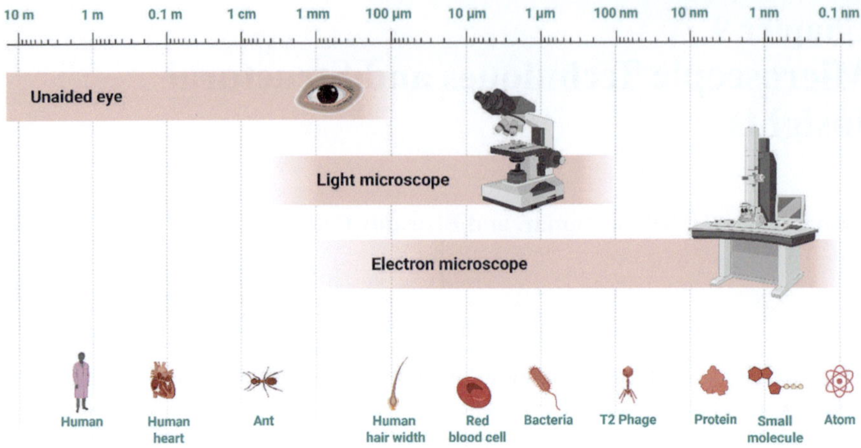

Fig. 9.1 Scale of visibility

This scale of visibility illustrates (Fig. 9.1) the observable size ranges of different biological or physical structures using various tools. The horizontal axis represents the size of objects, ranging from 10 m down to 0.1 nm, moving from left to right as the objects get smaller.

9.1.1 Mechanism of Image Formation in the Human Eye

The human eye works like a camera to form images. Light from an object enters the eye through the transparent outer layer (cornea), then passes through the aqueous humor, pupil (the black centre of the eye controlled by the iris), and the lens. The lens focuses the light onto the retina (thin layer of tissue at the back of the eye). The retina contains special photoreceptor cells that convert light into electrical signals. The image formed on the retina is inverted and reversed. These electrical signals travel through the optic nerve to the brain. At the optic chiasm, the nerve fibres from the inner half of each retina cross over to the opposite side of the brain. As a result, the right visual field is processed by the left visual cortex, and the left visual field is processed by the right visual cortex. The brain combines and decodes these signals to create a clear, upright image Fig. 9.2a.

While this explains how magnification works in the eye. When we look at an object with the naked eye, it makes a small angle at the eye, and the image formed on the retina is also small. But when we look at the same object through a magnifying lens, the object appears larger because the lens makes the object look like it's closer. This causes the object to form a larger angle at the eye, which results in a bigger image on the retina. This is called angular magnification. The more the angle increases, the bigger the object looks to us Fig. 9.2b.

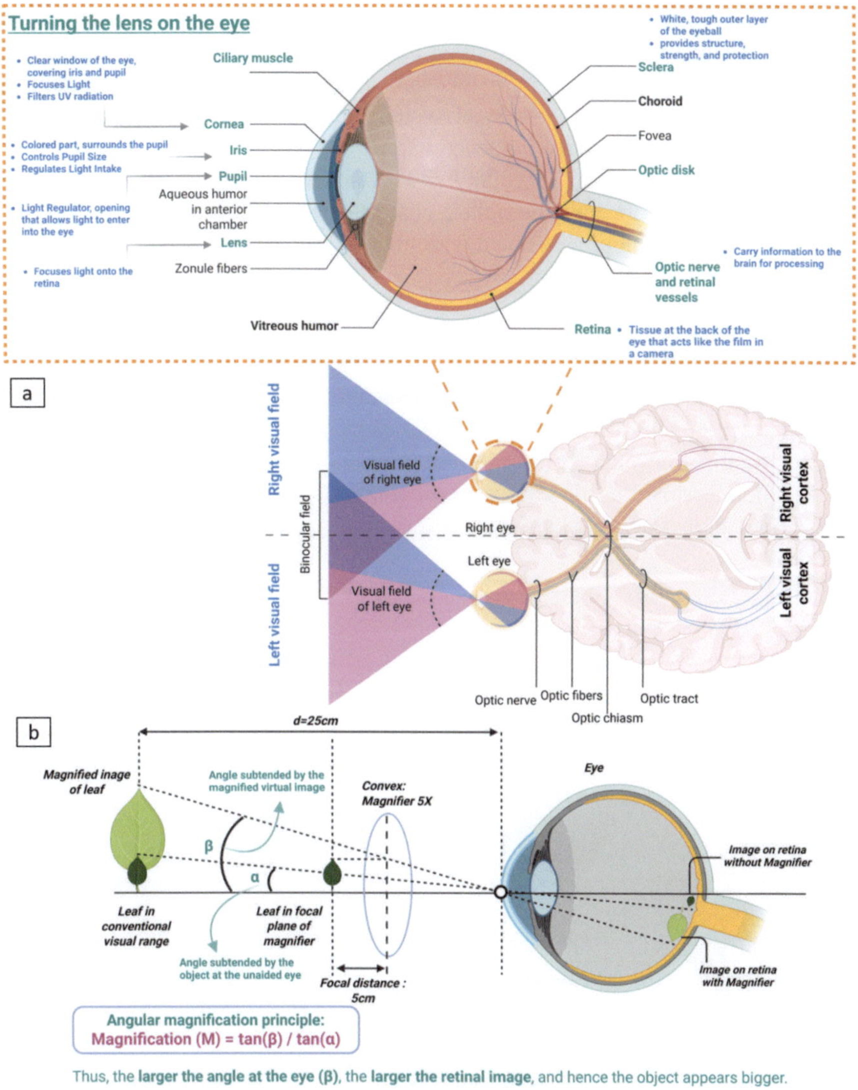

Fig. 9.2 Illustration of the human eye and brain pathways involved in image formation and visual processing. **a** Anatomy of the eye, showing structures responsible for focusing light and transmitting signals to the brain, and the visual pathway connecting the eyes to the visual cortex, demonstrating how visual fields are processed contralaterally. **b** Angular magnification principle depicting how the perceived size of an object increases with greater visual angle, enhancing the retinal image via optical aids like lenses

Microscopy is a powerful tool for uncovering the micro- and nano-structural details of polymers details that are often central to realizing their intended function Fig. 9.3. Whether it is providing mechanical strength, controlling permeability, or enabling responsiveness to external stimuli, the structure of a polymer at the microscopic scale can directly influence its macroscopic performance. This structure–function relationship is especially significant in engineered polymers, where the internal architecture must be tightly controlled to achieve application-specific goals. Basic terminologies associated with microscopy are explained in Table 9.1.

Why is microscopy of polymers important?

Polymers are inherently complex materials—they can be amorphous or semicrystalline, linear or crosslinked, homogeneous or composite, and their properties are often highly dependent on processing conditions. Microscopy allows us to visualize and quantify features like phase separation, crystallinity, domain size, filler distribution, and surface morphology, all of which influence mechanical, optical, thermal, and barrier properties [1]. In both research and industrial

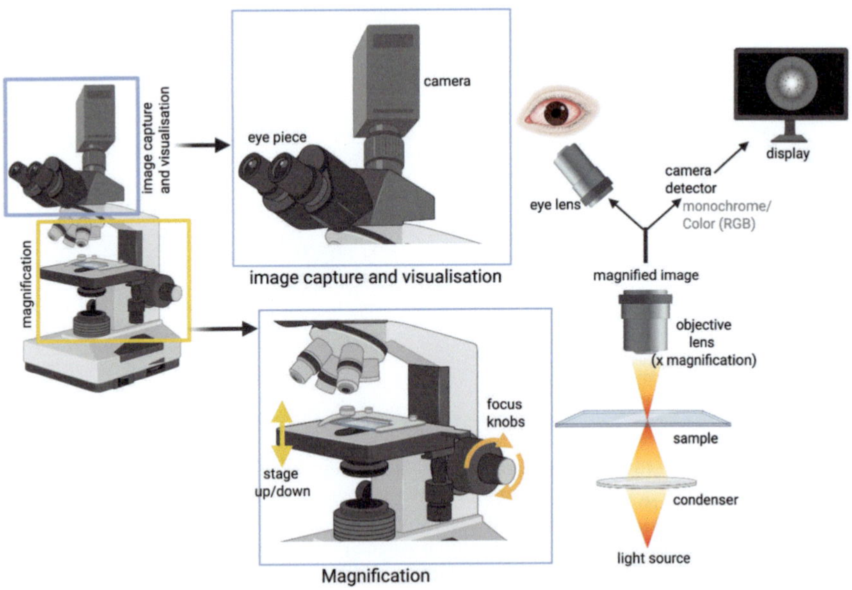

Fig. 9.3 Schematic of image formation and visualization in a compound microscope. The diagram illustrates the key components and processes involved in magnification and image capture. The light source illuminates the sample, which is focused by a condenser. The objective lens magnifies the image, which can then be viewed through the eyepiece or captured by a camera for digital visualization on a display. Stage position and focus knobs adjust sample positioning and image sharpness

Table 9.1 Terminology associated with microscopy

Term	Description	Important points
Magnification	The ratio of the size of the image to the size of the object	Optical magnification = Objective × Eyepiece
Resolution	The minimum distance between two points at which they can still be distinguished	Higher resolution enables finer detail; limited by wavelength and NA [Abbe Resolution: $d = \lambda/(2NA)$]
Contrast	The difference in light intensity between the object and background	Vital for observing transparent specimens
Sharpness	Sharpness is best described as the line contrast of resolved structures	Affects perceived clarity; related to resolution and focus precision
Depth of field	This is the thickness of the optical section of an object through which a sharp image is observed	Thicker depth allows more layers to appear in focus simultaneously
Numerical Aperture (NA)	A measure of the objective's ability to gather light and resolve fine detail	$NA = n \times \sin(\alpha)$; ['n' is the refractive index of the medium between the objective lens and the object, and 'α' is the half-angle of the cone of light entering the objective]
Field of View (FOV)	The diameter of the visible area seen through the microscope	Decreases with increasing magnification
working distance	The distance between the objective lens and the specimen	Smaller for higher magnification lenses

settings, microscopy bridges the gap between formulation and performance, enabling more informed design, processing, and failure analysis.

The choice of microscopy technique depends strongly on the polymer composition, structure, and chemical stability, as well as the specific information sought. For example, hydrogels, block copolymers, electrospun fibres, biodegradable polymers, and nanocomposites each pose distinct imaging challenges and require tailored strategies [2–4]. Selecting the right approach—whether it is optical microscopy for real-time observations [5, 6], electron microscopy for high-resolution surface and internal imaging [7, 8], or scanning probe techniques for nanoscale topology and mechanical mapping [9]—is key to extracting meaningful insights.

Beyond these conventional tools, emerging modalities such as acoustic microscopy [10] and opto-acoustic imaging offer new possibilities for characterizing the elastic and mechanical properties of biomaterials [11, 12] and synthetic materials [13–15] in a non-invasive manner. Likewise, infrared (IR) microscopy [16] and FTIR-based micro-spectroscopy [17] provide spatially resolved chemical information, enabling the mapping of functional groups, crystallinity gradients, or phase-separated regions within complex polymer blends. These techniques are particularly useful for systems where contrast in traditional imaging may be limited due to low electron density or optical transparency.

This chapter begins by establishing the importance of microscopic imaging in polymer science and then walks through the different techniques available, helping the reader align each method with the structural features of interest. By considering the diversity of polymer types and architectures discussed throughout this book, this chapter offers a practical guide to selecting the appropriate microscopy strategies for different material systems.

9.2 Introduction to Microscopy Techniques

The structural and functional complexity of polymers calls for characterization tools that can go beyond averaged, bulk measurements and reveal spatially resolved detail. Microscopy techniques offer that capability—allowing researchers to probe morphology, composition, and internal architecture at resolutions spanning from microns down to nanometres Fig. 9.4. In this context, microscopy is not a standalone solution, but a complementary asset to other well-established techniques used to evaluate polymer mechanical properties, chemical structure, crosslinking density, and degradation kinetics.

What sets microscopy apart is its ability to visualize heterogeneity. Polymers are rarely uniform; they are often semi-crystalline, phase-separated, or incorporate domains of varying stiffness, chemistry, or hydration [18]. These spatial variations can critically influence how a polymer performs under stress, interacts with cells or solvents, or degrades over time. As polymer science continues to push toward multifunctional, stimuli-responsive, and biologically interactive materials, microscopy's ability to provide high-resolution, spatially contextual insights has become increasingly valuable. In particular, applications in drug delivery, regenerative medicine, and soft robotics require precise control over polymer structure and microscopy is central to verifying and tuning that structure.

The vast structural diversity among polymers makes it clear that no single imaging method is universally applicable. Amorphous versus crystalline domains, hydrophobic versus hydrophilic components, and mechanically soft versus rigid architectures all demand different imaging considerations. Moreover, external processing and environmental conditions—such as pH, ionic strength, temperature, and shear stress—can modulate polymer morphology in subtle but functionally

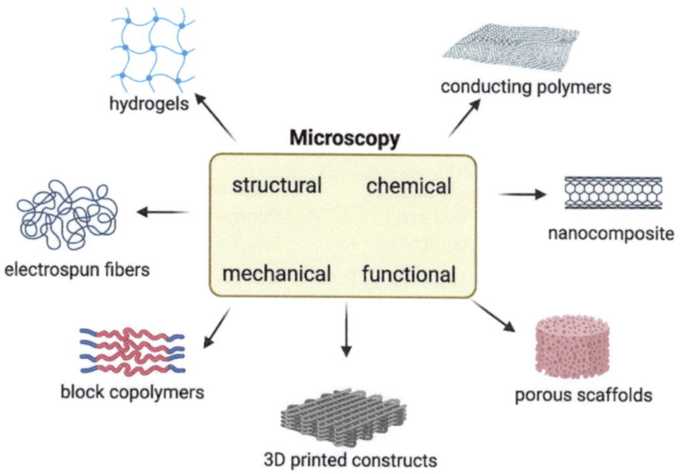

Fig. 9.4 Microscopy in polymer characterization

important ways. Therefore, it becomes essential to adopt an imaging strategy that matches both the physical nature of the polymer and the specific insight being sought.

Traditional bulk techniques such as spectroscopy or mechanical testing provide averaged signals, missing finer details like localized phase separation, microstructural defects, or interface behaviour [1, 19]. These features often have a disproportionate effect on material performance. For instance, nanodomain size and distribution in amphiphilic block copolymers can govern their self-assembly and release kinetics. Similarly, the crosslinking heterogeneity in hydrogels influences their mechanical integrity and cell-instructive behavior. As highlighted in recent studies [20], detailed structural imaging was crucial in correlating material design with biological outcomes, including cellular adhesion, scaffold remodeling, and therapeutic release. To aid the selection of the appropriate method, Table 9.2 summarizes the strengths, limitations, and optimal use cases for each category:

Choosing the right technique involves matching technical capabilities with the physical and chemical nature of the polymer under study. Key considerations include:

- The spatial scale of interest (microstructure vs. nanostructure)
- Desired contrast mechanism (morphological, mechanical, or chemical)
- Material properties (e.g., soft, hydrated, crystalline, thermosensitive).

Practical constraints (sample preparation, imaging environment, speed).

This chapter proceeds by detailing key microscopy approaches across five core sections: optical microscopy (Sect. 9.3), electron microscopy (9.4), scanning tunnelling microscopy (9.5), atomic force microscopy (9.6), acoustic microscopy

Table 9.2 Overview of microscopy techniques for polymer characterization

Technique	Resolution	Key features	Limitations	Best suited for
Bright-field optical	~200 nm	Simple, fast, in situ imaging	Low contrast for transparent samples	General morphology, fiber networks [21]
Confocal/ fluorescence	~100–300 nm	3D imaging, live-cell compatible	Requires fluorescent labeling	Hydrogels, labeled biointerfaces [22]
SEM	~1–10 nm	High-res surface imaging	Conductive coating, vacuum needed	Electrospun fibers, porous structures [23]
TEM	~0.1–1 nm	Internal structure, nanofillers	Extensive prep, thin slicing	Nanocomposites, crystalline domains (JOY)
AFM	~1–10 nm (z-axis)	Topography, stiffness, force mapping	Slow, small scan area	Soft films, biomaterials [24]
STM	Atomic	Electronic properties, atomic resolution	Conductive surfaces only	Conductive polymers, interfaces [25]
FTIR/Infrared microscopy	~2–10 μm	Chemical mapping, functional group analysis	Limited spatial resolution	Biopolymer composition, crystallinity gradients [17]
Acoustic microscopy	~1–5 μm	Mechanical imaging of soft materials	Limited chemical information	Hydrated scaffolds, gels [26, 27]
Opto-acoustic imaging	~10–200 μm	Deep imaging, contrast via absorption	Lower resolution, complex setup	Thick biological constructs, vascularized scaffolds [28]

(9.7) and futuristic trends in polymer microscopy (9.8). Each section highlights the principles, strengths, limitations, and real-world examples of how these tools can be applied in polymer research and development.

Ultimately, microscopy is more than a visualization tool—it is a strategic interface between molecular design and practical function. As polymer materials continue to evolve in precision and purpose, microscopy stands as a key enabler for advancing innovation across disciplines.

9.3 Optical Microscopy

9.3.1 Introduction to Optical Microscopy in Polymers

Optical microscopy encompasses a suite of techniques that image materials using light, offering key advantages for polymer characterization. Unlike electron or X-ray methods, optical approaches are largely non-destructive and can operate in real time under ambient or hydrated conditions. This makes them ideal for examining hydrated and soft biopolymers (e.g. hydrogels and extracellular matrix analogs) in their native state without vacuum drying or extensive sample prepration. For instance, super-resolution optical microscopes can probe polymer materials in situ, achieving spatial resolution approaching that of electron microscopy while preserving live or wet samples [5, 29, 30]. Such capability enables time-lapse studies of dynamic processes like polymer network swelling [31], degradation [32], or cell infiltration in polymer scaffolds [30, 33–36]. Optical methods thus fill a critical need by allowing repeated imaging of the same polymer specimen over time to capture evolving structure–property relationships that single end-point assays would miss.

In recent years, emerging demands in polymer science have pushed optical microscopy beyond its traditional limits. There is a growing need for depth-resolved 3D imaging of thick or opaque polymer constructs (tissues, bio-scaffolds, etc.), which has driven the development of techniques like laser-scanning confocal, multiphoton, and light-sheet microscopy for optical sectioning and deep penetration [19]. At the same time, researchers seek chemically specific, label-free imaging to map composition and molecular order in polymers without fluorescent dyes or stains. This has spurred advances in vibrational microscopy (infrared and Raman) [37] and quantitative phase imaging [19, 38], which provide inherent contrast based on chemical bonds or refractive index. Finally, polymer microstructures often contain features below the ~200 nm diffraction limit of light. To interrogate these, super-resolution microscopy techniques (e.g. STED, PALM, STORM) have been introduced, enabling nanoscale visualization of polymer morphology and tiny defects [19]. These trends have been highlighted in recent literature; for example, Coceancigh et al. [20] reviewed how state-of-the-art optical microscopies–from 3D and super-resolved imaging to label-free modalities–are revolutionizing synthetic polymer characterization. Thus, optical microscopy provides a versatile platform for polymer analysis that is continually extending its capabilities (3D, chemical, nanoscopic) to meet the complex characterization needs of modern polymer systems.

9.3.2 Structural Imaging of Polymers

Brightfield Microscopy: The simplest mode of structural imaging is brightfield microscopy, wherein transmitted or reflected white light produces contrast from inherent color, absorbance, or density differences in the polymer sample. Brightfield

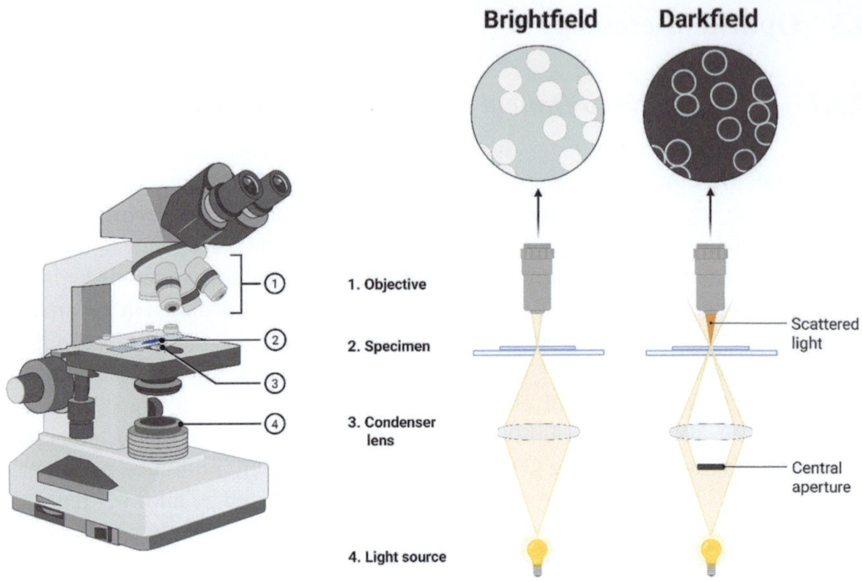

Fig. 9.5 Instrumentation and working principle of Bright and Dark field microscopy

is useful for examining pigmented or stained polymer films and bulk specimens for gross morphology (e.g. cracks, spherulites, fibers). However, many polymers are transparent and require contrast enhancement. Here, phase contrast and differential interference contrast (DIC) microscopy are invaluable. These techniques convert minute refractive index or thickness variations into intensity differences, allowing visualization of nearly transparent polymer matrices (such as hydrogels or thin films) without staining. Phase contrast and DIC [19] can reveal microstructural features like pores or inclusions in soft polymers in real time, though they provide qualitative images (intensity is a non-linear function of phase) and lack quantitative readouts Fig. 9.5.

Fluorescence and Confocal Microscopy: Fluorescence imaging is crucial for structural characterization when specific polymer components or regions can be tagged with fluorescent labels. By attaching dyes or quantum dots to polymer chains, crosslinkers, or functional groups, researchers can map the distribution of those targets with high specificity. Standard wide-field (epifluorescence) microscopy provides multi-color imaging of labeled polymer domains, but thicker specimens suffer from out-of-focus blur. Confocal laser scanning microscopy overcomes this by using a pinhole aperture to reject out-of-focus light, enabling sharp optical slices and 3D reconstructions of polymer structures. Confocal fluorescence imaging is widely used for electrospun fiber scaffolds and hydrogels [39]–for example, visualizing fiber alignment or pore networks by labeling fibers or embedding fluorescent nanoparticles. It offers sub-micron lateral resolution and ~0.5–1 μm axial resolution, sufficient to resolve micron-scale porosity or cell–polymer interactions in scaffolds. Notably,

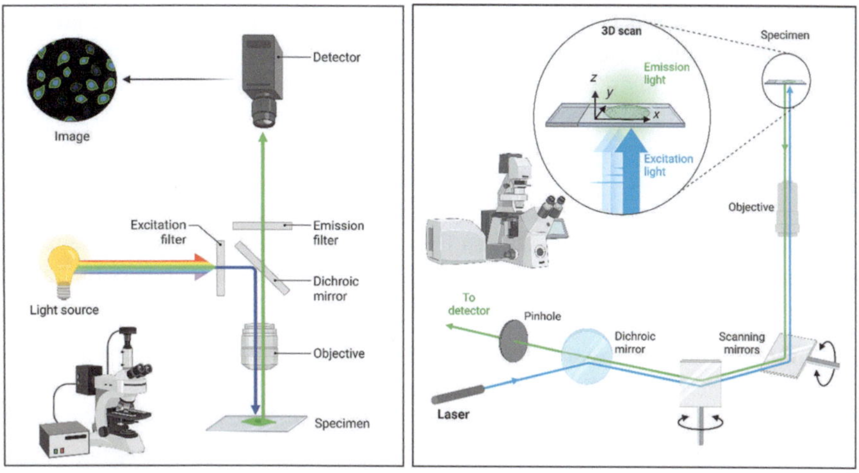

Fig. 9.6 Illustration of working principle Fluorescence and confocal microscopy

confocal microscopes can be combined with environmental controls (e.g. heating stages) to study polymer crystallization or phase separation in real time Fig. 9.6.

Beyond basic intensity imaging, advanced fluorescence modalities provide deeper structural insights. For instance, Förster resonance energy transfer (FRET) microscopy detects molecular-scale proximity (typically 2–10 nm) by non-radiative energy transfer between a donor–acceptor fluorophore pair [19]. This has been applied to polymers by labeling different components and observing FRET signals where those components intimately mix or interact [19]. A FRET-confocal approach, for example, has been used to probe the interface in polymer blends, yielding nanoscale information about component mixing while still mapping the sample at microscale resolution.

Another technique, fluorescence lifetime imaging microscopy (FLIM), maps the spatial distribution of the fluorescence decay time of probes, which is sensitive to the local environment (e.g. polarity, pH, or molecular binding). In polymers, FLIM can distinguish domains or detect solvent content and crosslink density [19]–for example, a fluorescent dye's lifetime might decrease in a densely crosslinked region due to quenching. In one study, FLIM on cell-seeded polymer scaffolds was able to discriminate autofluorescent cells from the scaffold matrix by their distinct lifetimes. Such lifetime contrast adds a functional dimension to structural imaging without needing multiple dyes.

Super-Resolution Optical Microscopy: To examine nanoscale polymer features far below the ~0.2 μm diffraction limit, we can now employ super-resolution (SR) optical microscopy. Techniques like stimulated emission depletion (STED) and single-molecule localization methods (PALM, STORM) have extended optical resolution to the 20–100 nm range [19, 40]. In polymer science, STED microscopy has been used to map the morphology of nanoscale polymer domains and defects. For example, modified STED methods achieved sub-diffraction imaging of conducting polymer nanoparticles (tens of nm in size) by suppressing unwanted two-photon-excited background fluorescence. STED has also been applied to electroluminescent polymer films, enabling visualization of local emissive intensity variations to detect nanoscale defects in devices. Similarly, PALM/STORM approaches, which localize sparse single molecules with nanometric precision, have been leveraged to image polymer chain organization and block copolymer microphase-separated structures at ~20 nm resolution. Super-resolution fluorescence modalities thus bridge the gap between optical and electron microscopy, allowing polymer researchers to "take a deep look" at nanoscale structure while still imaging under benign, in situ conditions.

Multiphoton and Nonlinear Microscopy: For thick polymer constructs and tissue-engineered materials, nonlinear optical techniques such as two-photon microscopy (TPM) are valuable. TPM [41] uses near-infrared femtosecond pulses to induce almost simultaneous two-photon absorption in fluorophores, confining excitation to the focal plane and enabling imaging hundreds of microns deep into scattering polymer matrices Fig. 9.7. This is ideal for 3D hydrogel scaffolds or organoids, where TPM can visualize embedded cells or structural features with less photodamage and deeper penetration than one-photon confocal imaging. Moreover, TPM can excite endogenous nonlinear signals like second harmonic generation (SHG) from non-centrosymmetric polymer structures. SHG microscopy detects frequency-doubled light generated by ordered polarizable assemblies without any fluorescent labels. It is famously used for collagen in extracellular matrices; similarly, SHG has been observed in engineered polymer fiber scaffolds that mimic collagen. For instance, in hybrid electrospun fibers composed of gelatin and poly-L-lactide, multiphoton imaging showed a strong SHG signal indicative of collagen-like triple-helical structure, whereas pure crosslinked gelatin fibers lost that SHG signal (due to disrupted molecular order). Incorporating poly-L-lactide preserved the SHG response in the hybrid fibers, revealing that the addition of a synthetic polymer maintained more aligned, crystalline domains in the scaffold. Thus, SHG provided a unique label-free readout of microscopic ordering in the polymer scaffold. Another nonlinear signal, two-photon-excited autofluorescence, can simultaneously visualize intrinsic signals (such as NADH/FAD in cells) in these scaffolds.

In addition to nonlinear microscopy, polarized light microscopy (PLM) is a classical structural tool, specifically for anisotropic or crystalline polymers. By placing the sample between crossed polarizers, PLM exploits birefringence: optically anisotropic polymer regions (e.g. crystalline lamellae or aligned fibers) rotate the polarization and appear bright against a dark background [19]. This technique quickly reveals features like spherulitic crystal textures in polymers and alignment in drawn films or electrospun mats. Modern quantitative PLM (qPLM) systems can further measure the retardance and orientation of birefringence, outputting maps of fiber orientation or degree of crystallinity for each image pixel. Overall, the assortment of optical modalities–from brightfield and DIC to confocal, multiphoton, and super-resolution microscopy–provides a toolkit to examine polymer morphology

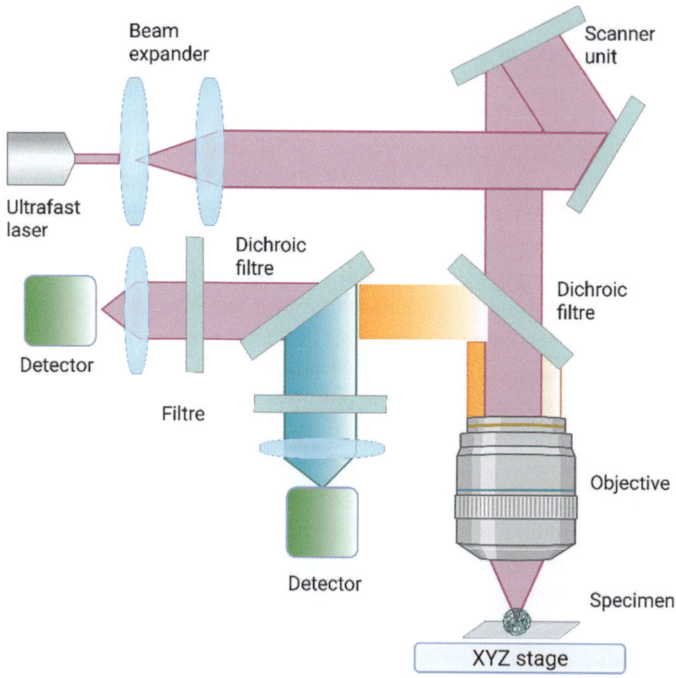

Fig. 9.7 Instrumentation and working principle of multiphoton microscopy

across length scales. These methods have been successfully used to study structures such as electrospun scaffolds, polymer blends, and semi-crystalline polymers, elucidating features like fiber alignment, phase separation, and crystalline domain architecture.

9.3.3 Label-Free and Phase-Based Imaging for Soft and Bioactive Polymers

While fluorescence techniques rely on added labels, many polymer systems (especially biomaterials and living polymer–cell constructs) benefit from label-free imaging to avoid perturbing their chemistry or mechanics. Label-free microscopy often leverages intrinsic optical properties of the sample–such as refractive index, thickness, or scattered light–to generate contrast.

Quantitative Phase Imaging: One major class is quantitative phase microscopy (QPM) [19], which encompasses methods like digital holographic microscopy (DHM), spatial light interference microscopy (SLIM), gradient light interference microscopy (GLIM), diffraction phase microscopy (DPM), and others. QPM techniques measure the optical path length delay introduced by the sample, producing maps of the optical phase shift which correspond to the sample's refractive index distribution or thickness. By capturing interferograms or intensity variations and computationally reconstructing the phase, QPM can map variations in polymer density or thickness without any stain [19, 38, 42]. This enables live imaging of phenomena like polymer swelling, degradation, or cell growth on polymers. For instance, phase imaging can monitor a hydrogel as it absorbs water by directly measuring changes in optical path length (and hence refractive index) over time–information not accessible by conventional brightfield. Modern QPM setups can be high-speed and high-resolution; some achieve lateral resolution near the diffraction limit (~0.2–0.3 μm) and millisecond-scale temporal resolution, making them suitable for dynamic studies of living cells within polymer scaffolds. Notably, optical diffraction tomography (ODT) extends QPM into 3D by acquiring phase images from multiple angles and reconstructing the volumetric refractive index. This yields true three-dimensional, label-free images of cells and polymer matrices, essentially providing a CT-like scan of the sample's internal structure. One implementation uses full 360° illumination ("holotomography") to achieve high-resolution 3D refractive index maps of unlabeled specimens. Such methods are now commercially available and have been used to examine live organoids and cell-laden hydrogels, visualizing internal cell–polymer interactions without fluorescence or physical sectioning.

Interference Reflection Microscopy: Another powerful label-free contrast arises from interference in reflection. Interference reflection microscopy (IRM) is essentially an optical analog of TIRF that highlights regions where a sample (such as a cell or thin polymer film) is extremely close to a reflective substrate. IRM illuminates the sample on a glass surface and captures the interference between light reflected from the sample–medium interface and the underlying glass interface [19]. Areas where a cell or polymer is in near-contact (within ~100 nm) of the glass cause destructive interference and appear dark. Originally developed to visualize cell adhesion footprints, IRM can also be applied to polymer films or bio-interfaces–for example, to monitor how a hydrogel pad contacts a glass slide or to detect protein adsorption patterns on a polymer-coated surface. IRM produces high-contrast images of attachment points, though it only probes very thin regions near the interface. In combination with fluorescence, IRM has been used to study immune cell interactions on functionalized polymer surfaces, correlating cell contact area dynamics with signals like calcium flux.

Light Scattering and Nanoparticle Imaging: For imaging nanoscale features and dynamics without labels, advanced scattering-based methods have emerged. Rotating coherent scattering (ROCS) microscopy uses a rapidly rotating oblique illumination beam to capture high-contrast scattering images of small structures [19]. By averaging scattering from many illumination angles in one frame, ROCS achieves ~150 nm lateral resolution at video-rate (tens of frames per second). It has been shown to visualize tiny, fast-moving objects such as cell membrane filopodia or virus-sized polymer nanoparticles near surfaces. Similarly, interferometric scattering (iSCAT) microscopy and its transmitted-light counterpart, COBRI (coherent brightfield), push sensitivity to the single-nanoparticle level [19]. These techniques detect the interference between light scattered by a nano-object and a reference beam (or unscattered light), allowing detection of objects as small as ~10–50 nm that would be invisible by ordinary microscopy. For example, iSCAT in reflection mode has been used to observe individual virus particles binding to a polymer-coated surface by capturing their faint scattering signal. In transmitted mode, COBRI uses a coherent laser in a brightfield configuration and achieves nanometer localization accuracy and microsecond temporal resolution, enabling tracking of single nanoparticles in 3D within transparent polymers. Such interferometric methods are well-suited for soft polymer nanostructures or nanoparticle–polymer systems: one can track the diffusion of unlabeled nanogels in a hydrogel or observe early stages of polymer colloid aggregation via scattering signals. Because scattered intensity does not rely on fluorophores (and is not limited by photobleaching), high illumination can be used to improve speed or precision–although excessive power may introduce background noise or heating. Overall, these label-free scattering microscopes open the door to interrogating polymer nanoparticles, colloids, and nanoscale phenomena with extreme sensitivity. Notably, interferometric approaches have achieved detection and tracking of single proteins in solution without any labels, underscoring their power for ultrahigh sensitivity.

Finally, many polymers and biomolecules possess intrinsic optical signals that can be harnessed for label-free imaging. Native autofluorescence from certain polymers (e.g. aromatic polymer backbones) or from biological components (like NADH and flavins in cells) provides built-in contrast. Autofluorescence microscopy [43] has been employed to monitor degradation of biomaterials or the metabolic activity of cells on polymer scaffolds. Though often considered a nuisance in labeled imaging, autofluorescence can report on polymer oxidation or crosslink formation (which introduce fluorescent chromophores) and on cell metabolic states in tissue-engineered constructs. In summary, phase and scattering-based modalities allow imaging of live cells, hydrated polymers, organoids, and ECM-mimicking materials entirely label-free. They quantitatively map physical properties (thickness, refractive index) or detect nanoscale objects via light scattering, thereby preserving the native chemistry and biomechanics of sensitive polymer systems during imaging.

9.3.4 Chemical and Functional Imaging of Complex Polymer Systems

Characterizing polymers often demands knowledge of their chemical composition, functional group distribution, and local chemistry (e.g. crosslink density or degradation products). Optical microscopy can be extended to provide chemical contrast through spectroscopic techniques. Fourier-transform infrared (FTIR) microspectroscopy [17] is a classic method for mapping chemical functionalities in polymers.

An IR microscope focuses mid-IR light onto a small region (on the order of 5–20 μm for diffraction-limited IR) and records the absorption spectrum. By raster-scanning the sample or using a focal plane array detector, one can build an IR absorption map where each pixel contains a full IR spectrum. Peaks corresponding to specific bonds (C=O, O–H, N–H, etc.) reveal the spatial distribution of chemical groups. FTIR microspectroscopy has been applied to polymer composites and laminates to, for example, examine the dispersion of functional additives or the degree of cure in different regions. Its strengths are label-free chemical specificity and the ability to identify unknown substances via spectral libraries. However, spatial resolution is limited by the long IR wavelengths (typically a few microns). Techniques like attenuated total reflectance (ATR) imaging can improve resolution (using an IR-crystal tip to probe a small region), and newer quantum cascade laser-based IR microscopes offer brighter sources for faster mapping. Still, for sub-micron chemical imaging, one turns to Raman scattering.

Raman Microspectroscopy: Raman microscopy uses visible or near-IR laser light to probe molecular vibrations in a polymer via inelastic light scattering (the Raman effect) [37]. A confocal Raman microscope can achieve ~0.5–1 μm lateral resolution, limited by optical diffraction of the excitation laser. By scanning the focused laser and collecting a full Raman spectrum at each point, one obtains a chemical map with sub-micron features. Raman peaks provide a fingerprint of polymer composition, sensitive to monomer structures, crystallinity, and even stress or molecular conformation. This technique has been widely used for polymer blends and composites–for instance, mapping the distribution of two immiscible polymers by their distinct Raman bands, or observing the diffusion of a monomer into a polymer matrix by the appearance/disappearance of characteristic vibrational peaks. It is also invaluable for tracking chemical reactions within polymers (e.g. monitoring a curing or crosslinking process by the decrease of an alkene C=C band and increase of an alkane C–H band over time). A strength of Raman microscopy is its ability to target specific bonds in complex environments (water is a weak Raman scatterer, so hydrated samples like hydrogels can be analyzed in situ). One limitation is that spontaneous Raman signals are inherently weak, making 2D mapping relatively slow (often seconds per spectrum). To address this, faster **coherent Raman** techniques have been developed.

CARS and SRS Microscopy: Coherent anti-Stokes Raman scattering (CARS) and stimulated Raman scattering (SRS) microscopy are nonlinear optical methods that amplify vibrational signals for rapid imaging. Both use two synchronized laser frequencies (pump and Stokes beams) tuned to a molecular vibration of interest, driving the vibrations coherently [37]. CARS detects the anti-Stokes light generated by the nonlinear four-wave mixing process, while SRS measures the intensity loss (or gain) in one of the beams due to stimulated excitation of the vibration. These methods can produce video-rate chemical images with sub-micron resolution, ideal for observing polymer processes in real time. CARS was among the first coherent Raman techniques applied in polymer contexts; for example, it can selectively visualize specific bonds like C–H stretches in polyolefins, allowing one to see polymer domain structures or lipid distributions in bio-polymer blends without labeling. SRS, which has gained prominence in the last decade, offers quantitative concentration maps of target bonds with linear response (and without the nonresonant background signal that complicates CARS). SRS microscopy has been used to distinguish different polymers or contaminants by their unique vibrational signatures. A recent application is in microplastic analysis: SRS imaging of environmental or biological samples can rapidly detect and identify individual microplastic particles by scanning for diagnostic bands (such as the C–H stretching region around 2850–2950 cm^{-1}). Researchers have even demonstrated 3D SRS imaging of nano- and micro-plastic particles inside aquatic organisms, differentiating polymer types by their Raman spectra. In biomedical polymers, SRS has been used to monitor drug release (tracking a drug's vibrational peak) and to map biochemical components in tissue-engineered constructs (e.g. distributions of lipids, proteins, and synthetic polymers) with high speed and sub-cellular resolution. Together, CARS and SRS bring high-speed chemically specific imaging capabilities, essentially turning vibrational spectra into real-time images.

Hyperspectral Imaging: Another approach to chemical imaging is optical hyperspectral imaging (HSI) in the UV–Vis–NIR range. In HSI, an image is acquired at many narrow wavelength bands (often tens or hundreds of channels), rather than the few channels of normal RGB or multi-color fluorescence imaging [44]. The resulting data cube can be analyzed for spectral signatures corresponding to different materials. For polymers, HSI has been used to detect and map photodegradation byproducts or additives that have distinct optical spectra. In one study, hyperspectral imaging in the UV–visible range visualized polymer weathering damage by capturing subtle color changes associated with chemical oxidation. Near-infrared HSI can identify different polymer types in mixed recycled plastics by their overtone absorption bands. While HSI may not achieve the molecular specificity of Raman or IR, it excels in rapidly classifying regions with differing optical properties and can be combined with multivariate analysis to infer chemical or structural variations.

Photoacoustic Microscopy: Finally, functional imaging in polymers extends to techniques that probe beyond static chemistry–for example, imaging mechanical or electrical function Fig. 9.8. One notable hybrid modality bridging chemical and functional contrast is photoacoustic microscopy (PAM). PAM detects ultrasonic waves generated by periodic laser excitation of light-absorbing molecules in the sample [28, 45]. Essentially, a pulsed laser heats regions containing chromophores (either intrinsic pigments or extrinsic labels like nanoparticles), causing thermoelastic expansion that is read out with an ultrasound transducer. In polymers, PAM can be used if the material or its inclusions absorb light (often in the optical or NIR range). The technique can achieve high resolution (optical diffraction-limited laterally, ~0.5–1 μm, and a few microns axially) at imaging depths of 1–3 mm, far deeper than pure optical microscopy due to the low scattering of ultrasound. Two modes exist: optically focused PAM (higher resolution, shallower imaging) and acoustically focused PAM (lower resolution, deeper reach).

In biomedical polymer constructs, PAM has been used to visualize microvasculature forming within a scaffold by detecting hemoglobin absorption in blood.

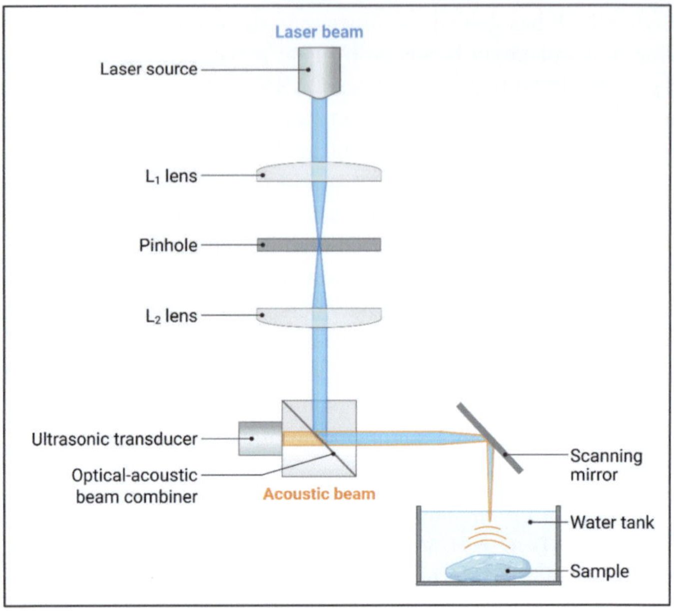

Fig. 9.8 Working principle of photoacoustic microscopy

The unique value of PAM is its ability to image through optically dense media and capture absorption-based contrast–for instance, one could monitor a dye-labeled drug's distribution in a thick polymer implant, or detect an absorbing polymer degradation product leaching out, even at millimeter depths. Recent advances have even pushed PAM's lateral resolution to the nanoscale (on the order of ~100 nm using specialized labels and nonlinear excitation schemes), blurring the line with super-resolution techniques. Although PAM requires coupling to an ultrasonic detector, its hybrid nature (optical excitation with acoustic detection) expands the reach of optical imaging into regimes of thickness and opacity that purely optical methods cannot penetrate.

9.3.5 Mechanical and Cross-Sectional Imaging of 3D Polymer Constructs

Three-dimensional polymer constructs (scaffolds, hydrogels, tissue-engineered cartilage, etc.) often exhibit spatial variations in mechanical and structural properties that are critical to their function. Optical imaging methods can non-invasively probe the internal architecture and mechanical state of such samples. Chief among these is **optical coherence tomography (OCT)** [46], an interferometric technique that provides cross-sectional images ("optical slices") of turbid samples analogous to ultrasound imaging, but with micrometer-scale resolution. OCT uses low-coherence near-IR light and interferometry to measure the echo time delay of backscattered light from within the sample, building up a depth profile. In polymer engineering, OCT is valuable for visualizing the internal layer structure and defects in materials up to a few millimeters thick. For example, an OCT image can reveal the layered morphology of a 3D-printed polymer scaffold, showing pore networks and strut architecture non-destructively. Typical OCT systems achieve spatial resolutions of around 5–15 μm and penetration depths of 2–5 mm in scattering tissue or polymer phantoms [47]. This mesoscopic resolution (better than medical ultrasound's ~100 μm, but coarser than optical microscopy's ~1 μm) hits a "sweet spot" for many tissue-like polymers: it can resolve multi-micron features such as hydrogel channels or scaffold fibers while surveying an entire construct. OCT additionally applies to polymer constructs to monitor degradation or swelling in a thick hydrogel over time, watch tissue ingrowth into a scaffold, or inspect an implant's integrity in situ.

Several OCT variants enhance specific contrast. **Polarization-sensitive OCT (PS-OCT)** detects changes in polarization state of the backscattered light, revealing birefringent features in the sample. In fibrous polymers or composites (and collagen-rich tissues), PS-OCT can map fiber orientation or zones of molecular alignment by their birefringence, effectively adding a 3D polarization map atop the structural image. This is useful for assessing alignment in anisotropic scaffolds or detecting stress-induced birefringence in loaded polymer components. **Doppler OCT** [19] (and the related OCT-angiography) measures motion within the sample by analyzing phase shifts between consecutive scans. These are predominantly used to image fluid flow (e.g. blood flow in vessels with OCT-A). In a polymer context, Doppler OCT could visualize the flow of fluid through porous scaffolds or microfluidic polymer devices, or detect moving cells or particles within a construct. **Optical coherence elastography (OCE)** builds on OCT to measure mechanical properties: it involves introducing a mechanical excitation (a vibration, compression, or acoustic pulse) and using OCT to track how the sample deforms or how mechanical waves propagate. From the observed motion, one can compute elastic moduli or viscoelastic parameters in a depth-resolved manner. For instance, by analyzing the speed of shear waves in a hydrogel of known density, the shear modulus (stiffness) can be mapped throughout the volume. OCE has enabled high-resolution elasticity mapping of hydrogels and tissues–one study used it to measure the stiffness of a thin hydrogel in a Petri dish *noninvasively*, by detecting ambient vibration responses via phase-resolved OCT. This technique is promising for evaluating gradients in crosslink density or stiffness in tissue-mimicking polymer constructs, or monitoring how a scaffold's stiffness evolves as it degrades or as cells deposit matrix. Notably, because OCT imaging is relatively fast, OCE can capture real-time changes and even perform time-lapse mechanical monitoring.

Beyond coherence-gated methods, other optical techniques probe mechanical properties on different scales. **Brillouin microscopy** is a spectroscopic technique that maps the viscoelastic properties of a material by measuring the frequency shift of light scattered from thermally excited acoustic phonons (i.e. spontaneous sound waves) in the sample [19]. When a focused laser beam interacts with a polymer, a tiny fraction of light undergoes Brillouin scattering, shifting in frequency by an amount related to the material's longitudinal modulus (and also its refractive index). By scanning the focus and collecting a Brillouin spectrum at each position, one obtains a 2D or 3D map of the sample's high-frequency elastic modulus. Brillouin microscopy can achieve sub-micron resolution in three dimensions (using a confocal collection setup). Its great advantage is that it is completely non-contact and label-free, directly sensing mechanical stiffness inside a material (even inside living cells) using only light. In polymer science, Brillouin imaging has been applied to hydrogels and bio-polymer tissues to measure local stiffness. For example, researchers have mapped the stiffness of bioprinted hydrogel constructs, showing variations corresponding to the printed lattice structure and differing crosslinker concentrations. In tumor-laden collagen gel models, Brillouin maps revealed a stiffening of the hydrogel in regions immediately surrounding growing tumor spheroids (consistent with local densification of the matrix). It should be noted that Brillouin measures a high-frequency (~GHz) modulus, which in hydrated materials is influenced by both the polymer network stiffness and the bulk modulus of the fluid in the pores. In fact, some studies have found that in very soft hydrogels Brillouin frequency shifts correlate more strongly with water content than with conventional elastic modulus, indicating a complex contrast mechanism. Even so, Brillouin microscopy provides a unique contrast mechanism to visualize mechanical heterogeneity within 3D polymer systems: stiffer or drier regions produce higher frequency shifts (brighter in a Brillouin map), whereas softer or more hydrated regions give lower shifts. It can thus identify features like a dense, stiff "shell" forming on a hydrogel or the softer core of a bioprinted construct–information difficult to obtain otherwise without physical sectioning or indentation. Confocal Brillouin imaging was first demonstrated in 3D cell cultures in 2015, and it has since emerged as a valuable tool for noninvasively mapping micro-mechanics inside materials.

9.4 Summary Table

To conclude, Table 9.3 summarizes the wide range of optical imaging modalities discussed, highlighting their typical resolution, contrast mechanism, examples of suitable polymer systems, and the unique insights each modality provides for polymer characterization.

9.5 Electron Microscopy (SEM and TEM)

Introduction to Electron Microscopy: Electron microscopy encompasses scanning electron microscopy (SEM) and transmission electron microscopy (TEM), which use focused electron beams to resolve structures at far smaller scales than light microscopy Fig. 9.9. These methods are ideal for dry, solid polymer samples

Table 9.3 Summary of optical microscopy modalities for polymer characterization

Imaging modality	Typical resolution	Polymer systems (Examples)	Value-added insights
Brightfield Microscopy	~0.2–0.5 μm lateral	Opaque or dyed polymer films, sections, fibers (e.g. stained tissue sections, coating, cracks)	Morphology and defects- size/shape of features (spherulites, cracks, fibers), polymer quality inspection
Phase Contrast Microscopy	~0.2–0.3 μm lateral	Transparent polymers films, hydrogels, live cells in polymers (e.g. cell-laden gels, microcapsules)	Internal structure of transparent samples (pores, inclusions) without staining. Allows real-time observation of swelling, gelation, cell motion, etc. (qualitative contrast)
Differential Interference Contrast (DIC)	~0.2–0.3 μm lateral	Thin polymer sections, fibers, surface textures (e.g. scaffold fiber topology, surface roughness of films)	Enhanced edge contrast for fine details (grain boundaries, tiny scratches). Produces 3D-like shadowed images showing texture and small features on polymer surfaces
Epifluorescence Microscopy (wide-field)	~0.2 μm lateral (high NA objective); ~~0.5–1 μm axial	Labeled polymers or composites (e.g. a polymer with a fluorescent dye, multi-component blends with each phase tagged)	Specific component mapping in multi-color; fast imaging of labeled features. Good for identifying phase domains or tracking labeled molecules (e.g. drug release) in polymers
Confocal Laser Scanning Microscopy	~0.2 μm lateral; ~ ~0.6 μm axial (with pinhole)	Thick or 3D polymer samples (e.g. electrospun fiber scaffolds with fluorophores, cells in hydrogels with fluorescent proteins)	3D imaging (z-stacks) of polymer architecture; eliminates out-of-focus blur. Allows reconstruction of pore networks or layer-by-layer structure, and co-localization of multiple labels in 3D
Fluorescence Lifetime Imaging (FLIM)	Spatial resolution same as confocal, nanosecond temporal resolution	Polymers with environment-sensitive dyes or autofluorescence (e.g. a hydrogel with a pH-sensitive dye, cells with NADH/FAD autofluorescence)	Microenvironment mapping: differentiates chemical or physical environments (pH, polarity, crosslink density) via lifetime differences. Can separate overlapping fluorophores by lifetime

(continued)

Table 9.3 (continued)

Imaging modality	Typical resolution	Polymer systems (Examples)	Value-added insights
Förster Resonance Energy Transfer (FRET) Microscopy	Diffraction-limited (depend on wide-field or confocal mode)	Polymer blends or block copolymers with each component labeled; biomaterials with ligand–receptor pairs	Nanoscale interaction detection: reveals mixing at molecular scale (distinguishes phase separation vs. miscibility) and crosslink formation (when labeled sites come together). Indicates molecular conformation changes
Super-Resolution Fluorescence (STED, PALM, STORM)	~20–100 nm lateral (STED ~ ~50 nm; PALM/STORM ~ ~20 nm); ~ ~50–150 nm axial (3D STED or biplane localization)	Nanoscale polymer structures (e.g. block copolymer domains, conducting polymer nanoparticles, nano-fibrous matrices labeled with fluorophores)	Nanoscale morphology resolved beyond diffraction limit: visualizes domains, defects, or polymer structures not seen by conventional optics. Enables in situ nano-characterization (including dynamic or in-solution imaging) as an alternative to electron microscopy
Two-Photon (Multiphoton) Microscopy	~0.3–0.5 µm lateral (lower than 1-photon); ~ ~1–2 µm axial	Thick, highly scattering samples; deep tissues and hydrogels (e.g. collagen or gelatin scaffolds, tissue-engineered constructs with cells)	Deeper 3D imaging (~0.5 mm) in turbid polymers with less photodamage. Can capture inherent signals like SHG or autofluorescence simultaneously, shows structure without dyes
Second Harmonic Generation (SHG) Microscopy	~0.3 µm lateral (like 2-photon); ~ ~1 µm axial	Nonlinear-optically active polymers or biopolymers (e.g. aligned collagen or silk fibers, poled electro-optic polymer films, certain crystalline polymers)	Collagen-like fiber detection and crystalline ordering visualizes fibrillar architecture (scaffold fiber alignment, polymer crystallites) that produce SHG signal. Indicates regions of high molecular order or polarity. Provides label-free contrast for fiber networks in tissues
Polarized Light Microscopy (PLM)	~0.2 µm lateral (diffraction-limited)	Anisotropic or crystalline polymers (e.g. polymer spherulites in thin film, stretched polymer fibers, liquid crystalline polymers)	Birefringence visualization: bright signals from oriented/crystal region-c, size, shape of spherulites, fiber alignment, and strain-induced orientation. Quick assessment of crystallinity in plastics and fibers

(continued)

Table 9.3 (continued)

Imaging modality	Typical resolution	Polymer systems (Examples)	Value-added insights
Quantitative PLM (qPLM)	~0.2 μm lateral	Highly birefringent biomaterials and fibers (e.g. collagen scaffolds, electrospun fiber mats, tendons or silk fibers—to measure alignment distribution)	Orientation mapping: outputs retardance and fiber orientation maps, enabling quantification of alignment and anisotropy in polymer fibers or composite laminates. Useful for quality control in fiber spinning (alignment consistency)
Quantitative Phase Imaging (QPI/QPM) (e.g. DHM, SLIM, DPM)	~0.2–0.3 μm lateral; ~ ~nm axial sensitivity in optical path length	Live cells in polymer scaffolds, microgel particles, transparent polymer structure (e.g. cell growth in a hydrogel, thin polymer film thickness uniformity, microcapsule)	Mass/thickness quantification: measures cell dry mass, polymer density, or swelling by phase imaging. Tracks dynamic processes (hydration, degradation) since it's non-invasive. Enables calculation of volumes or refractive indices of polymer microspheres, etc
Optical Diffraction Tomography (ODT)	~0.3 μm lateral; ~ ~0.3–0.5 μm axial (for 3D RI map)	3D polymer-cell constructs, organoids, large microgels (e.g. an organoid in hydrogel—3D RI distribution of cells and scaffold; porous scaffold architecture without sectioning)	Volumetric 3D label-free imaging: shows internal refractive index change matching to structure (cells, voids, matrix density). Digitally "slice" samples showing- layer structure, cell infiltration in scaffolds. 3D live imaging over time (cell division, migration in a polymer matrix)
Interference Reflection Microscopy (IRM)	Lateral ~ ~0.2 μm; probes ~ ~ < 0.1 μm axial gap at interface	Cell or protein adhesion on polymer coatings; polymer thin films on glass (e.g. a fibroblast footprint on a hydrogel-coated coverslip, protein multilayer on a biosensor surface)	Contact/gap visualization: highlights where the sample is within ~ ~λ/4 of the surface. Identifies adhesion zones, thin liquid layers, or dewetting patterns. In polymers, can monitor how a film thins or how a cell's contact area spreads on a substrate in real time

(continued)

Table 9.3 (continued)

Imaging modality	Typical resolution	Polymer systems (Examples)	Value-added insights
Rotating Coherent Scattering (ROCS) Microscopy	~150 nm lateral (high NA $\approx$ 0.9, λ ~ ~400 nm); real-time (~100 Hz framerates)	Fast dynamics at surfaces; small nanostructures (e.g. nanoparticles in films, cell membrane ruffles on scaffolds, quick events like polymer colloid aggregation)	High-speed label-free imaging: captures moving nano-objects (~100–200 nm) and fine filopodia with video-rate imaging. Useful for studying kinetics (e.g. virus–polymer interactions, protein adsorption) requiring high spatial and temporal resolution
Interferometric Scattering (iSCAT) Microscopy	~0.2 μm; localization precision ~ ~1–10 nm for single nanoparticles	Nanoparticles and nano-assemblies in polymers; single-molecule studies (e.g. 20–50 nm polymer nanoparticles diffusing in solution, a single virus binding to a polymer coating, unlabeled quantum dot in a polymer film)	Ultrasensitive detection: visualizes very small particles (<50 nm) that are invisible in ordinary microscopy by boosting their signal via interference. Allows single-particle tracking in polymers (e.g. nanoparticle diffusion in a gel) and label-free observation of nano events (e.g. nucleation of polymer domains)
Coherent Brightfield (COBRI) Microscopy	~0.2 μm resolution; ~ ~nm localization accuracy; ~ ~10 μs temporal resolution	Same as iSCAT but for transmitted light (particles anywhere in volume). E.g. gold or polymer nanoparticle moving in a hydrogel, tracking of polymer microswimmers in solution	Deep high-speed tracking: can follow single nanoparticle trajectories in 3D within transparent polymers at kHz frame rates. Great for studying transport properties (nanoparticle diffusion, active particle motion) in soft materials without fluorescent labels or slow imaging
Autofluorescence Imaging (label-free fluorescence)	~0.2 μm (lateral), ~ ~0.5–1 μm axial (depend on wide-field or confocal)	Intrinsically fluorescent polymers or tissues (e.g. elastin or collagen in decellularized matrix, polymer photo-oxidation products, cell metabolic cofactors in a graft)	Chemical fingerprint without dyes: indicates presence of certain functional groups/ degradation products (many polymers develop fluorescence upon oxidation). In cell-polymer constructs, can monitor cell metabolism or matrix formation. No fluorescence staining

(continued)

Table 9.3 (continued)

Imaging modality	Typical resolution	Polymer systems (Examples)	Value-added insights
FTIR Microspectroscopy	~5–15 μm; down to ~ ~3 μm with ATR accessories	Polymer films, fibers, laminates; chemical analysis of micro-regions (e.g. coating cross-sections to identify layers, polymer blend microspheres, protein/polymer hydrogel)	Chemical composition mapping: identifies functional groups and components in situ. Helps identify polymer in mixes, monitor reactions, or locating additives by their spectral signatures. Non-destructive and label-free (based on intrinsic molecular vibrations)
Raman Microspectroscopy	~0.5 μm lateral (with 532 nm excitation & high NA); ~ ~1–2 μm axial	Essentially all polymers (requires Raman-active bonds). E.g. polymer blend films or fibers (mapping one polymer vs. another), stressed plastics (peak shifts in polyethylene reveal stress)	Molecular fingerprint at μm scale: differentiates polymer types and crystallinity within a sample. Monitors in situ reaction (tracks peak changes with polymerization) and maps multi-component systems. Complementary to FTIR (also works with aqueous samples)
Coherent Anti-Stokes Raman Scattering (CARS)	~0.3 μm lateral (similar to one-photon confocal); ~ ~1 μm axial	Polymer domains with a known “marker” bond; typically single-band imaging. E.g. visualizing lipid-rich domains in a polymer scaffold via C–H stretch; mapping polymer chain orientation via polarized CARS	High-speed vibrational imaging: video-rate imaging possible for a specific bond, enables dynamic processes viewing (polymer crystallization, solvent diffusion).. Provides 3D sectioning capability (like multiphoton microscopy), chemically-selective visualization when one component has a unique Raman band
Stimulated Raman Scattering (SRS)	~0.3 μm lateral; ~ ~1 μm axial (confocal-like)	Complex polymer or biological samples requiring multiplex imaging. E.g. differentiating multiple microplastics by their spectra in a mixture; imaging distribution of drug vs. polymer in a drug-delivery implant	Quantitative chemical maps at speed: combines chemical specificity with video-rate scanning. Multi-color SRS map several functional groups in one scan. Ideal for label-free chemical imaging in tissues and composites (e.g. monitoring polymer degradation products or metabolite–polymer interactions in scaffolds)

(continued)

Table 9.3 (continued)

Imaging modality	Typical resolution	Polymer systems (Examples)	Value-added insights
Hyperspectral Imaging (HSI) (optical)	Varies—typically ~ ~0.5–5 µm (depends on optics and wavelength range)	Polymers with distinct optical spectra or for detecting degradation. E.g. mapping weathering on a painted polymer (yellowing extent), sorting different plastics in recycling by NIR spectral differences	Spatially-resolved spectroscopy: identifies materials or chemical states by spectral fingerprints at each location. Useful for quality control and failure analysis (e.g. visualize UV-degraded vs. intact polymer regions), and for examining spatial variation in additives (pigment or stabilizer distribution) via their spectra
Photoacoustic Microscopy (PAM)	Optical-resolution: ~ ~0.5–1 µm lateral (high NA objective); Acoustic-resolution: ~ ~10–50 µm lateral; u ~ ~1–3 mm depth	Absorbing polymer constructs or those with absorbing agents; vascularized tissue–polymer implants. E.g. a scaffold with embedded dye nanoparticles for contrast, or an engineered tissue with blood vessels (hemoglobin) in a polymer matrix	Deeper imaging of features: visualizes structures that absorb light (e.g. blood vessels in a scaffold) at depths beyond optical microscopy (~millimeters). Uses- viewing vascularization in implants, distribution of nanoparticle-based therapeutics, or polymer implant degradation (releasing colored byproducts). Combines high resolution and depth using optics with ultrasonics
Polarization-Sensitive OCT (PS-OCT)	~5–15 µm (same as OCT)	Birefringent polymer matrices or composites. E.g. collagen gel scaffolds (collagen is birefringent), fiber-reinforced polymer laminates (checking internal fiber orientation)	3D birefringence mapping: shows alignment or anisotropy inside the sample in 3D. Useful to assess if fiber alignment in a scaffold remains uniform through the thickness, or if a polymer's crystallinity varies by layer. Can indicate stress-induced birefringence inside plastic parts

(continued)

Table 9.3 (continued)

Imaging modality	Typical resolution	Polymer systems (Examples)	Value-added insights
Optical Coherence Elastography (OCE)	~5–15 μm (inherits OCT imaging resolution)	Soft polymers and tissues where spatial stiffness info is needed. E.g. gelatin or PEG hydrogels with stiffness gradients, layered cartilage repair scaffolds (stiff core, soft outer layer)	3D elasticity mapping: shows stiff or soft region, in a label-free manner. Can detect subtle changes (e.g. a gel region crosslinking over time becomes stiffer). Adds functional mechanical info atop the structural OCT image —critical for matching scaffold stiffness to native tissue or ensuring even crosslinking
Brillouin microscopy	~0.2–0.5 μm lateral; ~ ~1–5 μm axial (confocal Brillouin; trade-off with acquisition time)	Soft hydrated polymers, biomaterials, cells in materials. E.g. hydrogels with varying crosslink densities, cell-seeded collagen gels (mapping stiffness or density), polymer films with modulus gradients	**Non-invasive stiffness/viscosity imaging**: Reveals local longitudinal modulus (stiffness/compressibility) at sub-micron scale. Useful for detecting soft/stiff regions and tracking changes like gel stiffening or cell-driven matrix remodeling. Offers internal mechanical insights, complementing bulk rheology

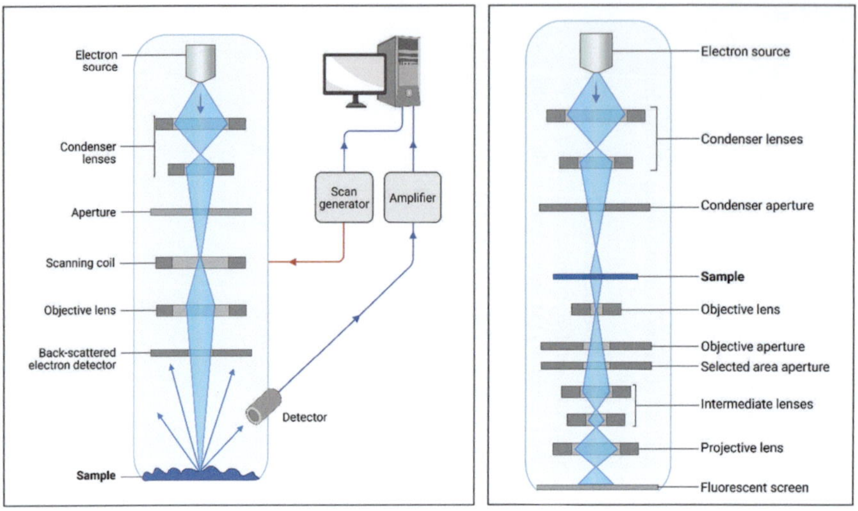

Fig. 9.9 Working principle of Scanning and Transmission electron microscopy

compatible with high vacuum [48]. SEM and TEM complement optical techniques by providing nanometer to sub-nanometer resolution and by enabling analytical measurements (e.g., elemental mapping via EDX or crystallinity via electron diffraction) beyond the reach of light microscopes [6]. They have become core tools for polymer characterization, revealing fine features that optical microscopy cannot discern [48].

Scanning Electron Microscopy (SEM): SEM (particularly field-emission SEM, FESEM) produces high-resolution images of a polymer's surface topology. Because polymers are insulating, samples often receive a thin conductive coating to prevent charging under the electron beam [48]. SEM operates under high vacuum, meaning samples must be dry; wet specimens would evaporate and deform in the chamber [48]. An environmental SEM (ESEM) permits imaging of moist samples by using a low-pressure gas to neutralize charge [49]. SEM systems can also integrate a focused ion beam (FIB) to slice into the sample and image cross-sections, enabling three-dimensional tomography of polymer microstructures [50]. SEM is widely used in polymer research–for example, to assess fiber alignment in electrospun scaffolds and to examine the morphology of 3D-printed or fractured polymers [51].

Transmission Electron Microscopy (TEM): TEM passes a high-energy electron beam through an ultrathin polymer specimen, projecting an image of the internal nanostructure. It offers higher resolution than SEM, resolving features down to the nanometer or sub-nanometer scale (e.g., crystalline lamellae in semicrystalline polymers or microphase-separated domains in block copolymers). Polymer samples are typically cut into ~50–100 nm thin sections (often using cryo-microtomy for soft polymers) and mounted on support grids [48]. To enhance contrast, heavy-metal stains like osmium tetroxide (OsO_4) or ruthenium tetroxide are often applied to bind specific polymer domains and darken them [2], though over-staining may distort morphology and hinder techniques like selected-area electron diffraction (SAED) [52]. TEM is especially useful for visualizing internal interfaces and inclusions: it can directly show the dispersion of nanoparticles or other fillers in a polymer matrix [48]. Electron diffraction patterns in TEM reveal polymer crystallinity and orientation. A variant, cryo-TEM, flash-freezes the sample to preserve hydrated polymer structures without drying artifacts [52].

Limitations and Complementarity: Despite their strengths, SEM and TEM have significant limitations. Both operate under vacuum and require sample preparation, meaning polymers must be examined in a fixed, dehydrated state. Delicate polymer networks (e.g., hydrogels or biological tissues) can be deformed by the drying, sectioning, or staining required for electron microscopy [5]. Moreover, the intense electron beam can damage polymers by breaking bonds or causing crosslinking, leading to artifacts during imaging [52]. Additionally, EM imaging covers only a small field in grayscale, so macroscale heterogeneity and live dynamics are not captured. For comprehensive characterization, SEM/TEM are often combined with other techniques: optical or X-ray imaging can survey bulk structure and in situ behavior, while electron microscopy provides high-resolution nanoscale detail [50].

9.6 Scanning Tunneling Microscopy (STM)

Scanning tunneling microscopy (STM) is a surface-sensitive imaging technique that enables atomic-scale visualization based on the principle of quantum tunneling. A conductive tip is brought within angstroms of a sample surface, and a tunneling current is measured as a function of voltage and distance. Because of this, STM can resolve individual atoms or molecules on a surface—far beyond the capabilities of conventional light microscopy and complementary to SEM and TEM, which offer topographic and internal structure imaging but lack direct probing of electronic density states [25].

STM requires samples to be conductive or semiconductive, making it particularly useful for imaging conjugated and conducting polymers, self-assembled monolayers, and nanostructures in organic electronics. One major application has been in elucidating the surface structure and chain alignment of conducting polymers poly(*p*-phenylene vinylene) and poly(3-hexylthienylene) [53]. STM can reveal molecular packing, stacking, and local disorder in these systems, all of which influence charge transport properties. Moreover, STM has been employed to probe local electronic structures, including the density of states, enabling direct correlations between molecular arrangement and electrical behavior in semiconducting polymers and organic thin films [54].

STM has also been used to study crystalline domains and phase separation at the molecular level in block copolymers and polymer blends. For instance, polymer crystallites on surfaces have been imaged to study nucleation and orientation at the molecular scale [54]. Despite these strengths, STM is rarely used for soft or hydrated polymer systems due to its sensitivity to surface cleanliness and requirement for ultrahigh vacuum (UHV) or cryogenic conditions. Such constraints restrict its use to flat, stable, and conductive samples [54].

Compared to AFM, which does not require electrical conductivity and can function in ambient or liquid conditions, STM provides higher spatial resolution and richer electronic information but at the cost of stricter environmental and material requirements. For conductive and functional polymer surfaces, however, STM remains unmatched in its ability to resolve and analyze structural and electronic features at the atomic level [55].

9.7 Atomic Force Microscopy (AFM)

Atomic force microscopy (AFM) maps surface topography and mechanical properties at nanometer resolution by scanning a sharp tip across the polymer surface and detecting tip-sample interactions. Unlike STM, AFM works under ambient, vacuum, or liquid conditions and does not require conductive samples, making it highly suitable for soft and hydrated polymers such as hydrogels, ECM-mimetic materials, and bio-functionalized films [24, 56].

AFM offers multiple operational modes. Contact mode directly traces surface contours but may damage soft polymers. Tapping mode oscillates the tip to reduce shear forces, providing surface morphology with minimal deformation. More advanced approaches like PeakForce Quantitative Nanomechanical Mapping (QNM) allow simultaneous imaging of modulus, adhesion, dissipation, and deformation, making them ideal for characterizing polymer blends or nanocomposites [57]. AFM enables nanoscale visualization of surface roughness, fibrillar architecture, and pore morphology. It is particularly valuable for mapping stiffness heterogeneities and crosslinking patterns in hydrogels and tissue-engineered scaffolds. Nanoindentation with force spectroscopy allows quantification of Young's modulus at specific sites, revealing phase differences in block copolymers or the mechanical gradient in bioengineered matrices [58]. Real-time monitoring of polymer swelling, degradation, or cell-material interactions is also possible using fluid cell setups.

AFM variants extend their analytical reach. Conductive AFM (C-AFM) maps local electrical conductivity in doped or semiconductive polymers. Kelvin Probe Force Microscopy (KPFM) measures surface potential, aiding studies on electronic devices or coatings. AFM-IR combines tip-based IR absorption with AFM mapping to yield nanoscale chemical composition, proving useful for analyzing polymer blend interfaces or local degradation [59]. Despite being surface-confined and relatively slow over large areas, AFM remains the gold standard for nanomechanical analysis of polymer surfaces. Its ability to operate under diverse environmental conditions and to integrate mechanical, electrical, and chemical contrast makes it an essential complement to optical and electron microscopy techniques [60].

9.8 Acoustic Microscopy in Polymer Characterization

Acoustic microscopy offers a unique, non-destructive window into the internal structure and mechanical properties of polymers, using high-frequency ultrasound waves rather than light or electrons. Unlike traditional optical or electron microscopy, which are primarily surface-sensitive, acoustic microscopy enables sub-surface and volumetric imaging based on mechanical impedance contrasts—providing insights that would otherwise remain hidden without physical sectioning.

At the core of this technique lies the principle of acoustic impedance, a material-specific property derived from the product of density and sound velocity. When ultrasound waves encounter regions of different impedance within a polymer sample—such as phase boundaries, defects, or inclusions—they are partially reflected and collected to construct high-resolution acoustic images. The most common variant, **Scanning Acoustic Microscopy (SAM)** [61–63], provides micron-scale resolution of internal features, while emerging tomographic approaches push toward 3D mechanical mapping. In contrast, **Photoacoustic Microscopy (PAM)** [45], covered in Sect. 9.3.4, combines optical excitation with acoustic detection to reveal absorption-based contrast deep within optically scattering polymer matrices.

What makes acoustic microscopy particularly relevant for polymer characterization is its ability to non-invasively probe internal interfaces and map mechanical heterogeneity in layered, hydrated, or thick samples—contexts where optical microscopy struggles with light scattering or shallow depth, and electron microscopy fails due to vacuum or conductivity constraints. For example, in 3D-printed constructs, SAM can assess adhesion quality between layers, while in hydrogels or soft scaffolds, it enables the visualization of internal swelling or porosity changes over time.

Acoustic microscopy effectively bridges the resolution–depth trade-off that limits other modalities. It operates in the mid-resolution regime (typically ~1–10 μm), offering deeper penetration than optical techniques (e.g., OCT or multiphoton microscopy) while achieving finer resolution than clinical tools like ultrasound or CT. This places it in a sweet spot for mesoscopic imaging critical for understanding how local mechanical variations influence bulk polymer function. When used in tandem with chemical, optical, and electron-based microscopies, acoustic imaging completes a holistic view of polymer systems—linking molecular-level interactions to bulk performance, and surface properties to internal architecture without compromising sample integrity or requiring invasive preparation.

9.9 Conclusion and Future Directions: Toward Integrative and Intelligent Microscopy

Microscopy has long been central to the characterization of polymers, from revealing gross morphology to resolving nanoscale structures. Yet, the field is now entering a transformative phase. As polymer materials grow more complex, soft, hydrated, and multifunctional, the demand is shifting from isolated imaging events to integrated, context-aware microscopy systems capable of capturing dynamic, multi-parametric information across scales. The future of microscopy will be defined not just by resolution or contrast, but by correlation, interpretation, and adaptability.

In this evolving landscape, each modality contributes a distinct but complementary perspective. Transformative chemical microspectroscopy techniques such as Raman, FTIR, CARS, and SRS, enable spatially resolved compositional mapping without labels. These are particularly critical for functional polymers, where chemical gradients, phase segregation, or drug-polymer interactions must be understood in situ. The pairing of such spectral data with high-resolution structural maps has given rise to correlative microscopy—a practical and conceptual advance allowing multiple forms of contrast to converge on a single sample. For instance, pairing AFM stiffness maps with Raman-derived chemical domains, or aligning fluorescence-based protein localization with SHG fiber orientation in scaffolds, provides a more holistic understanding of structure–function relationships in advanced materials.

Going forward, several frontiers are positioned to redefine microscopy:

- **Resolution**: Label-free nanoscopy techniques are narrowing the gap between super-resolution and intrinsic contrast, enabling high-fidelity imaging without exogenous agents.
- **Speed**: Ultrafast optical modalities—combined with smarter detection and denoising strategies—now allow for real-time observation of polymer deformation, swelling, and degradation.
- **Accuracy and Selectivity**: Advanced light–matter modeling, illumination engineering, and noise characterization are enhancing signal fidelity. Spatial and spectral selectivity are becoming critical for imaging discrete phases or features in chemically complex environments.
- **Automation and Interpretation**: AI-driven analysis is shifting from basic segmentation to **interpretive learning**, with machine learning models now capable of identifying emergent structures, phase transitions, or damage patterns in multi-modal datasets. These models not only accelerate data processing but also offer new ways to **discover latent features**, particularly when grounded in physical or chemical priors.

Most significantly, multi-modality and correlative microscopy will become the new standard. The idea is not to replace one method with another, but to create workflows where information from optical, electron, scanning probe, and spectroscopic systems feed into each other, each resolving ambiguities or extending the insight of the other. This is especially important in biomaterials and polymer–cell hybrid systems, where the material's function is often inseparable from its microstructural and biochemical evolution. In this future, microscopy is not just an image, but a narrative of material behavior, responsive to questions across biology, chemistry, mechanics, and design. As new polymers are engineered for regenerative medicine, soft robotics, smart packaging, and beyond, microscopy will remain the essential interface between material form and scientific understanding.

Acknowledgements We used the Biorender student license (BioRender.com) for the illustrations in this article.

Competing Interest The authors declare no competing interests.

References

1. Linda C, Sawyer DTG, Meyers GF (2008) Polymer microscopy. 3 ed. Springer New York, NY
2. Murr L (2009) Imaging systems and materials characterization. Mater Charact 60(5):397–414
3. Agarwal K, Sahay R, Baji A (2020) Tensile properties of composite reinforced with three-dimensional printed fibers. Polymers 12(5):1089
4. Agarwal K et al (2019) Biomimetic tough helicoidally structured material through novel electrospinning based additive manufacturing. MRS Adv 4(43):2345–2354
5. Chapman DV et al (2020) Optical super-resolution microscopy in polymer science. Prog Polym Sci 111:101312
6. Ghosh B et al (2021) Attenuation corrected-optical coherence tomography for quantitative assessment of skin wound healing and scar morphology. J Biophotonics 14(4):e202000357
7. Agarwal K et al (2021) Development of strong and tough electrospun fiber-reinforced composites. Electrospun polymers and composites, pp 287–313
8. Agarwal K et al (2018) Additive manufacturing enabled by electrospinning for tougher bio-inspired materials. Adv Mater Sci Eng 2018(1):8460751
9. Wang Y et al (2021) Scanning probe microscopy for electrocatalysis. Matter 4(11):3483–3514
10. Agarwal K et al (2023) Acoustic velocity estimation of salmon fish scales using acoustic microscope. In: 2023 IEEE international ultrasonics symposium (IUS). IEEE
11. Ghosh B et al (2024) Label-free acoustic and optical microscopy of live tumor spheroids in hydrogel for high-throughput 3D In-vitro drug screening. bioRxiv, p 2024.08.28.610181
12. Agarwal K et al (2024) 2P1–6 Acoustic properties of chitosan-GelMA composite hydrogel. In: Proceedings of symposium on ultrasonic electronics. Institute for Ultrasonic Elecronics
13. Agarwal K et al, Visualization of ultrasonic waves in piezoelectric materials
14. Ojha S et al (2024) Quantification of impedance and mechanical properties of Zeonor using scanning acoustic microscopy. Appl Acoust 221:109981
15. Shelke A et al (2024) 2E2–2 Mechanical properties estimation of PVDF polymer using scanning acoustic microscopy. In: Proceedings of symposium on ultrasonic electronics. Institute for Ultrasonic Elecronics
16. Bhowmik A et al (2022) Portable, handheld, and affordable blood perfusion imager for screening of subsurface cancer in resource-limited settings. Proc Natl Acad Sci 119(2):e2026201119
17. Bellisola G, Sorio C (2011) Infrared spectroscopy and microscopy in cancer research and diagnosis. Am J Cancer Res 2(1):1
18. Young RJAL, PA (2011) Introduction to polymers. 3rd Edition ed. CRC Press
19. Ghosh B, Agarwal K (2023) Viewing life without labels under optical microscopes. Commun Biol 6(1):559

20. Coceancigh H, Higgins DA, Ito T (2018) Optical microscopic techniques for synthetic polymer characterization. Anal Chem 91(1):405–424
21. Agarwal K et al (2020) Impact-resistant and tough helicoidally aligned ribbon reinforced composites with tunable mechanical properties via integrated additive manufacturing methodologies. ACS Appl Polymer Mater 2(8):3491–3504
22. Kundu S et al (2020) Surface ligand-controlled wavelength-tunable luminescence of gold nanoclusters: cellular imaging and smart fluorescent probes for amyloid detection. ACS Appl Bio Mater 3(7):4282–4293
23. Budiman AS et al (2022) Modeling impact mechanics of 3d helicoidally architected polymer composites enabled by additive manufacturing for lightweight silicon photovoltaics technology. Polymers 14(6):1228
24. Ghosh B et al (2021) Arecanut-induced fibrosis display dual phases of reorganising glycans and amides in skin extracellular matrix. Int J Biol Macromol 185:251–263
25. Rabe JP (1992) Molecules at interfaces: STM in materials and life sciences. Ultramicroscopy 42:41–54
26. Agarwal K et al (2024) Uncertainty analysis of Altantic salmon fish scale's acoustic impedance using 30 MHz C-Scan measurements. Ultrasonics 142:107360
27. Ojha S et al (2024) 1P5–13 Specific acoustic impedance mapping of shrimp scale using scanning acoustic microscopy. In: Proceedings of symposium on ultrasonic electronics. Institute for Ultrasonic Elecronics
28. Ogunlade O et al (2019) Monitoring neovascularization and integration of decellularized human scaffolds using photoacoustic imaging. Photoacoustics 13:76–84
29. Ghosh B et al (2024) Molecular histopathology of matrix proteins through autofluorescence super-resolution microscopy. Sci Rep 14(1):10524
30. Butola A et al (2024) Morphological imaging of 3D cultured kidney mesangial cells using gradient light interference microscopy. In: Digital holography and three-dimensional imaging. Optica Publishing Group
31. Liu S et al (2023) Optical mapping of the evolution of water content during the swelling of hydrophilic polymers. Chem Commun 59(5):599–602
32. Rosa D et al (2005) The use of optical microscopy to follow the degradation of isotactic polypropylene (iPP) subjected to natural and accelerated ageing. Polym Testing 24(8):1022–1026
33. Ghosh B, Agarwal K (2024) Kidney mesangium organotypic culture reveals co-stimulatory mechano-chemical interplay for fibrotic manifestation. Biophys J 123(3):405a
34. Kumar N et al (2023) Multilayered "SMART" hydrogel systems for on-site drug delivery applications. J Drug Delivery Sci Technol 80:104111
35. Ghosh B, Fenton KA, Agarwal K (2023) Mechano-chemical insights in diabetic kidney disease through 3D pathotypic model of mesangium. bioRxiv, p 2023.12. 10.571017
36. Butola A et al (2024) Label-free correlative morpho-chemical tomography of 3D kidney mesangial cells. arXiv preprint arXiv:2409.10971
37. Chalmers JM, Everall NJ (1995) The bole of vibrational spectroscopy-microscopy techniques in polymer characterisation. In: Macromolecular symposia. Wiley Online Library
38. Ghosh B et al (2024) 2P1–7 scanning acoustic microscope for visualizing 3D cell clusters embedded in hydrogel systems. In: Proceedings of symposium on ultrasonic electronics. Institute for Ultrasonic Elecronics
39. Kachwal V, Tan JC (2023) Stimuli-responsive electrospun fluorescent fibers augmented with aggregation-induced emission (AIE) for smart applications. Adv Sci 10(1):2204848
40. Valli J et al (2021) Seeing beyond the limit: a guide to choosing the right super-resolution microscopy technique. J Biol Chem 297(1)
41. Zhao Z et al (2023) Two-photon synthetic aperture microscopy for minimally invasive fast 3D imaging of native subcellular behaviors in deep tissue. Cell 186(11):2475–2491. e22
42. Ghosh B, Agarwal K (2024) Randomly oriented microgrooved hydrogel guides cellular motility, modulates speed, and governs directionality of cellular spread. bioRxiv, p 2024.09. 10.612339

43. Sarkar A et al (2016) Autofluorescence signatures for classifying lung cells during epithelial mesenchymal transition. RSC Adv 6(81):77953–77962
44. Ghosh B, Chatterjee J (2023) Advances in medical imaging for wound repair and regenerative medicine. Regenerative Medicine: Emerging Techniques to Translation Approaches. Springer, pp 57–76
45. Ranjan A, Agarwal K, Melandsø F (2023) Photoacoustic and ultrasound imaging of caridea for quantification of pigments and hardness measurement. In: 2023 IEEE International ultrasonics symposium (IUS). IEEE
46. Ghosh B et al (2021) Quantitative in situ imaging and grading of oral precancer with attenuation corrected-optical coherence tomography. Oral Oncol 117:105216
47. Ghosh B et al (2021) Structural mechanics modeling reveals stress-adaptive features of cutaneous scars. Biomech Model Mechanobiol 20:371–377
48. Alqaheem Y, Alomair AA (2020) Microscopy and spectroscopy techniques for characterization of polymeric membranes. Membranes 10(2):33
49. Martinez-Garcia FD et al (2022) A beginner's guide to the characterization of hydrogel microarchitecture for cellular applications. Gels 8(9):535
50. Wirth R (2009) Focused Ion Beam (FIB) combined with SEM and TEM: advanced analytical tools for studies of chemical composition, microstructure and crystal structure in geomaterials on a nanometre scale. Chem Geol 261(3–4):217–229
51. Furtado AS et al (2025) 3D-Printed PCL-based scaffolds with high nanosized synthetic smectic clay content: fabrication, mechanical properties, and biological evaluation for bone tissue engineering. Int J Nanomed, 53–69
52. Chen J (2021) Advanced electron microscopy of nanophased synthetic polymers and soft complexes for energy and medicine applications. Nanomaterials 11(9):2405
53. Bond S, Friend R, Howie A (1993) Scanning tunnelling microscopy of conjugated polymers. J Microsc 171(3):199–204
54. Gao H-J, Gao L (2010) Scanning tunneling microscopy of functional nanostructures on solid surfaces: manipulation, self-assembly, and applications. Prog Surf Sci 85(1–4):28–91
55. Bian K et al (2021) Scanning probe microscopy. Nat Rev Methods Primers 1(1):36
56. Lee D, Zhang H, Ryu S (2018) Elastic modulus measurement of hydrogels. Cellulose-based superabsorbent hydrogels, pp 1–21
57. Pittenger B, Erina N, Su C (2013) Mechanical property mapping at the nanoscale using PeakForce QNM scanning probe technique. Nanomechanical analysis of high performance materials. Springer, pp 31–51
58. Reggente M et al (2015) Atomic force microscopy techniques for nanomechanical characterization: a polymeric case study. Jom 67:849–857
59. Dazzi A et al (2012) AFM–IR: combining atomic force microscopy and infrared spectroscopy for nanoscale chemical characterization. Appl Spectrosc 66(12):1365–1384
60. Geisse NA (2009) AFM and combined optical techniques. Mater Today 12(7–8):40–45
61. Agarwal K et al (2022) 1Pb3–5 Visualization of ultrasonic waves in piezoelectric materials. In: Proceedings of symposium on ultrasonic electronics. Institute for Ultrasonic Elecronics
62. Singh H et al (2023) Digital refocusing in acoustic microscopy images using adversarial autoencoders. In: 2023 IEEE international ultrasonics symposium (IUS). IEEE
63. Khan MSA et al (2024) 2P2–13 coded signal scanning acoustic microscopy. In: Proceedings of symposium on ultrasonic electronics. Institute for Ultrasonic Elecronics

Komal Agarwal is a Principal Investigator at UiT The Arctic University of Norway, with expertise in polymer thin films and biomaterials for advanced functional applications. Her research spans piezoelectric and impact-resistant polymers, bio-inspired adhesives, and polymeric scaffolds for tissue engineering. She brings a strong innovation background with contributions to synthetic polymer design, biomimetic systems, and applied materials science.

Nikhil Kumar is a senior research fellow the Advanced Technology Development Centre, Indian Institute of Technology Kharagpur. He is specialized in fabrication of smart polymeric materials for diagnosis and therapeutics. Nikhil completed his Bachelor of Pharmacy and Master of Pharmacy in Pharmacology. His masters research work was focused on developing dressings for diabetic foot ulcers. His current research revolves around the design and application of smart polymeric biomaterials for advanced diagnosis and therapeutic interventions for cutaneous squamous cell carcinoma. He has published 12 peer-reviewed research articles. He is actively engaged in research focused on polymer characterization, with particular interest in the structural, thermal, and rheological analysis of advanced polymeric materials. Mr. Kumar has been involved in research projects that explore the application of spectroscopic techniques (FTIR, NMR), thermal analysis (DSC, TGA), to evaluate polymer performance under varied environmental conditions. Mr. Kumar is also developing a novel diagnostic tool that integrates multiple characterization techniques FTIR, Raman spectroscopy, AFM and nanoindentation to probe and differentiate pathological tissue conditions such as fibrosis and hyperkeratosis.

Biswajoy Ghosh is Principal Investigator at UiT The Arctic University of Norway, specializing in biomaterials and engineered organ systems. His interdisciplinary work integrates polymer-based materials in organ-on-chip platforms to study chronic kidney disease and cardiovascular biology. With over a decade of experience, he leads innovation-driven research in biomimetic tissue engineering and has contributed extensively to international publications and patents.

Chapter 10
Accelerated Stability Studies of Polymers

Chandra Khatua and Soumajyoti Ghosh

Abstract It has been a common practice to observe changes in the structural and chemical characteristics of polymers to comprehend their degradation process. The environmental conditions that a polymer is exposed to have a significant impact on its service life. However, the polymer's stability, including nanocomposites or polymer blends, governs its value. The characteristics of degradation can be investigated by using different appropriate analysis techniques. Therefore, the purpose of this chapter is to discuss the variables that may have a major impact on the degradation and accelerated stability testing of the polymer. This chapter discusses a variety of harsh conditions, such as artificial weathering, hot/wet conditions, and chemical exposure, with a primary focus on polymers and fiber-reinforced plastics (FRP).

Keywords Accelerated stability studies · Polymer degradation · Accelerated stability

10.1 Introduction

Polymers are an essential component of many common materials, from kitchenware to parts of electronic equipment. They are widely used in electronics, healthcare, construction, and packaging. The manufacturing and use of these materials have increased exponentially due to their convenience and adaptability. However, there are also serious environmental concerns due to this consumption boom. The service life of polymeric materials in the environment due to their intrinsic stability and resistance to degradation causes serious pollution problems. However, under different environmental factors, the qualities of the polymer may deteriorate over time. This deterioration can lead to failure and impair their performance in applications where consistency is important. Understanding how environmental factors influence polymer properties helps design better, environmentally stable, and durable

C. Khatua (✉) · S. Ghosh
Rubber Technology Centre, Indian Institute of Technology Kharagpur, Kharagpur, India
e-mail: chandra.khatua@gmail.com

S. Chattopadhyay and N. Kumar (eds.), *Polymer Characterization*,
https://doi.org/10.1007/978-981-95-2609-3_10

products. To determine the shelf life of a polymer under typical conditions, accelerated stability measurements occur by exposing the polymer to specific conditions that accelerate degradation processes. This eliminates the need for lengthy waiting times for shelf-life prediction and enables researchers to determine how long a polymer will retain its intended qualities in practical settings. This chapter focuses on accelerated degradation of polymers (thermosets, including adhesives, and thermoplastics) and fiber-reinforced plastics (FRPs). It takes into account a variety of harsh conditions, such as chemical (i.e., solvents, caustic, and acidic solutions), aqueous (liquid and vapour), and artificial weathering. Tests for static and creep rupture at both ambient and non-ambient temperatures are discussed.

10.2 Overview of Polymers' Degradability and Accelerated Stability

The stability of the polymer describes how a material's characteristics change after being exposed to heat, oxygen, moisture, sunlight (directly or indirectly), and chemicals. The changes in the polymer's properties after being used as the product are usually undesirable and lead to the deterioration of the product quality. There are connections between the terms "degradation" and "durability," as well as "long-term properties" and "end-of-life" of polymers. Long-term polymer disintegration occurs in a shorter period using techniques known as "accelerated polymer degradation," which speed up the degradation process of polymers. This is accomplished by subjecting the polymers to circumstances that accelerate the breakdown process, such as elevated temperatures, ultraviolet light, or exposure to specific chemicals. The external influences that can have a major effect on the deterioration process are chemicals, ozone, salt, contaminants (such as sulfur and nitrogen oxides), and microorganisms [1]. Additionally, other substances that are included in the polymer matrix, including dyes, pigments, processing additives, absorbers, and stabilizers, will be impacted by these environmental factors. Sometimes, it is subjected to the combined effect of several stimuli. Polymer degradation generally involves a change in molecular weight and distribution, a change in molecular structure, reduced ductility, embrittlement, cracking, color change, and a change in desirable properties. The polymer's deterioration generally begins at the interface between amorphous and crystalline phases and impacts their mechanical characteristics, thermal stability, lamellar thickness, and crystallinity [2]. Bio-based polymers are expected to perform well, still, how well their long-term qualities are preserved depends on how well their strengths are conserved. It is critical to choose appropriate degradation pathways to get rid of plastic garbage. To reduce the adverse effects on the environment, a plastic product's durability and end-of-life pathway must be balanced. To give such items second-life usage, the end-of-life of a product should be taken into consideration rather than the end-of-life of a material. Consequently, the product's value and service life will be improved. One easy way to make the above concept a reality is to move waste polymers to lower-grade applications where they will be integrated into other materials to increase their usefulness.

The factors that are affecting the polymer stability [3, 4]:

- **Polymer structure**: The stability of the polymer chain is influenced by its chemical bonds in the chain length, including the existence of functional groups.
- **Molecular weight**: Polymers with a higher molecular weight are usually more stable.
- **Additives**: To improve polymer stability, stabilizers such as UV absorbers and antioxidants can be added.
- **Environmental conditions**: The degradation of polymers can be greatly influenced by environmental factors like temperature, humidity, light exposure, and chemical exposure.

To predict a material's shelf life under typical storage conditions, accelerated stability testing involves exposing the polymer to a specific environment that quickens degradation processes, such as high temperatures or ultraviolet light. This minimizes the need for lengthy waiting times for stability determination and enables the determination of how long a polymer will retain its intended qualities in a practical environment. Accelerated stability is important for predicting the shelf life of polymer, efficiency of time, and economic viability. Accelerated testing methods enable more precise determination of shelf-life by predicting how long a polymer will be stable and functioning under typical conditions. It offers a significantly faster way of assessing a polymer's degradation behaviour than waiting a long time for it to break down naturally. Accelerated approaches can drastically lower the cost and resources required to evaluate polymer stability by cutting down on testing time. Accelerated stability determination is important in different industries like pharmaceuticals, cosmetics, food packaging, and other industries. In pharmaceutical industries, to ascertain the shelf life of drugs, they are exposed to accelerated stress conditions. Accelerated stability studies also evaluate the stability of chemicals and formulations used in cosmetics and assess the longevity and stability of materials used in food packaging.

10.3 Environmental Factors and Functions of Accelerated Stability

The degradation behavior of polymers strongly depends on different environmental influences, including physical, chemical, and biological. Environmental factors like heat, light, moisture, chemicals, etc., play a role in the degradation of their properties [2] (Table 10.1). Therefore, understanding these issues is crucial for selecting the appropriate polymers and designing durable products for specific applications.

It is important to know the stability of the polymer before using the end product. Various analysis methods can determine their stability. Accordingly, the durability of polymers will be increased by employing stabilizers, selecting resilient components,

Table 10.1 Different environmental factors affect polymeric degradation

Environmental factors	Types of degradation
Sunlight	Photodegradation
Heat	Thermal degradation
Atmospheric agent	Oxidative degradation
Humidity	Hydrolytic degradation
Microorganisms	Biodegradation
Light + Oxygen	Oxidative photodegradation
Heat + Oxygen	Thermoxidative degradation
Light + Humidity	Photohydrolytic degradation

and streamlining production procedures. Additionally, proper handling and storage prolong the life of these materials.

Accelerated testing entails subjecting the polymer to certain environmental conditions that speed up the degradation rate. The outcomes of accelerated studies are analysed to determine the rate of degradation and to extrapolate the findings to real conditions.

- Temperature: Chemical reactions that cause polymer degradation can be accelerated by higher temperatures.
- Humidity: In certain polymers, higher humidity can faster the hydrolysis reactions.
- UV Radiation: Certain polymers can degrade and undergo photodecomposition when exposed to UV light.
- Additional Environmental Factors: Accelerated degradation may also involve other environmental factors like mechanical stress, oxygen exposure, or different pH levels.

Accelerated stability of the polymer is mainly determined by the Arrhenius equation. It explains the effect of temperature on the rate of reaction. In Eq. 10.1, the standard Arrhenius equation is shown (Eqs. 1 and 2). According to this equation, the speed of the reaction increases about 2–3 times for every 10° rise I temperature.

$$k = Ae^{-E_a/RT} \tag{10.1}$$

$$log\frac{k}{RT} = logA - \frac{E_a}{2.303} \tag{10.2}$$

A = Arrhenius Constant, E_a = Energy of activation, RT = Ideal gas constant, Arrhenius Constant is the frequency of molecular collisions occurring between molecules.

The other is the humidity-modified Arrhenius equation (Eq. 3), which is important for humidity-sensitive polymers. Equation 10.3: humidity-adapted Arrhenius equation, which is an extended form of the standard Arrhenius equation by the factor B(RH).

$$lnk = lnA - \frac{E_a}{RT} + B(RH) \tag{10.3}$$

ln k: specification limit or conversion time, *ln A*: collision frequency, Ea: activation energy, RT: 1.986 cal/deg B: humidity sensitivity factor, RH: equilibrium relative humidity.

Typical values of B are between 0 and 0.10, where 0 designates low moisture sensitivity and 0.10 refers to high moisture sensitivity. E_a values are between 10–45 kcal/mol. The degradation depends on E_a value and increases with E_a value.

There are some limitations to the determination of accelerated polymer stability as it depends on assumptions, complex degradation mechanisms, and different environmental factors.

- Accelerated tests are predicated on assumptions that may not always be true, such as the deterioration process being mostly temperature-dependent.
- Simple temperature-dependent kinetics may not adequately capture due to the complexity of some degradation processes.
- Not every response depends on temperature. Arrhenius equation of stability studies only considers the temperature. Some are also influenced by humidity, oxygen, ultraviolet light, etc.

10.3.1 Heat exposure

Heat exposure to the polymer at elevated temperatures breaks down the chemical bonds into smaller parts that do not recombine when cooled. It causes physical, mechanical, and electrical properties of the polymer. Thermal oxidation and chain scission are two primary mechanisms through which heat impacts the integrity of polymers [5]. Oxidation processes take place when polymers are heated to high temperatures in the presence of oxygen [6]. Free radicals are formed as a result of these interactions, and they propagate chain reactions that degrade the polymer. Chain scission can result in breaking the bonds inside the polymer backbones due to high heat. This process causes deterioration of the molecular weight and physical characteristics like elasticity and tensile strength. Elevated temperatures considerably accelerate chemical reactions and lead to bond breakage and thermal degradation [6]. For the overall degradation process, a lengthy induction time is required, during which minimal degradation is observed. The mechanical properties of the polymer have significantly deteriorated due to a sharp rise in degradation at the end of this process. Temperature affects this induction period, which can be reduced at higher temperatures. The serviceable lifetime of the polymer can often be considered as the induction period of the deterioration process. The Arrhenius equation is used to correlate degradation rates at various temperatures in order to predict the degradation rate under typical conditions. Several processes are used to minimize thermal degradation, such as adding stabilizers and radical scavengers and controlling oxidation in an inert environment. Structural stability, additives, and environmental conditions determine the understanding of polymers to thermal degradation [7]. In general, polymers with more stable structures, like cross-linking or aromatic rings, are more heat resistant. Polymers with strong bonds like C-F in Teflon showed better thermal resistance. Some additives, including UV stabilizers and antioxidants, are used to slow down the deterioration process and improve the thermal stability of polymers [8]. The extent of thermal degradation is significantly influenced by temperature, exposure time, and the presence of oxygen. Oxygen diffuses in adhesively bonded joints through the adhesive, which can accelerate at higher temperatures. Metallic adherends act as a barrier to the gas as they are impervious to oxygen. FRPs, on the other hand, degrade at a significantly faster rate because of their permeability to ambient gases, and the fibre matrix interface is also weakened at the elevated temperatures. Temperature can cause additional physical changes in polymers, thermal expansion being one of the most common. As long as the temperature changes don't cause any harm, thermal expansion is a reversible process. In FRPs, residual stresses can be caused by differences in the thermal expansion of the fibre and matrix, which can reduce the efficiency of the fiber-matrix bond and cause thermo-mechanical degradation during thermal cycling. Low temperatures can cause embrittlement of plastics, debonding in adhesive joints, and fiber-matrix interfacial cracking in FRPs. At below-freezing temperatures, environmentally deteriorated polymeric materials may break quickly. When multicomponent systems are exposed to cryogenic temperatures, complete interfacial debonding may happen due to the combined effects of embrittlement and differential thermal contraction. The design of polymers, especially used in high-temperature applications like the automotive or aerospace industries, requires an understanding of how heat impacts polymer deterioration. With this information, engineers can choose the right polymers, fabricate the components that reduce thermal risks, and design novel materials with improved heat resistance.

10.3.2 Light Exposure

When polymers are exposed to sunlight and some artificial lights, light-induced homolytic fission of chemical bonds (photolysis) and photo-oxidation occur. Photodegradation is caused by visible light and ultraviolet (UV) light, which destroy chemical bonds of polymer (polypropylene and polyethylene terephthalate), eventually resulting in cracking, chalking, color changing, and the deterioration of the physical properties [9]. UV radiation from sunlight has a higher binding energy than the C-H and C–C bonds, which causes the molecular chains to break, lowering their molecular weight and mechanical characteristics [10]. Chain scission results from oxidation reactions that are started by light energy (Fig. 10.1). The light must be absorbed by the polymeric system. Thus, the commencement of any photochemical reaction requires the presence of chromophoric groups in the polymer [11]. Engineering polymers like fluoropolymers exhibit excellent UV resistance. Impurities

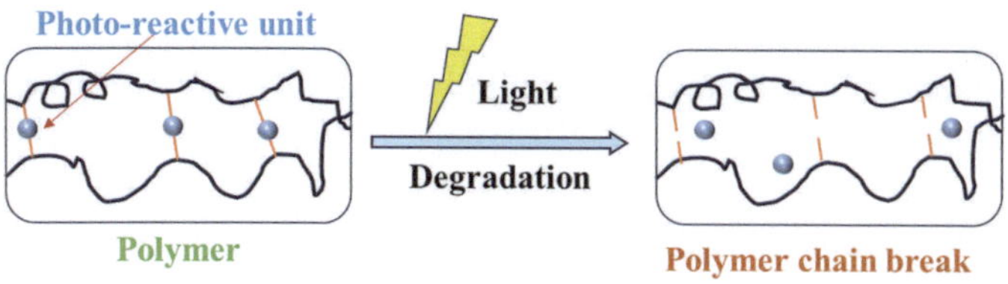

Fig. 10.1 Degradation of polymer under exposure to light

present in the polymer cause photodegradation. Additives, pigments, metallic catalysts, and residues of photoinitiators are the external impurities [12]. Internal impurities, including peroxides, hydroperoxides, and ketone chromophores, are formed during the processing of the polymer. These impurities absorb more solar light and cause photodegradation [10].

Different types of stabilizers are used in combination:

- UV screeners,
- UV absorbers,
- Singlet oxygen, ($^1O_2$), quenchers,
- Free radical scavengers,
- Hydroperoxide decomposers.

The photooxidation of polymers has been explained by two different mechanisms. In the first, singlet oxygen reacts directly with the substrate, whereas in the second, radicals are created and then react with oxygen. The stability of the polymer under light exposure is important in coating applications [13].

The light source, which should ideally mimic sun radiation, is the most important component of any accelerated stability testing system. The three primary types of artificial light sources that are utilised are fluorescent UV, xenon-arc, and carbon-arc. Three wavelengths can be used to categorise UV light, as explained below:

- UV-A radiation-which makes up 6% of all solar energy that reaches Earth, is the least damaging to polymers (315–400 nm).
- UVB, which makes up 0.1% of all sun energy that reaches Earth, are more harmful to polymers (280–315 nm).
- The earth's atmosphere filters out UV-C, which is the most damaging to polymers (<280 nm).

10.3.3 Moisture Exposure

When water or high humidity reacts with functional groups of the polymer, such as esters and amides (polyesters, polyamides), hydrolytic breakdown results. Water permeates the polymer when it comes into contact with an aqueous medium. It results in swelling, breaks down the intermolecular hydrogen bonds, and ultimately, the hydrolysis of the bonds [14]. Moisture exposure at higher temperatures can lead to the hydrolysis of the polymer. Hydrolytic cleavage of functional groups can happen in both the main chain and side chain of the polymer. Moisture changes the characteristics of polymers and accelerates bond cleavage. Sometimes, hydrolytic degradation and microbial degradation can take place simultaneously in aquatic media [15]. In that case, biocides are the common antimicrobial degradants that can be used to prevent microbial degradation. Though it is not desirable for some applications like packaging, this degradation process is useful for controlled drug delivery systems [16, 17].

10.3.4 Chemical Exposure

Only processes that are triggered by chemical reagents (such as acids, bases, solvents, oxidizing agents, reactive gases, etc.) are referred to as chemical degradation (polycarbonate, polyurethanes) [18]. Depending on the chemical agent and the structure of the polymer, a polymer may become degraded by side group reactions, crosslinking (construction of new bonds between chains), or chain scission (the breaking of polymer chains) [19]. Modifications in mechanical characteristics result from chemical degradation. Oxidative degradation occurs when exposed to oxygen. Ozone gas exposure leads to ozonolysis. These degradations usually lead to discoloration, hardening, and changes in surface properties. The extent of degradation depends on the structure of the polymer. A combination of antioxidants is used for the stability of reactive gas exposure. Solvolysis degradation occurs in the presence of acid, base, or catalyst. Understanding of chemical degradation is essential for determining the lifetime of pipes and seals. The different factors that affect on chemical degradation of polymeric substances are given below [20].

- Chemical structure of the polymer
- Types of chemical exposure
- Duration of chemical exposure

10.4 Key Aspects of a Polymer Stability Study

10.4.1 Thermal Stability

The thermal stability of the polymer is an important property that regulates how a polymer behaves at high temperatures. Polymers with higher thermal stability can maintain their structure at high temperatures, resist decomposition, and function effectively under thermal stress. The chemical composition and structure of the polymer greatly influence this attribute. Therefore, understanding of thermal stability of a polymer makes it easier to choose the right polymer for an application like aerospace, automotive, and electronics where high-temperature resistance is crucial such as electronics, automotive, aerospace, etc. ISO 2578 provides a standard process for evaluating the thermal stability of polymers exposed to high temperatures over extended periods. Accelerated stability tests for thermal endurance were carried out in air without the impact of external factors like applied stress. The reciprocal of the relevant absolute temperature and the logarithm of the time are considered to have a roughly linear connection (Arrhenius equation). The process gives ideas to the user about the maximum withstanding temperature for a specified time before deterioration of a specific property. The stability of plasticisers in polymers is covered by ISO 176 and ASTM D1203, which both quantify the plasticisers lost as a result of absorption on activated charcoal.

1. Extended or even brief exposure to high temperatures will diminish an adhesive's short-term qualities and may cause irreversible chemical and physical alterations. For accelerated stability testing, a temperature-controlled chamber is recommended to assess mechanical properties (such as stiffness and strength) at high temperatures.
2. Low temperatures can be tested using similar fundamental ideas as elevated temperatures. The size of the environmental chamber and level of insulation may have a significant impact on the testing temperatures. To avoid moisture condensation and ice development on the test specimen and test equipment, dry nitrogen gas should be used in the test chamber. Stainless steel should be used in the construction of the test instrument to prevent corrosion products from accumulating on its surface.
3. The freeze–thaw conditioning methods outlined in ASTM D7792 are intended to assess and compare the impact of 100 freeze–thaw cycles on pultruded materials in a controlled laboratory setting. The average freeze–thaw cycle range is 23 °C to −20 °C, with a total cycle time of about 2–5 h. The test chamber's capacity to heat and cool will determine the maximum cycle rate (cycle duration). Throughout the testing process, it is advised that samples be kept sealed inside polyethylene bags to reduce environmental contact and stop moisture loss from the samples.

10.4.2 Chemical Stability

The chemical stability of the polymer endures the change from chemical contact. This problem with exposure to corrosive fluids needs to be assessed to measure the resistance of the polymer's sustainability. It is important to determine the susceptibility of a polymer to chemical degradation and possible defects in its composition or application conditions by analyzing the changes in chemical structure and properties in response to various chemicals. The chemical stability studies are important for polymer selection for specific application and their quality control, development of new materials, and potential impact of polymer on the environment.

To ascertain the level of chemical degradation (i.e., chemical resistance) as a function of exposure duration, the environmentally conditioned material is put through a series of tests. Mechanical characteristics like strength, hardness, and flexural modulus, as well as dimensional stability (such as swelling), appearance, and weight change (such as gloss, colour, crazing, retention, fibre prominence, loss of surface, blister formation, etc.) are all taken into consideration when evaluating chemical resistance. ISO 877–1, 2, and 3, ASTM D1435, ASTM D4364, and ASTM G24 provide standard test methods for the weathering of polymers in intense sunlight, bright daylight utilising Fresnel mirrors. Polymer exposure in a marine environment is outlined by ISO 15314.

10.4.3 Mechanical Stability

The knowledge of the mechanical stability of the polymer provides the basis for predicting the behavior of the polymer under different load conditions [21]. The basic phenomenon involved when the polymer is subjected to high shear forces leads to the breakdown of the molecule. Mechanical degradation of the polymer occurs mainly in the solid state or the presence of solvent or moisture [22].

The accelerated test program examines its moisture absorption characteristics and simultaneously mechanical properties at equilibrium. The temporal and spatial distributions of stresses and strains within a bonded structure can be ascertained by measuring mechanical properties at intermediate moisture levels (based on average moisture contents). Strain and displacement can be measured using a variety of contact and non-contact methods, including strain gauges, crosshead movement, electronic speckle pattern interferometry (ESPI), video extensometers, and linear voltage displacement transducers (LVDTs). The measured strain will be an average strain, except for strain mapping methods (such as ESPI and digital image correlation, or DIC). Furthermore, strain is only recorded at the gauge location.

10.4.4 Light Stability

The light stability of the polymer is another crucial feature. It causes the loss of physical properties and discoloration. The photons mainly found in sunlight, such as infrared radiation, visible light, and ultraviolet light, affect the stability of the material. Therefore, investigation of light stability is important for applications like coating, drug delivery, etc. [23, 24].

- Carbon-arc lamps: Almost all polymer degradation is caused by UV light with wavelengths less than 320 nm. Carbon-arc lamps are nevertheless used in industry today, and ASTM D1499 and ISO 4892–4 standardise its application for polymers.
- Xenon lamps: Although xenon-arc lamps are far more costly than other light sources, they provide a better spectrum simulation of natural sunlight. UV wavelengths shorter than those present in solar radiation are in the xenon-arc spectra,

but these are readily eliminated by optical filters. To avoid the specimen overheating, the high quantities of infrared radiation emitted by xenon lamps must also be eliminated. Excellent spectral reproduction of the entire solar spectrum is possible with a xenon lamp with the proper optical filter (such as a daylight filter), which replicates both the UV and the sunlight. Nowadays, xenon lamps are thought to be the best option when the entire spectrum of the sun is needed. Standardised test protocols with xenon lamps are found in ASTM D265 and ISO 4892–2.

- Fluorescent UV lamps: Fluorescent tubes, compared to xenon-arc lamps, are far less expensive and can replicate the solar spectrum. The long wavelength spectra from a low-pressure mercury vapour source are re-emitted to create the radiation from a fluorescent bulb. The light emitted by the tubes may have a different spectral distribution depending on the type of tube utilised. UV-B tubes can be used to expedite the tests quickly. They expedite the deterioration process because they have a significant UV peak at about 313 nm. UV-A fluorescent lamps produce spectra that are more similar to solar radiation because they emit light at a considerably higher wavelength. Q Panel's UVA-340 lamps are widely utilised for testing because of their UV spectrum (295–350 nm), which is quite similar to that of sun radiation. ASTM D4329 and ISO 4892–3 provide standard test techniques for the use of fluorescent tubes.

10.4.5 Microbial Stability

Bacterial degradation of the polymer by the microbes is very important to analyse [25, 26]. Enzymes secreted by living organisms, such as bacteria and fungi, rupture the polymer chains [25]. The mechanism of microbial degradation is illustrated in Fig. 10.2. Several factors affect the polymeric biodegradation [27] (Fig. 10.3). Sometimes, hydrolytic degradation and biodegradation occur simultaneously in aquatic media. It is particularly very important to investigate biomedical applications. It is important for the biopolymers that are produced from natural sources, either chemically synthesized from biodegradable material or entirely biosynthesized from living organisms [28]. The main degradation mechanism of interest in vivo for synthetic bioresorbable polymers or natural biopolymers is chemical degradation by hydrolysis. The process of hydrolysis forms a weak linkage of the polymer chain after reacting with water molecules, resulting in the bond breaks, shortening the chain, and ultimately leading to deterioration. Anhydride, ortho-ester, ester, carbonate, and amide bonds are among the bonds that are subject to passive hydrolytic breakdown. The stability of the biopolymer depends on the temperature, light, moisture, pH, microbial activity, exposed surface area, etc. It is particularly very important to study for biomedical applications.

Chemical pigments and additives are typically susceptible to microbial attack, which typically happens on exposed surfaces as a result of the additives' oxidation.

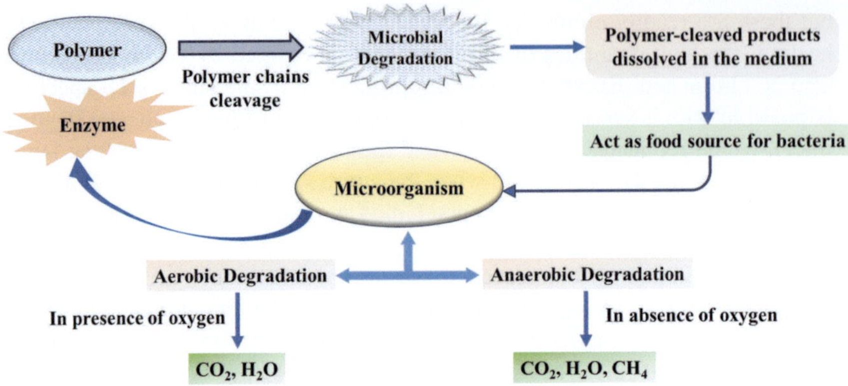

Fig. 10.2 Mechanism of microbial degradation of the polymer

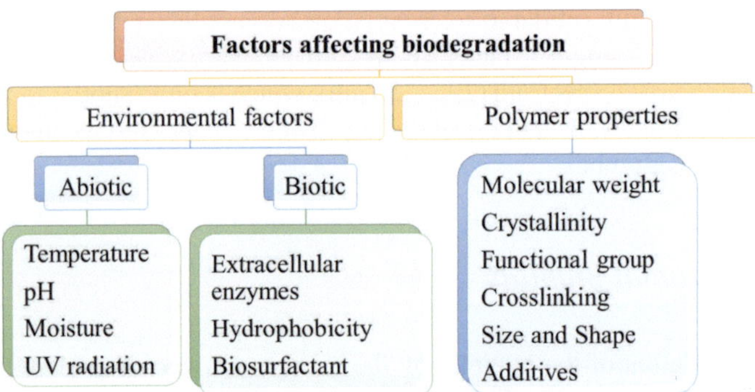

Fig. 10.3 Factors affecting the polymeric microbial degradation

Polymers with high water resistance are more resilient to microbial attack. Microorganisms develop more quickly in warm, humid settings than in cold, dry ones, therefore, geographic location and seasonal impacts are crucial. Microbial testing involves exposing items that are conducive to microbial development. The degree and length of microbial attack will depend on the weather and the angle of exposure to sunlight.

Microbial stability: Analysis by following some established standards protocol, like ISO 846 for microbial testing of plastics.

10.5 Mitigation Strategies for Polymer Degradation

There are different strategies to lessen the polymer degradation. The lifespan of polymers will be improved by employing stabilizers, selecting resilient components, and optimizing production procedures [29, 30]. Appropriate handling and storage of polymer products also influence their prolonged service life.

- Stabilizers and additives
  - Antioxidants: By scavenging free radicals or breaking down peroxides, they prevent oxidative deterioration (hindered phenols, phosphites).
  - UV stabilizers: Block or absorb UV rays to avoid photodegradation (benzotriazoles, benzophenones).
  - Hydrolysis stabilizers: Decrease the rate of hydrolytic breakdown by reacting with water molecules or hydrolyzable groups (carbodiimides, epoxides).
- Material selection and design
  - Select polymers that are naturally resistant to particular environmental conditions (fluoropolymers, silicones).
  - To reduce exposure to degrading chemicals, products should be designed with protective coatings or barriers.
- Processing and manufacturing optimization
  - Reduce production deterioration by optimizing processing conditions (lower temperatures, inert atmosphere).
  - To increase the polymer's resistance to external influences, and regulate cooling rates and residual stresses.
- Storage and handling practices
- To reduce exposure to heat, moisture, and light, keep polymers in a cold, dry, and dark place.
- To prevent polymers from becoming contaminated by chemicals while being stored and transported, use the proper packing materials (moisture barrier films, desiccants).

10.6 Conclusions

Accelerated stability studies offer a useful technique for predicting the long-term performance of polymers for different applications by subjecting the polymers to controlled stress conditions and examining the degradation behaviour. Choosing the appropriate polymer for a given application requires an understanding of these variables and how they interact with different types of polymers. Polymers are essential in many applications, but a major problem is their performance and service life under different environmental conditions. When combined with sophisticated analytical methods, a thorough understanding of polymeric degradation pathways enables scientists and engineers to enhance polymer technology and produce materials that will satisfy future needs.

Acknowledgements We have used bioRender.com (student licence) for making illustration.

Conflict of Interest Authors declare no competing conflict of interest.

References

1. Hawkins WL (1984) Polymer degradation. In: Polymer degradation and stabilization. Springer, pp 3–34
2. Göpferich A (1996) Mechanisms of polymer degradation and erosion. Biomater: Silver Jubil Compend 117–128
3. Sazali N, Ibrahim H, Jamaludin AS, et al (2020) Degradation and stability of polymer: a mini review. In: IOP conference series: materials science and engineering. IOP Publishing, p 012048
4. Jouenne S (2020) Polymer flooding in high temperature, high salinity conditions: selection of polymer type and polymer chemistry, thermal stability. J Pet Sci Eng 195:107545
5. David C (1975) Thermal degradation of polymers. In: Comprehensive chemical kinetics. Elsevier, pp 1–173
6. Madorsky SL, Straus S (1959) Thermal degradation of polymers at high temperatures. J Res Natl Bur Stand A Phys Chem 63:261
7. Pielichowski K, Njuguna J, Majka TM (2022) Thermal degradation of polymeric materials. Elsevier
8. Ray S, Cooney RP (2018) Thermal degradation of polymer and polymer composites. In: Handbook of environmental degradation of materials. Elsevier, pp 185–206
9. Rånby B (1989) Photodegradation and photo-oxidation of synthetic polymers. J Anal Appl Pyrolysis 15:237–247
10. Woo RSC, Chen Y, Zhu H et al (2007) Environmental degradation of epoxy–organoclay nanocomposites due to UV exposure. Part I: Photo-degradation. Compos Sci Technol 67:3448–3456
11. Rabek JF (2012) Polymer photodegradation: mechanisms and experimental methods. Springer Science & Business Media
12. Pospíšil J, Pilař J, Billingham NC et al (2006) Factors affecting accelerated testing of polymer photostability. Polym Degrad Stab 91:417–422
13. Makki H, Adema KNS, Peters EAJF et al (2014) A simulation approach to study photo-degradation processes of polymeric coatings. Polym Degrad Stab 105:68–79
14. Roy S (2011) Moisture-induced degradation. In: Long-term durability of polymeric matrix composites. Springer, pp 181–236
15. Fan XJ, Lee SWR, Han Q (2009) Experimental investigations and model study of moisture behaviors in polymeric materials. Microelectron Reliab 49:861–871
16. Viana G, Costa M, Banea MD, Da Silva LFM (2017) A review on the temperature and moisture degradation of adhesive joints. Proc Inst Mech Eng, Part L: J Mater: Des Appl 231:488–501
17. Aniskevich A, Bulderberga O, Stankevics L (2023) Moisture sorption and degradation of polymer filaments used in 3D printing. Polymers (Basel) 15:2600
18. Clifford PJ, Sorbie KS (1985) The effects of chemical degradation on polymer flooding. In: SPE international conference on oilfield chemistry? SPE, p SPE-13586

19. Cimatu KLA, Ambagaspitiya TD, Premadasa UI et al (2022) Polymer-solvent interaction and conformational changes at a molecular level: Implication to solvent-assisted deformation and aggregation at the polymer surface. J Colloid Interface Sci 616:221–233
20. Zou D, Nunes SP, Vankelecom IFJ et al (2021) Recent advances in polymer membranes employing non-toxic solvents and materials. Green Chem 23:9815–9843
21. Caruso MM, Davis DA, Shen Q et al (2009) Mechanically-induced chemical changes in polymeric materials. Chem Rev 109:5755–5798
22. He G, Liu Y (2019) Theoretical modeling of the mechanical degradation of polymer composites due to moisture/water absorption and damage progression. SAE Technical Paper
23. Dos Santos KAM, Suarez PAZ, Rubim JC (2005) Photo-degradation of synthetic and natural polyisoprenes at specific UV radiations. Polym Degrad Stab 90:34–43
24. Jin C, Christensen PA, Egerton TA et al (2006) Rapid measurement of polymer photo-degradation by FTIR spectrometry of evolved carbon dioxide. Polym Degrad Stab 91:1086–1096
25. Mohan K (2011) Microbial deterioration and degradation of polymeric materials. J Biochem Technol 2:210–215
26. Banerjee A, Chatterjee K, Madras G (2014) Enzymatic degradation of polymers: a brief review. Mater Sci Technol 30:567–573
27. Brito J, Andrianov AK, Sukhishvili SA (2022) Factors controlling degradation of biologically relevant synthetic polymers in solution and solid state. ACS Appl Bio Mater 5:5057–5076
28. Kliem S, Kreutzbruck M, Bonten C (2020) Review on the biological degradation of polymers in various environments. Materials 13:4586
29. Mokobia K, Jonathan EM, Oyiborhoro G, et al (2024) Environmental degradation of polymer-based composite materials: challenges and mitigation strategies. In: TMS annual meeting & exhibition. Springer, pp 1218–1236
30. Zatoń M, Rozière J, Jones DJ (2017) Current understanding of chemical degradation mechanisms of perfluorosulfonic acid membranes and their mitigation strategies: a review. Sustain Energy Fuels 1:409–438

Chapter 11
Advanced Techniques for Polymer Characterization

Nikhil Kumar, Komal Agarwal, Shashi Kumar, and Biswajoy Ghosh

Abstract This chapter offers a comprehensive exploration into the combination of various analytical techniques to provide a thorough understanding of polymer properties and behaviours. By integrating diverse technologies employed in prior chapters, this section aims to allow the strengths of each method to achieve a more nuanced and comprehensive characterization of polymers. Through the synergistic application of spectroscopic, chromatographic, microscopic, and thermal analysis techniques, among others, this chapter delves into the intricate details of polymer structure, morphology, composition, and thermal properties. Each section provides a detailed exploration of the respective topic, contributing to a holistic understanding of advanced polymer characterization methods and their applications.

Keywords Correlative techniques · LC–MS · Spectroscopy-microscopy · Multidimensional chromatography

11.1 Introduction

The characterization of polymers has evolved significantly over the past decades, driven by the growing complexity of modern polymeric materials and their diverse applications across various fields such as aerospace, electronics, biomedical devices, and energy storage (Fig. 11.1). While traditional characterization techniques such as

N. Kumar
Advanced Technology Development Centre, Indian Institute of Technology, Kharagpur, India

K. Agarwal · B. Ghosh (✉)
Department of Physics and Technology, UiT-The Arctic University of Norway, Tromsø, Norway
e-mail: biswajoy.ghosh@uit.no

S. Kumar
Department of Chemical Engineering, National Institute of Technology, Rourkela, Odisha, India

S. Chattopadhyay and N. Kumar (eds.), *Polymer Characterization*,
https://doi.org/10.1007/978-981-95-2609-3_11

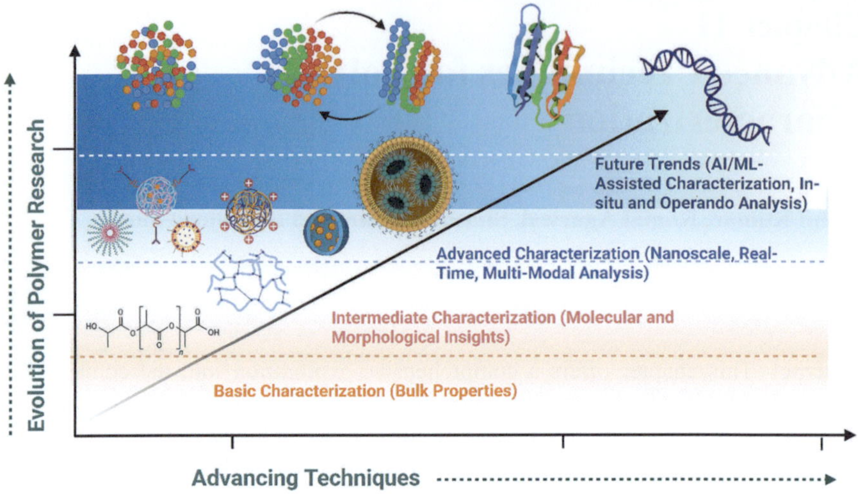

Fig. 11.1 Evolution of polymer research from basic to advanced characterization, highlighting a shift toward AI/ML-assisted, real-time, and nanoscale analysis

basic spectroscopy, microscopy, and thermal analysis laid the foundation for understanding polymer properties, the need for deeper, molecular-level, and application-specific insights has led to the development and adoption of advanced characterization techniques [1]. These techniques are designed to probe not only the bulk properties of polymers but also their molecular architecture, micro- and nano-scale morphology, dynamic behaviour, and interactions with other materials. These sophisticated methods enable the elucidation of structural, mechanical, thermal, chemical, and functional attributes of polymers with high precision and sensitivity (Fig. 11.2).

Furthermore, the advent of hybrid and correlative techniques, combining multiple analytical tools, has expanded the ability to simultaneously obtain complementary data, thus offering a better understanding of complex polymer systems. Computational methods such as molecular dynamics simulations [2, 3] and machine learning-based data analysis [4] are increasingly being integrated with experimental techniques to predict and interpret polymer behaviour under various conditions. This chapter focuses exclusively on advanced polymer characterization methodologies, offering insights into how these techniques contribute to cutting-edge polymer research. Particular attention is given to modern developments, interdisciplinary approaches, and future prospects that are shaping the landscape of polymer science.

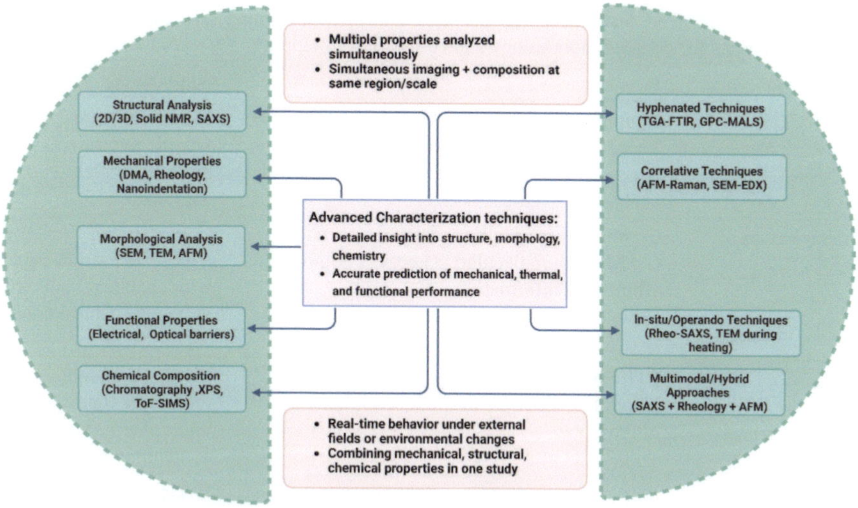

Fig. 11.2 Advanced characterization technique

11.2 Hyphenated/Correlative Techniques

Hyphenated or correlative techniques represent a significant advancement in polymer characterization, allowing researchers to combine the strengths of two or more analytical methods into a single, integrated workflow. This approach offers comprehensive insight into the physical, chemical, morphological, and thermal properties of polymers—often in real time and at micro- or nano-scale resolution (Table 11.1). This integration leads to multi-dimensional analysis, offering a more complete and accurate understanding of complex polymer systems. By combining techniques like TGA-FTIR [5], AFM-Raman [6], and Microscope-based spectroscopies, researchers can observe not only what changes occur in a polymer but also why and how those changes relate to the molecular structure or external conditions. These methods reduce experimental time and sample requirements while enhancing the reliability of results. For instance, TGA-FTIR allows for real-time monitoring of thermal degradation and identifies the nature of evolved gases, which is critical for understanding polymer stability. AFM-Raman provides both nanoscale surface topography and chemical composition, ideal for analysing phase separation or nanoparticle dispersion in polymer blends. Microscope-FTIR and Microscope-Raman techniques offer spatially resolved chemical insight, making them highly effective for investigating interfaces, multilayer structures, or contaminant distribution. These techniques are especially powerful because they enable multi-modal data collection from the same sample area, improving data accuracy and interpretability. They also help in correlating structure–property relationships, understanding degradation mechanisms, and evaluating the spatial distribution of components in complex polymer systems.

Table 11.1 Hyphenated techniques for polymer analysis combining structural, thermal, and chemical insights for detailed morphology and composition mapping

Techniques	Principle	Inference	Application
TGA-FTIR	Coupling of Thermogravimetric Analyzer with FTIR to analyze evolved gases	Thermal stability; decomposition products	Identification of volatile byproducts during polymer degradation [5]
AFM-Raman	Combines topography (AFM) with molecular vibrational information (Raman)	Surface morphology + chemical composition	Correlation of nanoscale morphology with chemical structure in blends, nanofillers [6]
Microscope-FTIR	FTIR spectroscopy integrated with optical microscope	IR spectra from micro-regions	Chemical mapping of multilayer films, microphase separation, contaminants [7, 8]
Microscope-Raman	Raman spectroscopy combined with high-resolution optical microscopy	Spatially resolved chemical fingerprint	Phase distribution, crystallinity, molecular orientation in micro-structured polymers [9–11]
GPC-MALS	Gel permeation chromatography with multi-angle light scattering detector	Absolute molecular weight, size, distribution	Determination of polymer molar mass, branching, and aggregation behavior [12, 13]
SEM–EDX	Scanning electron microscopy coupled with energy-dispersive X-ray spectroscopy	Surface morphology + elemental composition	Filler dispersion, elemental mapping in polymer composites and coatings [14, 15]

11.3 In-Situ and Operando Techniques

In-situ and operando techniques are cutting-edge analytical approaches used to study polymers under realistic or working conditions, providing real-time insights into structural, chemical, or morphological changes as they happen (Table 11.2). These techniques bridge the gap between static characterization and dynamic behaviour, enabling researchers to understand how polymers respond to temperature, stress, electric fields, solvents, or mechanical strain during actual processing or use.

- In-situ techniques: Measurements taken during a physical or chemical process (e.g., heating, swelling, crystallization) to monitor changes in real time.
- Operando techniques: Go to a step further by measuring performance and structure simultaneously under operational conditions (e.g., during catalysis, battery cycling, mechanical deformation).

Significance:

Table 11.2 Summary of in-situ and operando techniques for real-time polymer analysis, enabling dynamic monitoring of structural, chemical, and morphological changes during processing or external stimuli

Techniques	System coupled with	Key insights	Application
In-situ FTIR/Raman	Heating/stretching/exposure cells	Monitors chemical changes during reactions or aging	Crosslinking, photooxidation, curing, and crystallization [16]
In-situ SAXS/WAXS	Temperature/mechanical loading	Real-time structural evolution at nano/micro-scale	Crystallization kinetics, phase transitions, orientation under strain [17, 18]
Operando NMR	Rheometer or temperature stage	Tracks molecular mobility and phase transitions dynamically	Rheological behavior, polymer dynamics during processing [19]
In-situ AFM	Thermal or liquid cell	Morphological changes during solvent swelling or annealing	Surface restructuring, drug release from polymer matrices [20]
Operando SEM/TEM	Heating or electrical bias stages	Monitors nanoscale morphology and phase transitions live	Nanocomposite deformation, conductive polymer behavior under bias [21]
In-situ UV–Vis/Fluorescence	Light/thermal environment	Follows chromophore behavior or degradation mechanisms	Photostability, dye-polymer interaction, stress response [22]

- Reveal dynamic behaviour often missed by static methods.
- Enable real-time correlation of structure–property relationships.
- Crucial for processing–performance linkage, especially in smart materials, biopolymers, and energy devices.
- Reduce artifacts from sample handling by studying polymers in native or functional states.

11.4 Multimodal Hybrid Approaches

Multimodal hybrid approaches involve the strategic combination of multiple analytical techniques either sequentially, simultaneously, or interactively to generate a comprehensive dataset that bridges gaps left by individual methods (Table 11.3). These approaches go beyond simple hyphenation, often incorporating cross-platform correlations across physical, chemical, thermal, and mechanical properties. They are especially valuable when dealing with complex, multi-phase, nanostructured, or

Table 11.3 Multi-technique combinations for detailed polymer analysis of structure, chemistry, mechanics, and dynamics across diverse applications

Technique combination	Purpose/Strength	Applications
AFM + Raman + Nanoindentation	Morphology + chemical mapping + mechanical properties	Characterization of toughened polymers, nanocomposites
SEM–EDX + FTIR Microscopy	Morphological + elemental + chemical analysis	Coatings, composites with multiple fillers
TGA + DSC + MS + FTIR	Thermal behaviour + decomposition products + mass analysis	Degradation profiling of additives in complex polymers
Rheology + SAXS + NMR	Viscoelastic properties + structural changes + molecular dynamics	Processing–structure–function analysis
XPS + ToF–SIMS + AFM	Surface chemistry + molecular fragments + topography	Surface functionalization and depth profiling
Confocal Raman + Fluorescence microscopy	Chemical identity + active component localization	Biodegradable polymers, drug delivery matrices

functional polymers, where a single technique is insufficient to resolve all relevant structural or compositional details.

11.5 Multidimensional Chromatography

Multidimensional chromatography refers to the coupling of two or more different chromatographic separation stages in one analysis, to achieve a much higher resolving power than is possible with a single separation alone [23]. In polymer analysis, this technique is motivated by the inherent *molecular heterogeneity* of polymers. A given polymer sample is often distributed in multiple ways–for example by molar mass, chemical composition, functionality (end-groups), and molecular architecture (branching). Traditional *bulk characterization methods* (such as DSC or TGA for thermal behavior, and DMA or rheology for mechanical properties) generally provide only averaged or overall responses of a polymer material. They cannot resolve individual components or distributions within complex polymer mixtures. In contrast, chromatographic separations physically isolate different molecular species. By combining two orthogonal separations (e.g. one based on size and another based on chemical affinity), multidimensional chromatography can disentangle overlapping distributions and reveal detailed correlations between polymer properties that single-dimensional methods would miss [24]. In essence, each dimension adds its own selectivity, multiplying the total peak capacity and yielding far greater separation power than one-dimensional chromatography alone. This provides a powerful advantage

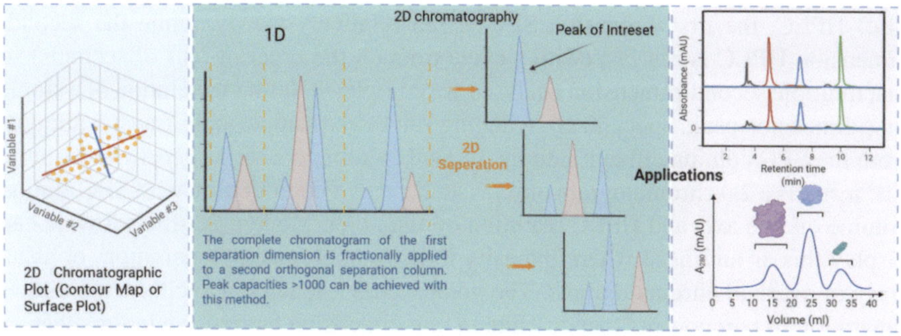

Fig. 11.3 Overview of 2D chromatography enabling enhanced separation and analysis of complex mixtures, with applications in resolving overlapping peaks and characterizing molecular size and structure

for characterizing polymers with complex microstructures that are not adequately described by bulk measurements or one-dimensional separations (Fig. 11.3).

11.5.1 Two-Dimensional SEC–HPLC Coupling

To overcome the limitations of one-dimensional SEC, researchers often combine SEC with a second, complementary liquid chromatography (HPLC) mode–creating a two-dimensional liquid chromatography (2D-LC) approach [25, 26]. In a typical SEC–HPLC coupling, one dimension separates by size (SEC) while the second dimension separates by some chemical property such as polarity, composition, or functionality. The two separation mechanisms are chosen to be as orthogonal as possible, so that each targets a different aspect of polymer heterogeneity. By doing this, one can obtain a 2D map [25] of the polymer where one axis corresponds to molecular weight and the other to composition (or another property). This reveals correlations, for example, whether high-molecular-weight chains also have higher comonomer content, which cannot be seen in a 1D chromatogram.

There are different strategies to implement 2D separations. One common workflow is an off-line (or stop-flow) two-dimensional analysis: fractions are collected across the first-dimension elution, then each fraction is analyzed in turn on the second dimension. For instance, one could first perform SEC and collect slices of the eluate corresponding to various molecular weight ranges, then inject each slice onto an HPLC column that separates by chemical composition. The outcome would be, say, the composition distribution at several discrete molecular weight cuts, which can be reconstructed into a continuous 2D distribution. This off-line approach is flexible (one can adjust fraction ranges, concentration, etc.) but can be labor-intensive. Increasingly, online or comprehensive 2D-LC (LC $\times$ LC) is used, wherein the effluent of the first column is continuously modulated into the second column. In comprehensive

SEC–HPLC, the first-dimension SEC is run relatively slowly, while the second-dimension HPLC is run fast (with a cycle time on the order of tens of seconds) so that multiple second-dimension analyses are completed during the elution of a single first-dimension peak. Specialized switching valves and sample loops transfer aliquots from the SEC column effluent into the second column in a serial, repetitive fashion. The result is a 2D chromatogram often visualized as a contour plot, with SEC elution volume on one axis and HPLC retention on the other. Modern 2D-LC software can display this as an image where intensity indicates polymer concentration, or as an array of overlaid chromatograms. The greater peak capacity of such a setup allows separation of components that co-elute in one dimension but resolve in the other.

SEC–HPLC combinations for polymer analysis generally pair SEC with one of the following second dimensions:

Interaction Polymer HPLC (LAC):

Liquid adsorption chromatography (LAC), also called interaction chromatography, separating polymers by their chemical affinity to a stationary phase. This can be done in normal-phase (NP) mode for polar polymers or in reversed-phase (RP) mode for polar additives, etc., depending on the solvent. In polymer analysis, a common goal is to separate copolymers by chemical composition distribution (CCD). For example, gradient HPLC on a polar column can separate poly(acrylate) chains by their ratio of methyl acrylate to ethyl acrylate, since the polarity (and thus retention) changes with composition. By coupling this to SEC, one can determine how composition and molecular weight are related. In fact, for copolymer analysis (block, graft, etc.), one effective scheme is composition separation by LAC in the first dimension, followed by SEC in the second dimension to measure MWD within each composition slice. This way, each spot in the 2D map corresponds to polymer chains of a specific composition and a specific molecular weight. Researchers have applied such 2D-LC to complex copolymers——for instance, analyzing styrene–acrylate copolymers where LAC (under gradient or critical conditions) separates chains by acrylate content, and SEC reveals the molecular weight distribution of each composition class. The result is a complete characterization of the copolymer's heterogeneity.

Liquid Chromatography at Critical Conditions (LC–CC):

This is a special mode of polymer HPLC where conditions are tuned such that one type of polymer segment has *no chromatographic retention* (the "critical" point of adsorption). At critical conditions for a given polymer backbone, separation is highly sensitive to other molecular features–e.g. a different comonomer or an end-group–while differences in the backbone length (molecular weight) cause minimal retention differences (Fig. 11.4). LC-CC can therefore separate polymers by end-group type or by the presence of a particular comonomer block, independently of size [26].

Coupling LC-CC with SEC is very powerful for block copolymers or end-functional polymers. For example, one can set the first dimension to critical conditions for the major polymer component so that separation is achieved according to the amount of a second component (or an end-group); then use SEC in the second dimension to find the molecular weight of those fractions. This approach has been

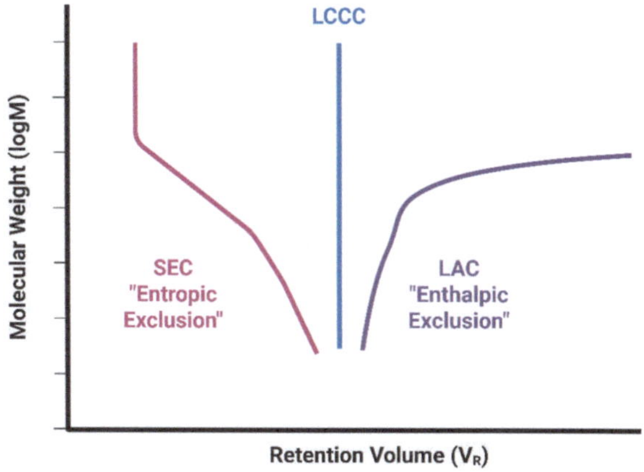

Fig. 11.4 SEC and LAC use entropic and enthalpic exclusion, while LCCC allows separation independent of molecular weight

used to analyze, for instance, block copolymers of styrene and butadiene: by running the first dimension at the critical point of polystyrene, the separation depends only on the polybutadiene content of each chain, and SEC then characterizes the molecular weight of each fraction of differing composition [27]. Such 2D data reveal how the block length distributions are linked with overall chain length–information not attainable by SEC or HPLC alone.

Other modes (e.g. ion-exchange, chiral LC): Less commonly, SEC can be coupled with ion-exchange chromatography (to separate by charge or ionic functionality in polyelectrolytes), or with chiral stationary phases (to resolve stereochemical sequence distributions in polymers). These specialized combinations have been explored in research for very specific polymer systems. The general principle remains the same: an orthogonal second dimension adds insight that SEC by itself cannot provide.

The applications of SEC–HPLC 2D-LC are numerous. They have enabled detailed characterization of compositionally complex polymers, such as multi-block copolymers and polymer blends, by producing a "fingerprint" that shows each distinct population of polymer molecules. For example, multidimensional LC was crucial in one case to analyze an ethylene/α-olefin copolymer with a broad MWD and CCD–SEC alone gave a broad unimodal MWD, and FTIR suggested varying comonomer content, but only a 2D-LC (TREF or LAC for composition × SEC for size) could map how shorter chains were enriched in comonomer while longer chains were mostly ethylene. In another case, researchers studying graft copolymers used 2D-LC to separate ungrafted homopolymer, grafted chains with varying graft frequency, etc., providing quantitative insight into a complex reaction mixture that would appear as a single broad peak by SEC. These examples underscore that by leveraging orthogonal separation mechanisms, multidimensional chromatography can deconvolute multi-faceted polymer distributions that are impossible to resolve otherwise.

SEC–GC Coupling (LC–GC for Polymers)

Another powerful 2D technique is the coupling of SEC with gas chromatography (GC). In SEC–GC, the first-dimension SEC (in liquid phase) separates the high-molecular-weight polymer from volatile or semi-volatile low-molecular-weight components, and then these lighter components are transferred to a GC for further separation and identification. The primary use of SEC–GC is for analyzing additives, residuals, or degradation products in polymer samples. This approach essentially extends the idea of using SEC as a sample prep step: after SEC, the polymer itself (which is non-volatile and high-boiling) can be excluded, and the small molecules of interest are sent to the GC, which offers high resolution separation and detection (often by MS).

One implementation is off-line GPC–GC: the polymer sample is first run through an SEC column, and the fraction eluting at the total permeation volume (which contains the oligomers, monomers, plasticizers, etc., but excludes the bulk polymer) is collected. This collected solution is then injected into a GC (or GC–MS). In this way, one can directly analyze, for example, the mixture of additives present in a polyolefin. This method has been used to determine antioxidants and stabilizers in polyethylene: the SEC step removes the interfering polymer, and the GC–MS step identifies each additive by its unique retention time and mass spectrum. One study by authors in [9] demonstrated simultaneous determination of polymer MWD (by the SEC portion) and additive content (by GC) for a polyolefin sample, by splitting the flow–the large polymer was measured by an SEC detector, while the small-molecule portion was routed to a GC for analysis. Such a setup provides a comprehensive picture of a material, showing, for instance, that a particular antioxidant is preferentially concentrated in the low-molecular-weight polymer fractions.

On-line coupling is also possible: specialized interfaces (often a large-volume injector or an evaporative interface) can continuously transfer eluent from an SEC to a GC. For example, an on-line SEC–GC system can evaporate a portion of the SEC effluent and inject it into the GC at set time intervals. This has been utilized in the analysis of complex petrochemical resins and polymer additives–sometimes termed LC–GC or GPC–GC–to automate the process. The challenge in on-line coupling is solvent compatibility (SEC uses a liquid solvent that must be removed or minimized before GC) and timing (GC runs must be fast enough or coordinated with the SEC output). Nonetheless, commercial on-line GPC–GC interfaces have been developed for rapid screening of polymer additives and impurities.

Applications of SEC–GC include detecting residual monomers or oligomers in polymers (for quality control of food-contact plastics, where any residual vinyl chloride or styrene monomer must be quantified at ppm levels), analyzing plasticizers and phthalates that leach from PVC (GPC can isolate the plasticizer fraction for GC analysis), and studying thermal degradation products. For instance, when polymers are aged at high temperatures, they may generate a host of small-molecule byproducts; SEC–GC allows one to first separate out any intact polymer and then use GC–MS to identify the degradation products in the low-MW fraction. Another intriguing application is pyrolysis-GPC-GC: the polymer is first partially degraded (thermally or chemically) to generate fragments, which are size-separated by SEC, and each fraction is then analyzed by GC–MS to deduce structural information as a function of parent chain size. This has been used in research to understand how branching or cross-linking in a polymer influences the distribution of fragments upon pyrolysis. While specialized, it highlights the creative combinations of SEC and GC for tackling polymer analysis problems.

Overall, SEC–GC coupling extends the reach of chromatography for polymers beyond what either SEC or GC could do alone: SEC handles the non-volatile, high-MW matrix and prevents it from fouling the GC, while GC provides superb separation and identification of low-MW compounds that SEC cannot resolve (due to co-elution or insufficient resolution in the low-MW end). It effectively bridges the gap between macromolecular analysis and traditional small-molecule analysis, which is especially useful in fields like polymer additives, environmental microplastics (where one might need to identify additives within tiny polymer particles), and forensic polymer analysis.

11.6 Advanced Microscopic Imaging

Advanced microscopic imaging techniques, including Scanning Electron Microscopy (SEM) [28, 29], Transmission Electron Microscopy (TEM), and Atomic Force Microscopy (AFM) [30], are essential for detailed polymer characterization. SEM primarily investigates surface morphology, TEM reveals internal structures at atomic resolution, and AFM provides nanoscale 3D topographical and mechanical property information. Each modality has distinct principles, strengths, limitations, and specific applications in polymer research. Table 11.4 below gives a general summary of these techniques.

However, this general breakdown does not give a detailed description of the advantages each of these modalities provides. Moreover, new advancements in optical

Table 11.4 Comparison of SEM, TEM, and AFM techniques highlighting differences in principles, resolution, sample preparation, environment, imaging modes, and typical applications

Feature	SEM	TEM	AFM
Principle	Scans surface with electrons	Transmits electrons through sample	Measures forces between tip and surface
Resolution	1–10 nm	0.1–0.2 nm (atomic)	Sub-nanometer (atomic)
Sample prep	Conductive coating often needed	Ultra-thin samples required	Minimal preparation
Environment	Vacuum	Vacuum	Air, liquid, or vacuum
Imaging mode	Surface topography	Internal structure	3D surface topography
Applications	Surface analysis, materials science	Atomic-level internal structure	Surface properties, nanomechanics

imaging have enabled advanced optical microscopy imaging applications in polymer characterization. Optical microscopy, encompassing brightfield [31], phase contrast [32, 33], infrared [34, 35], fluorescence [36–38], confocal [39], and super-resolution techniques [40], offer a non-destructive and real-time approach to imaging polymers. It is especially valuable for examining hydrated and soft polymers under ambient conditions. Optical methods enable the study of polymer morphology, dynamic processes like swelling or degradation, and even chemical composition through vibrational imaging techniques like Raman and infrared microscopy. Emerging methods such as multiphoton microscopy and quantitative phase imaging extend optical microscopy's capability to three-dimensional and label-free analysis. Additionally, acoustic microscopy techniques, such as Scanning Acoustic Microscopy (SAM) and Photoacoustic Microscopy (PAM), provide non-invasive imaging of internal structures and mechanical properties of polymers by utilizing high-frequency ultrasound waves or optically induced acoustic signals, offering deeper penetration into thick or hydrated samples compared to optical or electron microscopy [41–46]. For comprehensive details on the applications, principles, strengths, and limitations of these imaging techniques in characterizing synthetic polymers and biopolymers, readers are encouraged to refer to Chapter 9.

11.6.1 Evolved Gas Analysis (EGA) with Spectroscopic Methods

Evolved Gas Analysis (EGA) is a technique used to analyze the gases evolved during the thermal decomposition or degradation of a material. It is often coupled with spectroscopic methods such as Fourier Transform Infrared Spectroscopy (FTIR) or Mass Spectrometry (MS) to identify and quantify the chemical composition of the evolved gases. EGA is widely used in polymer science, materials research, and environmental analysis to study the thermal degradation behavior of synthetic and biopolymers.

Basic Working Principle and uses of EGA: EGA operates by heating a sample in a controlled atmosphere (e.g., nitrogen, air, or oxygen) while continuously analyzing the gases released during the thermal decomposition process. The key steps in EGA are (1) Sample Heating: The sample is placed in a furnace or thermal analyzer (e.g., TGA or DSC) and subjected to a temperature program (heating, cooling, or isothermal conditions). As the temperature increases, the sample undergoes thermal decomposition, releasing gases. (2) Gas Collection and Transfer: The evolved gases are collected and transferred to a spectroscopic analyzer (e.g., FTIR or MS) through a heated transfer line to prevent condensation. (3) Gas Analysis: The spectroscopic analyzer identifies and quantifies the chemical composition of the evolved gases. FTIR provides insights into functional groups and molecular structures, while MS identifies molecular weights and fragmentation patterns, displaying results as temperature or time functions, showing the evolution of specific gases during the thermal decomposition process.

For instance, EGA-TOFMS combined with principal component analysis (PCA) and Kendrick mass defect (KMD) analysis revealed molecular-scale structural changes during thermo-oxidative degradation in polymer nanocomposites. This approach effectively identified ion series and chemical structures evolving from the degradation of triblock copolymer thermoplastic elastomers, highlighting structural alterations and stability improvements due to specific filler-matrix interactions [47].

EGA provides detailed insights into the thermal degradation behavior and chemical composition of both synthetic and biopolymers. It identifies the specific gases released during thermal decomposition, such as CO_2, H_2O, CH_4, NH_3, and volatile organic compounds. It elucidates degradation mechanisms and pathways including chain scission, crosslinking, and oxidation. For example, TG/FTIR and Pyrolysis/GC–MS analyses revealed the detailed thermal degradation pathway of organically modified montmorillonite (OLS), identifying evolved gases like alkanes, alkenes, CO_2, and H_2O and providing insights into decomposition reactions including secondary reactions catalyzed by clay structures [48].

Applications of EGA: EGA is widely used in both research and industry for various applications. In polymer characterization, it helps study the thermal degradation behavior of synthetic and biopolymers, as well as identifying the chemical composition of evolved gases. For instance, TG/FTIR/MS effectively characterized evolved products from polyimide composites, identifying main decomposition products like H_2O, CO_2, CO, CH_4, and aromatic compounds. This analysis provided insights into the degradation mechanism, revealing carbon–nitrogen and carbon–carbon bond cleavage followed by rearrangement and condensation reactions [48]. For quality control, EGA is used to ensure the thermal stability and performance of polymer products and to verify the presence and effectiveness of additives. In formulation development, it assists in optimizing polymer formulations to improve thermal stability and studies the effects of fillers, stabilizers, and flame retardants. Environmental analysis applications include analyzing emissions from polymer degradation or combustion and studying the environmental impact of polymer waste. Additionally, EGA is valuable in biopolymer analysis, where it helps analyze the thermal degradation of proteins, polysaccharides, and other biopolymers, as well as studying their denaturation and decomposition behavior.

Limitations and advantages of EGA: Interpretation of EGA data can be complex due to overlapping gas peaks, especially in multi-component systems. It only analyzes evolved gases, not solid or liquid residues, and experimental conditions can significantly influence results. The results can be sensitive to experimental conditions, such as heating rates, sample pan types, and the atmosphere (e.g., nitrogen or air) used during the analysis. However, EGA offers several advantages, including the provision of detailed chemical information about the evolved gases during thermal decomposition, enabling the identification of specific degradation products.

11.7 Nanoindentation

Nanoindentation involves pressing a very small indenter (often with a tip radius on the order of nanometers) into the surface of a material and measuring the force and displacement during the indentation process. The load–displacement data obtained are used to determine mechanical properties such as hardness and elastic modulus. It can also provide insights into a material's creep behavior, fracture toughness, and viscoelastic response (Fig. 11.5).

Basic Working Principle and Uses: In nanoindentation, typically a diamond indenter with a defined geometry (e.g., a Berkovich tip) is pressed onto the surface of the material. This process produces an elastic–plastic deformation, and by measuring the load and displacement curves during loading and unloading, the Oliver-Pharr

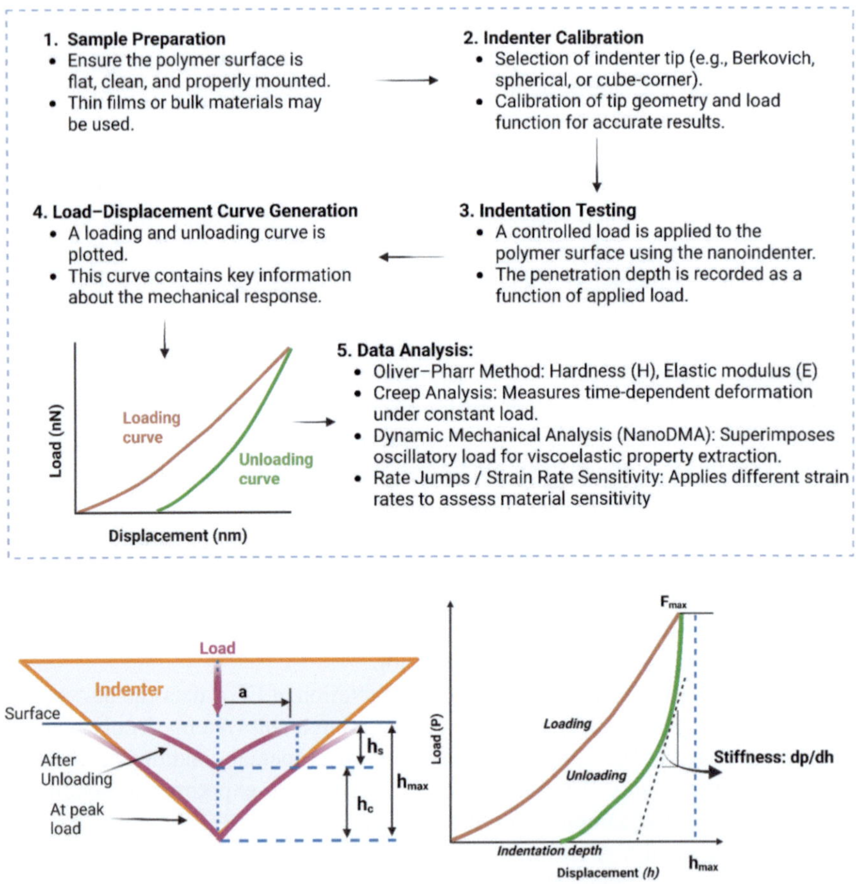

Fig. 11.5 Workflow of nanoindentation testing for polymers, detailing sample prep, indentation, load–displacement curve generation, and mechanical property analysis like hardness and modulus

method is used to estimate hardness (H) and elastic modulus (E). The hardness is calculated as the maximum load divided by the projected contact area, and the elastic modulus is derived from the reduced modulus, which incorporates both the material and indenter properties [49].

Nanoindentation can precisely measure local mechanical properties of materials, making it highly suitable for surface coatings, thin films, nanoparticles, and biomaterials. For synthetic polymers, nanoindentation helps measure local mechanical properties, assess surface treatments, and evaluate aging effects. In biopolymers, it aids nanoscale mechanical characterization, tissue mechanics studies, hydrogel characterization, and thin-film analysis.

Specific Examples from Polymer Characterization: Nanoindentation has been effectively employed to investigate the surface mechanical properties of polymers such as polyethylene terephthalate (PET), polycarbonate (PC), polystyrene (PS), and poly(methyl methacrylate) (PMMA). These studies have helped to clarify the impact of surface treatments, filler additions, and aging effects on the polymer's mechanical performance. For instance, nanoindentation studies on PET films revealed variations in hardness and elastic modulus with different surface treatments, demonstrating how mechanical properties can be tailored for specific applications such as packaging materials [49].

Applications in Polymer Science: Nanoindentation is valuable for characterizing thin polymer films, coatings, and polymer composites, especially for assessing mechanical behavior relevant to surface integrity and durability. For example, nanoindentation studies on graphene and diamond-like carbon films demonstrated exceptionally high hardness and favorable H/E ratios, indicating excellent resistance to wear and mechanical deformation [49]. Additionally, nanoindentation has proven useful in biomechanical studies, evaluating the mechanical properties of human tooth enamel and fingernails, significantly enhancing our understanding of biomaterial resilience and biocompatibility [49]. For example, nanoindentations were performed on healthy and fibrotic tissues to determine their stiffness in health and disease [30, 50].

Limitations and Advantages: Nanoindentation offers high spatial resolution, minimal sample preparation, and suitability for small volume samples. However, it is sensitive to surface roughness, requires careful calibration, and primarily measures surface properties, which may not fully represent bulk material characteristics. Interpretation of data can be complex due to surface artifacts such as pile-up effects, where material extrusion around the indentation affects accuracy in estimating contact area and thus hardness values [49].

Overall, nanoindentation remains a crucial technique in materials science and polymer research, allowing detailed exploration of surface mechanical properties and facilitating advanced material development and application-specific optimizations.

11.8 Computational Modelling and Simulation in Polymer Characterization

As computational simulations and modeling have emerged as an invaluable tool in guiding the experiments across various scientific domains, their study has become essential staple in the advanced characterization of polymers systems. These molecular simulations have become a key to understand the details of macromolecules. These advanced techniques provide atomistic details into the structure and property relationships of polymer systems, and help in understanding how polymers interact under different conditions respectively [23]. By simulating the polymeric structures, dynamics and structural interactions over time with respect to different factors such as temperature, pressure, and chemical surroundings can be observed. These computational methods not only deepen the researchers understanding of how different attributes influence polymer properties, but also allow the scientists to predict and optimize the polymer systems for real world conditions. *Molecular Dynamics (MD) simulations* and *Density Functional Theory (DFT)* calculations are well accepted and primarily used approaches in polymer characterization. Both techniques offer valuable information, span in depth, and complementary insights into polymer materials at different levels of analysis.

11.8.1 MD Simulations in Polymer Systems

With the advent of computers, MD simulations have become powerful tools to study dynamical behaviour of 3-dimensional structures of polymers. It simulates the motion of the molecules, revealing how molecules change and respond to different environments by solving the principles of classical mechanics [24]. These simulations allow the researchers to visualize the structures at atomic level and provide molecular insights to materials and biological macromolecules [51]. In context to polymer science, MD simulations are primarily used to understand various phenomena, such as conformational rigidity and flexibility of polymer units, self-assembly behaviour (micelle, vesicles) of polymers chain, crystallization processes, and mechanical properties (Young's modulus, tensile strength) under deformations. MD simulations offer real-time insights into the motion and interactions of polymer molecules, enabling scientists to visualize the complex processes viz. folding/unfolding pathways [51, 52], processing, and phase transitions of polymer chains. One of the major advantage associated with MD simulations is to study large and complex polymeric structures such as block copolymers, and nanocomposites. Additionally, MD simulations also provide the details into thermodynamic and mechanical properties of these materials, which are essential to understand how polymeric systems behave under different operating conditions, leading to the design of novel polymers with enhanced properties for specific industrial applications.

11.8.2 DFT Calculations in Polymer Systems

Both DFT calculations and MD simulations are used to study and model polymer systems, however differences occur in methodology and the detailed information they provide. While MD models the time-dependent behaviour of molecules, on the other hand, DFT calculations are quantum mechanical atomistic simulations method that allow the investigation the electronic structures and associated properties on the basis of first principles (ab initio) by approximating the Schrodinger equation [53]. DFT methods are powerful tools to predict the behaviour of new materials. DFT methods reduce the cost and development time by down-selection of variety of materials without performing experiments into the lab. DFT methods are extensively employed to understand and examine the properties of both organic and inorganic compounds. In polymer science, these are utilized as a tool to investigate different properties for polymer structures such as band gap, band structure, vibrational frequencies, charge distribution, and interactions with other fields that are important for materials having applications in solar cells and semiconductors. DFT methods aid in the understanding how these polymer materials behave and work under different operating conditions. Moreover, DFT analyses correlate the simulations data with wet lab data to get insights into the actual physics of the effects associated with these materials/devices, which are arduous to explain with other simple models. Apart from the molecular insights into the electronic structures, DFT calculations also provide crucial information on reactive properties of polymers chains, revealing valuable information about the phenomena such as stability, transitions states, transition energies, and charge transport. Further, organic macromolecules having applications in light emitting diodes and in solar cells, DFT calculations tailor the electronic properties by providing detailed analysis of structural factors viz. chain length, molecular packing effect, functional groups.

11.8.3 Recent Developments in Computational Polymer Characterization

The integration of quantum mechanical calculations (DFT) and MD simulations allow researchers to study large and complex systems with high degree of accuracy by simulating both dynamics and electronic properties of structures simultaneously. Apart from these methods, quantum mechanics and molecular mechanics (QM/MM) methods are also employed to understand various aspects of complex polymer systems. QM/MM methods are divided into two regions: one is the QM region (chemical reactions or electronic properties of interest) used for reactive fragment with high accuracy and the surrounding environment is taken care by the MM region with greater efficiency [26]. The synergy between these computational methods provides a detailed insights into the characteristics of polymers, where MD simulations focus on structural changes and thermodynamics properties, and DFT (QM) calculations

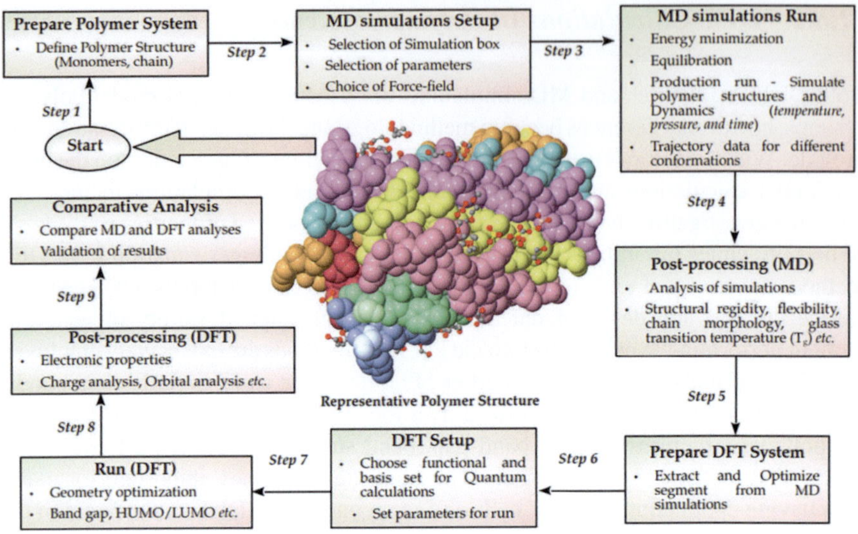

Fig. 11.6 Workflow integrating molecular dynamics (MD) and density functional theory (DFT) simulations for polymer analysis—covering system setup, simulation, post-processing, and comparative evaluation of structural and electronic properties

focus on information crucial for chemical reactions and bond formation/breaking of polymeric macromolecules, respectively. Additionally, with the modern era of technology advancements, advances in high performance supercomputing facility and GPU-based acceleration systems have made it possible to simulate large (~ upto millions of atoms) and complex systems over longer timescales. This nano to micro seconds time scales simulations have enabled the researchers to model complex polymer structures in an environment that closely resembles to real world conditions. Figure 11.6 shows how MD simulations and DFT calculations complement each other in polymer characterization, illustrating basic steps involved in carrying out the MD simulations and DFT calculations.

11.8.4 Emerging Techniques in Computational Polymer Science

The field of computational polymer science is growing rapidly and is guided by variety of of new methods. These emerging methods are capable enough to reduce the computational time, and enhance the speed and accuracy of molecular simulations for polymer systems. These techniques include the methods namely machine learning (ML), artificial intelligence (AI), enhanced sampling, and multi-scale modeling methods. These advanced algorithms and computational molecular simulation methods have addressed the problem of studying and modeling of macromolecules for polymer characterization. These methods study systems under different environmental conditions, and optimize the materials design aiming for specific industrial applications.

11.9 Artificial Intelligence and Machine Learning in Polymers Simulations

Artificial intelligence (AI) has emerged as a powerful and transformative tool of the modern era. It provides a variety of analytical and predictive algorithms, which can be used to examine the properties of polymeric structures. Also, AI is being utilized to assist in modeling and designing of materials for specific applications in materials science. Machine learning (ML) method, which is a subset of AI, has become prevalent due its predictive nature, simple integration, cheaper computational cost, and high efficacy across material science. Due to these characteristics, it is relatively easy to understand and model the behaviour of new or uncharacterized polymer structures. ML methods have been used for mechanical and optical properties prediction, rational material design [27], and automating materials characterization from molecular descriptors/structures. Further, MD simulations have largely benefited by its integration with ML algorithms to determine the potentials used to optimize the force-fields. This optimization has allowed the researchers to sample large polymer conformations with reduced computation time and with high accuracy. In this field of polymer science, another application associated with ML is in the modeling of novel polymeric materials, where ML algorithms are vastly exploited to screen large datasets of polymers for desired industrial applications. Apart from many tools available in the basket of ML, their adoption has been significantly affected. There are many reasons for this hindrance, few are difficulty of expressing complex polymer macromolecule structure details into computer-readable format, and the lack of sufficient data to train and test ML models [54]. Many promising approaches are being developed to tackle these obstacles and then research community can later be benefited from the rational polymer design, polymer property prediction with any combination of ML tools with minimal human invention for material science.

11.9.1 Multi-Scale Modeling for Polymer Systems

The understanding of structure–property and processing-performance relationship of polymer materials is a challenging task. Modeling of industrial grade polymers with molecular details pose a significant challenge due to the system large size and long simulations run that are usually out of the conventional computational power. Also, precise estimation of qualitative and quantitative properties for polymers modeling and simulation is crucial. In this regard, multi-scale modeling methods offers information between the fine and coarse ensembles of molecules to study polymer materials properties across different temporal and length scales. These techniques capture the information on properties from molecular level to macroscopic level [29]. Due to the combined computational and analytical features, these techniques are efficient to address the existing problems of accuracy and applicability of extreme simulations times in materials. For instance, while MD simulations provide molecular insights into the dynamic behaviour of polymer units, on the other hand, continuum models capture the bulk behaviour. As mentioned in the above text that multi-scale modeling techniques capture the information from atomic to microscopic level, integration of MD, DFT, and continuum modeling techniques are much needed for polymer materials which are affected by multiple scales. These advanced techniques allow the researchers to simulate system close to real-world conditions which can be helpful for polymer design and process optimization.

11.9.2 Enhanced Sampling Techniques in Polymer Simulations

Enhanced sampling techniques are one of the simulation techniques that explore the classical/regular MD simulations for high dimensional space and with effective sampling for conformational space [28]. In context of polymers materials, these techniques are particularly helpful, where transition between different energy states occurs. These methods overcome the drawbacks of regular MD simulations by the exploration of rare events. Replica Exchange Molecular Dynamics (REMD), Metadynamics, and umbrella sampling methods are few, which are used to get rid of existing challenges. While MD simulations take longer time scales to capture materials properties, on the other hand, these enhanced techniques allow the users to escape local minima by accelerating transition between different metastable states. This leads to sampling of more different polymers conformations. Enhanced sampling methods are particularly employed to study polymers folding/unfolding transitions.

11.9.3 Conclusions

Computational modeling and simulation are important for structure–property relationship of polymer materials. While conventional methods such as MD simulations and DFT calculations provide information on dynamics of motion of molecules and electronic structures respectively, quantum computing, ML, and enhanced sampling techniques enables the prediction of materials properties for designing of new polymers with enhanced characteristics, respectively. With computational advancement, these emerging techniques will largely contribute to modeling and optimization of polymers materials for industrial applications.

Acknowledgements We have used bioRender.com (student licence) for making illustration.

Conflict of Interest Authors declare no competing conflict of interest.

References

1. Gajarska Z, Brunnbauer L, Lohninger H, Limbeck A (2022) Advanced polymer characterization. laser-induced breakdown spectroscopy in biological. Forensic Mater Sci 253–281. https://doi.org/10.1007/978-3-031-14502-5_10/FIGURES/10
2. Prathab B, Subramanian V, Aminabhavi TM (2007) Molecular dynamics simulations to investigate polymer–polymer and polymer–metal oxide interactions. Polymer (Guildf) 48:409–416. https://doi.org/10.1016/J.POLYMER.2006.11.014
3. Shi R, Qian HJ, Lu ZY (2023) Coarse-grained molecular dynamics simulation of polymers: structures and dynamics. Wiley Interdiscip Rev Comput Mol Sci 13:e1683. https://doi.org/10.1002/WCMS.1683;PAGE:STRING:ARTICLE/CHAPTER
4. Sha W, Li Y, Tang S, et al (2021) Machine learning in polymer informatics. InfoMat 3:353–361. https://doi.org/10.1002/INF2.12167; PAGEGROUP:STRING:PUBLICATION
5. Wilkie CA (1999) TGA/FTIR: an extremely useful technique for studying polymer degradation. Polym Degrad Stab 66:301–306. https://doi.org/10.1016/S0141-3910(99)00054-3
6. Cosas Fernandes JP, Castro LDC, Mareau VH et al (2018) New insights on the compatibilization of PA6/ABS blends: a co-localized AFM-Raman study. Polymer (Guildf) 146:151–160. https://doi.org/10.1016/J.POLYMER.2018.05.012
7. Yoshie N, Asaka A, Yazawa K et al (2003) In situ FTIR microscope study on crystallization of crystalline/crystalline polymer blends of bacterial copolyesters. Polymer (Guildf) 44:7405–7412. https://doi.org/10.1016/J.POLYMER.2003.09.024
8. L'Abee R, Li W, Goossens H, Van Duin M (2008) Application of FTIR microscopy in combinatorial experimentation on polymer blends. Macromol Symp 265:281–289. https://doi.org/10.1002/MASY.200850532
9. Markwort L, Kip B (1996) Micro-Raman imaging of heterogeneous polymer systems: general applications and limitations. J Appl Polym Sci 61:231–254. https://doi.org/10.1002/(SICI)1097-4628(19960711)61:2%3c231::AID-APP6%3e3.0.CO;2-Q
10. Ma Y, Du Y, Chen Y et al (2020) Intrinsic Raman signal of polymer matrix induced quantitative multiphase SERS analysis based on stretched PDMS film with anchored Ag nanoparticles/Au nanowires. Chem Eng J 381:122710. https://doi.org/10.1016/J.CEJ.2019.122710
11. Frick A, van Vliet WA, Žukauskaitė A et al (2024) Volumetric 3D-printed piezoelectric polymer films. Adv Mater Technol 9:2301469. https://doi.org/10.1002/ADMT.202301469;WGROUP:STRING:PUBLICATION
12. Hanton SD, Liu XM (2000) GPC separation of polymer samples for MALDI analysis. Anal Chem 72:4550–4554. https://doi.org/10.1021/AC000095N/ASSET/IMAGES/LARGE/AC000095NF00008.JPEG
13. Kissin Y V. (1995) Molecular weight distributions of linear polymers: detailed analysis from GPC data. J Polym Sci A Polym Chem 33:227–237. https://doi.org/10.1002/POLA.1995.080330205; PAGE:STRING:ARTICLE/CHAPTER
14. Fay F, Linossier I, Langlois V et al (2005) SEM and EDX analysis: two powerful techniques for the study of antifouling paints. Prog Org Coat 54:216–223. https://doi.org/10.1016/J.PORGCOAT.2005.05.005

15. Vandi LJ, Truss R, Veidt M et al (2013) Fluorine mobility during SEM-EDX analysis: a challenge for characterizing epoxy/fluoropolymer interfaces. J Phys Chem C 117:17933–17941. https://doi.org/10.1021/JP403314T/ASSET/IMAGES/LARGE/JP-2013-03314T_0008.JPEG
16. Cornel J, Lindenberg C, Mazzotti M (2008) Quantitative application of in situ ATR-FTIR and raman spectroscopy in crystallization processes. Ind Eng Chem Res 47:4870–4882. https://doi.org/10.1021/IE800236V/ASSET/IMAGES/LARGE/IE-2008-00236V_0014.JPEG
17. Schneider K, Zafeiropoulos NE, Stamm M (2009) In situ investigation of structural changes during deformation and fracture of polymers by synchrotron SAXS and WAXS. Adv Eng Mater 11:502–506. https://doi.org/10.1002/ADEM.200800323; SUBPAGE:STRING:ABSTRACT;JOURNAL:JOURNAL:15272648;WGROUP:STRING:PUBLI(
18. Pepin J, Gaucher V, Rochas C, Lefebvre JM (2019) In-situ SAXS/WAXS investigations of the mechanically-induced phase transitions in semi-crystalline polyamides. Polymer (Guildf) 175:87–98. https://doi.org/10.1016/J.POLYMER.2019.04.073
19. Cattaneo AS, Villa DC, Angioni S et al (2015) Operando electrochemical NMR microscopy of polymer fuel cells. Energy Environ Sci 8:2383–2388. https://doi.org/10.1039/C5EE01668A
20. Ono Y, Kumaki J (2018) In situ real-time observation of polymer folded-chain crystallization by atomic force microscopy at the molecular level. Macromolecules 51:7629–7636. https://doi.org/10.1021/ACS.MACROMOL.8B01428/SUPPL_FILE/MA8B01428_SI_007.AVI
21. Marceau H, Kim CS, Paolella A et al (2016) In operando scanning electron microscopy and ultraviolet–visible spectroscopy studies of lithium/sulfur cells using all solid-state polymer electrolyte. J Power Sources 319:247–254. https://doi.org/10.1016/J.JPOWSOUR.2016.03.093
22. Abyan M, De Caro D, Suzanne FF (2009) Suspensions of organic microcrystals produced in the presence of polymers: Diversity of UV/Vis Absorption and fluorescence properties according to the preparation conditions. Langmuir 25:1651–1658. https://doi.org/10.1021/LA803549U/SUPPL_FILE/LA803549U_SI_001.PDF
23. Pasch H (1999) Analysis of complex polymers by multidimensional techniques. Invit Lecture Phys Chem Chem Phys 1:3879–3890. https://doi.org/10.1039/A902365H
24. Marriott PJ, Chin ST, Maikhunthod B et al (2012) Multidimensional gas chromatography. TrAC, Trends Anal Chem 34:1–21. https://doi.org/10.1016/J.TRAC.2011.10.013
25. Kilz P (2017) Two-dimensional chromatography of copolymers
26. Lynen F, Wicht K, Kajtazi A, et al (2023) Possibilities and limitations of aqueous temperature responsive liquid chromatography in comprehensive 2D–LC. LC-GC N Am 41:248–253. https://doi.org/10.56530/LCGC.NA.VM4090L5
27. Olesik SV (2004) Liquid chromatography at the critical condition. Anal Bioanal Chem 378:43–45. https://doi.org/10.1007/S00216-003-2319-X/FIGURES/2
28. Ghosh B, Chatterjee S, Dhakar R et al (2021) Arecanut-induced fibrosis display dual phases of reorganising glycans and amides in skin extracellular matrix. Int J Biol Macromol 185:251–263. https://doi.org/10.1016/J.IJBIOMAC.2021.06.093
29. Ghosh B, Chatterjee J, Paul RR, et al (2024) Molecular histopathology of matrix proteins through autofluorescence super-resolution microscopy. Sci Rep 14(1):14. https://doi.org/10.1038/s41598-024-61178-0
30. Ghosh B, Mandal M, Mitra P, Chatterjee J (2021) Attenuation corrected-optical coherence tomography for quantitative assessment of skin wound healing and scar morphology. J Biophotonics 14:e202000357. https://doi.org/10.1002/JBIO.202000357; WGROUP:STRING:PUBLICATION
31. Ghosh B, Chatterjee J (2023) Advances in medical imaging for wound repair and regenerative medicine. Regen Med: Emerg Tech Transl Approaches 57–76. https://doi.org/10.1007/978-981-19-6008-6_4/FIGURES/7
32. Mandal M, Ghosh B, Anura A et al (2016) Modeling continuum of epithelial mesenchymal transition plasticity. Integr Biol 8:167–176. https://doi.org/10.1039/C5IB00219B
33. Mandal M, Ghosh B, Rajput M, Chatterjee J (2020) Impact of intercellular connectivity on epithelial mesenchymal transition plasticity. Biochim Biophys Acta (BBA) —Molecular Cell Research 1867:118784. https://doi.org/10.1016/J.BBAMCR.2020.118784

34. Bhowmik A, Ghosh B, Pal M, et al (2022) Portable, handheld, and affordable blood perfusion imager for screening of subsurface cancer in resource-limited settings. Proc Natl Acad Sci U S A 119:e2026201119. https://doi.org/10.1073/PNAS.2026201119/SUPPL_FILE/PNAS.2026201119.SM03.MP4
35. Ghosh B, Bhandari A, Mandal M, et al (2021) Quantitative in situ imaging and grading of oral precancer with attenuation corrected-optical coherence tomography. Oral Oncol 117. https://doi.org/10.1016/j.oraloncology.2021.105216
36. Chatterjee R, Ghosh B, Mandal M et al (2021) Pathophysiological relationship between hypoxia associated oxidative stress, Epithelial-mesenchymal transition, stemness acquisition and alteration of Shh/ Gli-1 axis during oral sub-mucous fibrosis and oral squamous cell carcinoma. Eur J Cell Biol 100:151146. https://doi.org/10.1016/J.EJCB.2020.151146
37. Sarkar A, Barui A, Ghosh B et al (2016) Autofluorescence signatures for classifying lung cells during epithelial mesenchymal transition. RSC Adv 6:77953–77962. https://doi.org/10.1039/C6RA16866C
38. Ghosh B, Agarwal K (2024) Kidney mesangium organotypic culture reveals co-stimulatory mechano-chemical interplay for fibrotic manifestation. Biophys J 123:405a. https://doi.org/10.1016/j.bpj.2023.11.2483
39. Ghosh B, Agarwal K (2024) Randomly oriented microgrooved hydrogel guides cellular motility, modulates speed, and governs directionality of cellular spread. bioRxiv 2024.09.10.612339. https://doi.org/10.1101/2024.09.10.612339
40. Ghosh B, Agarwal K (2023) Viewing life without labels under optical microscopes. Commun Biol 6(1):12. https://doi.org/10.1038/s42003-023-04934-8
41. Ranjan A, Agarwal K, Melandso F (2023) Photoacoustic and ultrasound imaging of caridea for quantification of pigments and hardness measurement. IEEE Int Ultrason Symp, IUS. https://doi.org/10.1109/IUS51837.2023.10308042
42. Ojha S, Agarwal K, Khan MSA, et al (2024) 1P5–13 Specific acoustic impedance mapping of shrimp scale using scanning acoustic microscopy. Proc Symp Ultrason Electron 45:65. https://doi.org/10.24492/USE.45.0_1P5-13
43. Ojha S, Agarwal K, Shelke A, Habib A (2024) Quantification of impedance and mechanical properties of Zeonor using scanning acoustic microscopy. Appl Acoust 221:109981. https://doi.org/10.1016/J.APACOUST.2024.109981
44. Agarwal K, Ojha S, Dalmo RA et al (2024) Uncertainty analysis of Altantic salmon fish scale's acoustic impedance using 30 MHz C-Scan measurements. Ultrasonics 142:107360. https://doi.org/10.1016/J.ULTRAS.2024.107360
45. Shelke A, Ojha S, Ghosh B, et al (2024) 2E2–2 mechanical properties estimation of PVDF polymer using scanning acoustic microscopy. Proc Symp Ultrason Electron 45:83. https://doi.org/10.24492/USE.45.0_2E2-2
46. Agarwal K, Ojha S, Agarwal C, et al (2024) 2P1–6 Acoustic properties of chitosan-GelMA composite hydrogel. Proc Symp Ultrason Electron 45:93. https://doi.org/10.24492/use.45.0_2P1-6
47. Watanabe R, Nakamura S, Sugahara A et al (2024) Revealing molecular-scale structural changes in polymer nanocomposites during thermo-oxidative degradation using evolved gas analysis with high-resolution time-of-flight mass spectrometry combined with principal component analysis and Kendrick mass defect analysis. Anal Chem 96:2628–2636. https://doi.org/10.1021/ACS.ANALCHEM.3C05269/ASSET/IMAGES/LARGE/AC3C05269_0006.JPEG
48. Xie W, Pan WP (2004) Thermal characterization of materials using evolved gas analysis. J Therm Anal Calorim 65:669–685. https://doi.org/10.1023/A:1011946707342
49. Bartali R (2024) Surface mechanical proprieties of materials: a brief review. Next Materials 3:100120. https://doi.org/10.1016/J.NXMATE.2024.100120
50. Ghosh B, Mandal M, Mitra P, Chatterjee J (2021) Structural mechanics modeling reveals stress-adaptive features of cutaneous scars. Biomech Model Mechanobiol 20:371–377. https://doi.org/10.1007/S10237-020-01384-7/FIGURES/5
51. Long C, Dong Z, Liu X et al (2022) Simultaneous enhancement in processability and mechanical properties of polyethylenes via tuning the molecular weight distribution from unimodal

to bimodal shape. Polymer (Guildf) 258:125287. https://doi.org/10.1016/J.POLYMER.2022.125287
52. Brooijmans T, Okhuijsen RA, Oerlemans GMM et al (2019) Heterogeneity analysis of polymeric carboxylic acid functionality by selective derivatization followed by size exclusion chromatography. Anal Chim Acta 1072:87–94. https://doi.org/10.1016/J.ACA.2019.04.051
53. Iglesias-Montes ML, Soccio M, Luzi F, et al (2021) Evaluation of the factors affecting the disintegration under a composting process of poly(lactic acid)/Poly(3-hydroxybutyrate) (PLA/PHB) Blends. Polymers 13:3171. https://doi.org/10.3390/POLYM13183171
54. Kobayashi N, Arimoto H, Nishikawa Y (2000) Simultaneous determination of polymer average molar mass and molar mass distribution, and concentration of additives by GPC–GC. J Microcolumn Sep 12:501–507. https://doi.org/10.1002/1520-667X(2000)12:9%3c501::AID-MCS4%3e3.0.CO;2-L

Chapter 12
Reverse Engineering and Case Studies

Shashi Kumar, Nikhil Kumar, and Santanu Chattopadhyay

Abstract Polymers are the essential part of modern life, found in everything from household products to advanced engineering materials. In situations when there is no proper documentation, understanding their structure, composition, and manufacturing history becomes extremely important. Reverse engineering, through careful mechanical breakdown, is a comprehensive method used to uncover the mechanical, molecular, morphological, and thermal properties of finished polymer products. This chapter explores the concept and significance of reverse engineering in understanding polymer materials and developing correlations between their structure and properties. To study polymer composition, morphology, additives, and performance, it uses widely applied analytical techniques such as FTIR, DSC, TGA, GPC, SEM, NMR, and XRD. Real-world examples, including the reconstruction of vehicle parts, multilayer food packaging, and the modification of safety glass, highlight the practical value of reverse engineering in ensuring regulatory compliance, improving reliability, and enhancing design quality. Overall, this method is a valuable tool for material replacement, advancement, and performance improvement across various industries.

12.1 Introduction

Polymers are everywhere in our daily lives ranging from plastic bottles to strong materials used in the airplanes. These materials are engineered to meet specific jobs such as chemical, thermal, and mechanical requirements. Ever wonders what really happens when we see a finished polymer product and try to decipher how it is made

S. Kumar
Department of Chemical Engineering, National Institute of Technology, Rourkela, Odisha, India

N. Kumar
Advanced Technology Development Centre, Indian Institute of Technology, Kharagpur, India

S. Chattopadhyay (✉)
Rubber Technology Centre, Indian Institute of Technology, Kharagpur, India
e-mail: santanu@rtc.iitkgp.ac.in

S. Chattopadhyay and N. Kumar (eds.), *Polymer Characterization*,
https://doi.org/10.1007/978-981-95-2609-3_12

or what is its composition, when there is no documentation are available to it. To understand this, we move backward from the final finished product to its raw material step and try to find out how it is made. This process is called *reverse engineering* [1, 2].

12.1.1 Definition and Concept

Reverse engineering has become a key tool to understand the fragments and materials composition inside a product. With reference to polymers, this approach delves into what type of polymers they are, what other additive substances are mixed, and how they have processed or their orientation are [3]. In other words, we can say that reverse engineering is a process of disintegrating a finished polymer product to unravel its molecular attributes, chemical and physical properties, and manufacturing history of building blocks. This approach not just identifies the concerned materials or its properties, but it provides a comprehensive history from polymer design to the performance. This technique allows the research community to develop a thorough understanding of polymer functionalities, its peculiar features, and how it might have been processed. For instance, an aluminum foil (a simple plastic food packaging wrapper) seems to be simple at its first glance. However, its reverse engineering unveils that it is a multilayered product with a particular polymer material. Other daily life examples include shampoo bottles, phone case, sneaker sole, plastic pens etc. These examples illustrate how reverse engineering is not just only for the industries, but it is an integral part of analysis and product development in everyday products. Overall, this approach provides deep insights not only on what the material is, but also on why it is composed that way only.

12.1.2 Importance of Reverse Engineering in Understanding Polymer Structure–Property Relationships

Reverse engineering plays a critical role in understanding polymer products and therefore, the structure–property relationship has become crucial to material sciences [1]. Sometimes, polymers look the same from the outset or on surface, but small differences such as their monomeric units, their ordered structures, and different types of additives mixed, can leads to a very different behaviour. By analyzing different polymer products for their rigidity, flexibility, heat resistance and durability, the scientists and engineers can understand why a polymer behaves in a certain way and how these critical determinants can lead to some performance and behaviour in real-world applications. For instance, a plastic part of bottle breaks or cracks while exposed in the sun, reverse engineering can identify the reasons whether it was not made from proper material or was not having enough ultra-violet protection. In another example, a plastic toy can be tested and analyzed to check whether it contains harmful chemicals or not, which can be taken into considerations in the next make of product. Understanding of these insights can lead to the making of superior quality products. With respect to this, reverse engineering uncovers the information between behavior and the performance. This in-depth knowledge and understanding of the above approach are beneficial for several reasons and presented in the text to follow.

i. Product performance: The particular companies can learn from the improvement of the existing products to make better materials.
ii. Troubleshoot failure analysis: When a certain product fails in the market, reverse engineering helps in troubleshooting weaknesses and unsuitable materials.
iii. Materials replacement: When a material do not work and perform as intended, or material is too expansive or unavailable due to some supply chain management issue, reverse engineering can help in guiding the development of substituent materials with similar properties.
iv. Research and development: Depending upon the history of materials that has worked and performed, reverse engineering can aid in designing new and innovative structural changes in the existing materials.
v. Safety and regulatory compliance: Reverse engineering helps in analyzing and identifying the materials/chemicals for unknown or imported product and ensure that it follows the safety and regulatory guidelines set by the government.

12.1.3 Overview of Common Methodologies in Polymer Reverse Engineering

Reverse engineering in polymers utilizes a wide variety of analytical techniques to characterize a number of chemical, structural, and associated thermal properties [1, 4]. In addition to these, the information on additive mixing, or fillers is also provided. There are several common methods which provide a complete picture of materials manufacturing, troubleshoot failure analyses [2], and materials substitution. In the series of methods, first includes Fourier Transform Infrared Spectroscopy (FTIR), which gives the information on chemical structure and functional groups of the polymers. Differential Scanning Calorimetry (DSC) tells about the melting and glass transition temperatures. Thermogravimetric Analysis (TGA) provides insights into the thermal stability and weight loss of polymers. Further, Gel Permeation Chromatography (GPC), Scanning Electron Microscopy (SEM), and Nuclear Magnetic Spectroscopy (NMR) are used to gain insights into molecular weight distribution, microstructure properties, and details on molecular structure, respectively [1]. Integration of these methods can help the researchers in understanding of how polymers are constructed, what their compositions are, and how to restructure or improve. These insights can act as a guideline for the development of new materials or similar materials with enhanced properties [5].

12.2 Analytical Techniques for Reverse Engineering

There are several analytical techniques that are available to deal with the reverse engineering process. Each one of the following methods sheds light on decoding how a polymer product was made and processed. The different methods [1, 2, 4, 6] utilized to gain insights on polymers composition, and structural properties are presented in the text to follow.

1. Fourier Transform Infrared Spectroscopy (FTIR): FTIR (discussed in Chap. 8) is a fast-analytical technique utilized to identity the polymers functional groups or backbones based on the information of absorption peaks in the infrared region. In other words, we can say that this technique is responsible to provide the chemical fingerprints of the polymers.
2. Differential Scanning Calorimetry (DSC): DSC (discussed in Chap. 6) is common method used to get insights into various thermals properties such as melting point temperatures, glass transition temperatures and crystallinity behaviour. These parameters are helpful in determining the processing conditions.
3. Thermogravimetric Analysis (TGA): Weight loss, filler contents and thermal degradation of polymers with respect to temperature and time are evaluated with TGA technique (discussed in Chap. 6).
4. Solvent extraction and Gas Chramatography-Mass Spectroscopy (GC–MS): Identification of low molecular weight antioxidants and stabilizers in plastics is done using a suitable solvent followed by GC–MS analysis. This technique is very useful in detecting the chemical or stabilizers which are harmful and against the standards set by the government (discussed in Chap. 7).
5. Gel Permeation Chromatography (GPC): GPC method is used to get the weight distribution of polymers. These insights are critical to mechanical properties of the polymers (discussed in Chap. 6).
6. Microscopy: Microscopy techniques like scanned electron microscopy (SEM) and optical microscopy reveal the deformation behaviors, surface defects and morphology in polymer materials (discussed in Chap. 9).
7. X-ray Diffraction (XRD): With respect to polymers, XRD reveals the information on the crystalline phases (discussed in Chap. 4).
8. Nuclear Magnetic Resonance (NMR) Spectroscopy: NMR analysis helps to get the detailed information on monomers ratio, polymeric chains, sequence of polymer units and also on the molecular weight analysis (discussed in Chap. 8).
9. Energy Dispersive X-Ray Spectroscopy (EDS): Another analytical technique in this series is EDS, which is utilize to get insights on quantification of elemental composition of polymers units, especially additives mixed, or fillers.

These above-mentioned methods are generally used in combination, and helps scientist to utilize the reverse engineering to get comprehensive understanding for large and complex polymer systems (Fig. 12.1).

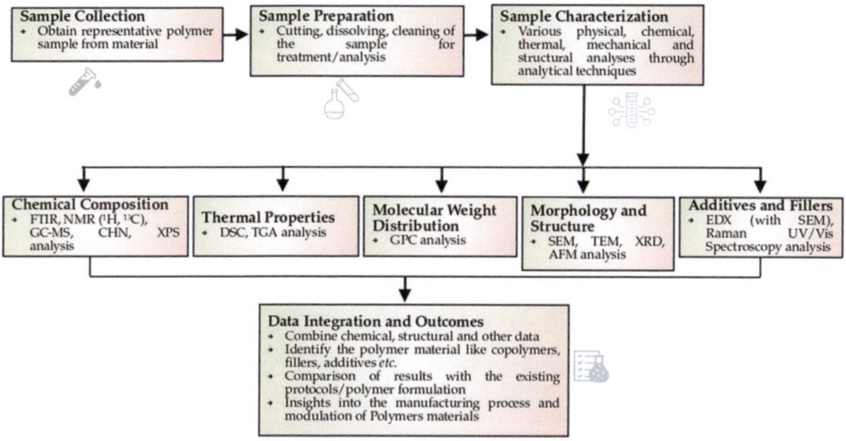

Fig. 12.1 Polymer characterization workflow showing steps from sample collection to analysis of properties and data integration for material identification and evaluation

12.3 Case Studies

The following three real-world case studies shows how the technique of reverse engineering of polymers is used in daily practical lives.

12.3.1 Case Study 1: Reverse Engineering of Broken Car Bumper

Situation Statement

The Hyundai car company received a car in damaged conditions and after analyzing, the concerned engineer finds out that car's bumper is totally damaged and need to replace. But, it was reported to the engineer that this car model is no longer in production and replacement of bumper is impossible.

Observations and Solution to Situation

The engineer has agreed to replace the whole front part of the car, but the customer was not ready for this deal. Therefore, after a careful observation, the engineer finally decided to carry out reverse engineering for the plastic car bumper. He has investigated several analytical techniques such as FTIR, and DSC etc. to know about the type, melting point, and structure rigidity of the polymer. It was found that this bumper was made up of polypropylene. Further, its behaviour under deformation was also analyzed with the SEM technique.

Conclusions

With this detailed information, the engineer was successful to select, and reconstruct the right plastic (polymer) with molding techniques. The final product recreated met the quality and standards of the original product. Finally, the customer has got his car in same shape and has avoided the expensive replacement of the bumper.

12.3.2 Case Study 2: Reverse Engineering for Multilayered Food Packaging Film [7]

Situation Statement

A food packaging company wants to develop a packaging film with similar properties and functionalities as a competitor one in the market to keep the food fresh for longer duration of time.

Process and Solution to the Situation

The thin food film was sliced into extremely small parts for microscopic examination using a process called *Microtomy*. These thin sections were then further analyzed using FTIR, XRD, and GPC to get details on identification of polymers, crystallinity behaviour, and factors critical for food preservation like oxygen, moisture, respectively.

Conclusions

Based on the observations in case of the above process, the engineer was able to successfully characterize and identify the different materials used in multilayer thin film (a competitor film in the market). From the analytical characterization, it was found that the polypropylene and ethylene vinyl chloride was used in the manufacturing the outermost layer of thin film, important for mechanical rigidity and central layer, crucial for the oxygen barrier, respectively. Further, special polymer layers called tie-layers were also utilized to stick the incompatible materials to stick together. Using this information provided in above text, engineers were able to make multilayered thin film for food packing replicating the competitor film's structure and different layer thicknesses. In this process, the final step includes the performance test of the constructed film to ensure food preservation capabilities. In a nutshell, we can say the reverse engineering allowed the food packing company to compete in the market with an alternative, and less costly multilayered film.

12.3.3 Case Study 3: Reverse Engineering of Polycarbonate Glasses

Situation Statement

Polycarbonated materials are used to make safety glasses and it was found that they developed cracks after being used in the field practices, raising the high safety concerns.

Process and Analysis

The observations revealed that the cracks developed in the safety glasses are mainly due to the ultraviolet oxidation. With exposure to sun for a prolonged period of time, polymer chains on the surface of glasses break down, leads to the formation of these surface cracks. It was also found that the micro-cracks and degradation in the glass polycarbonated materials were severe in the areas associated with direct sunlight. Analytical techniques like FTIS, TGA, SEM was conducted to identify the functional groups or any chemical changes due to oxidation, behavior of molecular degradation, and characteristics associated with brittle collapse, respectively. After getting these insights on the microcracks of the safety glasses, and to address the issue, the engineer team introduced ultraviolet stabilizer for the modification of the polymer material. This modification in polymer material was done so as to lessen the harmful effect of ultraviolet rays, which are responsible for crack formation on the glass surface. Finally, mechanical deformation test was done to ensure the resistance to cracks formation and longer life span under real-world conditions. In this way, reverse engineering helped the concerned company in restoring their product reliability and meeting the expected safety norms, aids in avoiding safety risks.

12.4 Conclusions

Reverse engineering is an indispensable tool in material sciences. This approach helps researchers and engineers to uncover materials for their structure–property relationships. Utilizing several analytical techniques, we can identify the structural, morphological, and design properties of polymer materials. The approach of reverse engineering is useful especially in enhancing the product quality, minimizing the troubleshoot failure, and placing the restructured materials in the competitive market with less expensive alternative. In total, the process of reverse engineering plays an integral role in guiding innovations, and enabling smart strategy useful for industries.

Conflict of Interest Authors declare no competing conflict of interest.

References

1. Gupta S, Mukhopadhyay R, Baranwal K (2013) Reverse engineering of rubber products: Concepts, tools, and techniques
2. Forrest M (2019) Rubber analysis: characterisation, failure diagnosis and reverse engineering
3. Chemical reverse engineering of polymers: a strategic ally for your innovation processes.—PolymerExpert. https://polymerexpert.fr/en/analysis-and-expetise/reverse-engineering-a-competitive-tool/. Accessed 15 May 2025
4. Campbell D, Pethrick RA, White JR (2017) Polymer characterization: physical techniques. CRC Press
5. Patra TK (2021) Data-driven methods for accelerating polymer design. ACS Polymers Au 2:8–26. https://doi.org/10.1021/ACSPOLYMERSAU.1C00035
6. Yang R (2018) Analytical methods for polymer characterization. CRC Press
7. Widjaja E, Today MG-M, 2011 undefined Reverse engineering of multi-layer films. Elsevier

MIX
Papier aus verantwortungsvollen Quellen
Paper from responsible sources
FSC® C105338

If you have any concerns about our products,
you can contact us on
ProductSafety@springernature.com

In case Publisher is established outside the EU,
the EU authorized representative is:
Springer Nature Customer Service Center GmbH
Europaplatz 3, 69115 Heidelberg, Germany

Printed by Libri Plureos GmbH
in Hamburg, Germany